# An Introduction to Electrical Science

## Second Edition

Heavily updated and expanded, this second edition of Adrian Waygood's textbook provides an indispensable introduction to the science behind electrical engineering.

While fully matched to the electrical science requirements of the 2330 levels 2 and 3 Certificates in Electrotechnical Technology from City & Guilds (Electrical Installation), the main purpose of this book is to develop an easy understanding of the how and why within each topic. It is aimed at those starting careers in electricity and electronics, as well as any hobbyists, with an array of new material to reflect changes in the industry.

New chapters include:

- Electrical drawings
- Practical resistors
- Measuring instruments
- Basic motor action
- Practical capacitors
- Basic transformer theory
- The electricity supply industry
- … and more

The author details the historical context of each main principle and offers a wealth of examples, images and diagrams, all whilst maintaining his signature conversational and accessible style. There is also a companion website, with interactive multiple choice quizzes for each chapter and more, at www.routledge.com/cw/waygood

**Adrian Waygood** is a retired lecturer and instructional designer, having had a career in the armed forces, technical institutes and the defence industry in the UK, Canada and the Middle East. These include the Royal Navy, the Royal Navy of Oman, the Northern Alberta and British Columbia Institutes of Technology, the Higher Colleges of Technology (United Arab Emirates) and British Aerospace. He has also managed government trades apprenticeship programmes in Canada, and has consulted on instructional-design methodology at technical-teacher training colleges in India and Indonesia. He holds a Higher National Certificate in Electrical and Electronics Engineering, together with a Master's Degree in Education Technology.

# An Introduction to Electrical Science

Second Edition

Adrian Waygood

Routledge
Taylor & Francis Group

LONDON AND NEW YORK

Second edition published 2019
by Routledge
2 Park Square, Milton Park, Abingdon, Oxon OX14 4RN

and by Routledge
711 Third Avenue, New York, NY 10017

*Routledge is an imprint of the Taylor & Francis Group, an informa business*

© 2019 Adrian Waygood

The right of Adrian Waygood to be identified as author of this work has been asserted by him in accordance with sections 77 and 78 of the Copyright, Designs and Patents Act 1988.

First edition published by Routledge 2013

*British Library Cataloguing-in-Publication Data*
A catalogue record for this book is available from the British Library

*Library of Congress Cataloging-in-Publication Data*
Names: Waygood, Adrian, author.
Title: An introduction to electrical science / Adrian Waygood.
Description: Second edition. | Abingdon, Oxon ; New York, NY : Routledge,
    [2018] | Includes bibliographical references and index.
Identifiers: LCCN 2018008929 (print) | LCCN 2018012113 (ebook) | ISBN
    9781351190428 (Adobe PDF) | ISBN 9781351190411 (ePub) | ISBN 9781351190404
    (Mobipocket) | ISBN 9780815391814 (pbk.) | ISBN 9780815391821 (hardback)
    | ISBN 9781351190435 (ebook)
Subjects: LCSH: Electrical engineering—Textbooks. | Electrical
    engineering--Equipment and supplies—Textbooks. | Electricity—Textbooks.
Classification: LCC TK146 (ebook) | LCC TK146 .W39 2018 (print) | DDC
    621.3--dc23
LC record available at https://lccn.loc.gov/2018008929

ISBN: 978-0-8153-9182-1 (hbk)
ISBN: 978-0-8153-9181-4 (pbk)
ISBN: 978-1-351-19043-5 (ebk)

Typeset in Times
by Apec CoVantage, LLC

Visit the companion website: www.routledge.com/cw/waygood

# Contents

Foreword ................................................................................................................................. xv

Introduction ........................................................................................................................ xvii

**1  SI system of measurements** ......................................................................................... 1
   Objectives............................................................................................................................ 1
   Introduction ........................................................................................................................ 1
   Base units ............................................................................................................................ 1
   Derived units ...................................................................................................................... 2
   Multiples and sub-multiples .............................................................................................. 3
   SI conventions .................................................................................................................... 4
   Proposed new definitions for SI base units ...................................................................... 4
   Definitions of electrical SI units........................................................................................ 5
   Misconceptions.................................................................................................................... 5
   Review your learning .......................................................................................................... 6

**2  Electrical schematic and wiring diagrams**................................................................. 7
   Objectives............................................................................................................................ 7
   Introduction ........................................................................................................................ 7
   Schematic diagrams............................................................................................................ 8
   Wiring diagrams ................................................................................................................ 11
   Other types of electrical diagram .................................................................................... 12
   Creating electrical drawings with CAD applications...................................................... 13
   Review your learning ........................................................................................................ 13

**3  The 'electron theory' of electricity** ........................................................................... 14
   Objectives.......................................................................................................................... 14
   Introduction ...................................................................................................................... 14
   Bohr's model of the atom ................................................................................................ 16
   Electrons and electricity .................................................................................................. 17
   Displacement current in insulators or dielectrics .......................................................... 21
   'Electron flow' versus 'conventional current' flow ........................................................ 22
   Practical conductors and insulators ................................................................................ 22
   Summary ............................................................................................................................ 23
   Misconceptions.................................................................................................................. 24
   Review your learning ........................................................................................................ 25

**4   Electric current** .................................................................................... 26
   Objectives ............................................................................................. 26
   Measuring electric current ...................................................................... 26
   Drift velocity of charge carriers ............................................................. 29
   Summary ............................................................................................... 29
   Misconceptions ...................................................................................... 30
   Review your learning .............................................................................. 30

**5   Potential and potential difference** ......................................................... 31
   Objectives ............................................................................................. 31
   Introduction ........................................................................................... 31
   Electric fields ......................................................................................... 31
   Creating a potential difference through charge separation ......................... 33
   Frictional contact (triboelectricity) ......................................................... 34
   Thermoelectricity ................................................................................... 35
   Cells and batteries ................................................................................. 36
   Electric current – misconceptions .......................................................... 37
   Voltage drop .......................................................................................... 38
   Potential v potential difference (voltage) ................................................. 38
   Summary ............................................................................................... 41
   Review your learning .............................................................................. 42

**6   Resistance** ........................................................................................... 43
   Objectives ............................................................................................. 43
   Resistance ............................................................................................. 43
   'A.C. resistance' due to the 'skin effect' .................................................. 47
   Resistance of conductor insulation .......................................................... 48
   General effect of temperature upon resistance ......................................... 49
   Misconceptions ...................................................................................... 49
   Review your learning .............................................................................. 50

**7   Resistors** ............................................................................................. 51
   Objectives ............................................................................................. 51
   Introduction ........................................................................................... 51
   Types of resistor .................................................................................... 51
   Resistor power ratings ........................................................................... 54
   Resistance values of resistors ................................................................ 54
   Identifying the resistance of a resistor ..................................................... 55
   Alpha-numeric resistance code ............................................................... 56
   Review your learning .............................................................................. 57

**8   Conductors and cables**................................................................................................ 58

Objectives.......................................................................................................................... 58

Introduction ...................................................................................................................... 58

Conductors ........................................................................................................................ 58

Insulation........................................................................................................................... 60

Protective sheath .............................................................................................................. 63

Review your learning ....................................................................................................... 63

**9   Effect of temperature change upon resistance** ..................................................... 64

Objectives.......................................................................................................................... 64

Introduction ...................................................................................................................... 64

Temperature coefficient of resistance............................................................................ 65

Negative temperature coefficient of resistance ............................................................ 68

Constantan and similar alloys ......................................................................................... 68

Summary ............................................................................................................................ 68

Review your learning ....................................................................................................... 69

**10   Ohm's Law of constant proportionality**.................................................................. 70

Objectives.......................................................................................................................... 70

Relationship between resistance, voltage and current.................................................. 70

Ohm's Law ......................................................................................................................... 71

Repeating Ohm's experiments in the laboratory .......................................................... 72

Further worked examples ................................................................................................. 75

Summary ............................................................................................................................ 76

Misconceptions about Ohm's Law .................................................................................. 76

Review your learning ....................................................................................................... 76

**11   Series, parallel and series-parallel circuits** ............................................................ 77

Objectives.......................................................................................................................... 77

Introduction ...................................................................................................................... 77

Understanding sense arrows............................................................................................ 78

Kirchhoff's Laws................................................................................................................ 79

Series circuits .................................................................................................................... 82

Parallel circuits.................................................................................................................. 85

Series-parallel circuits ..................................................................................................... 87

Voltage drops along cables .............................................................................................. 91

Review your learning ....................................................................................................... 92

**12   Electrical measurements**............................................................................................ 93

Objectives.......................................................................................................................... 93

Introduction .................................................................................................... 93
Analogue instruments ..................................................................................... 94
Analogue multimeters ..................................................................................... 97
Digital measuring instruments ......................................................................... 99
Review your learning ....................................................................................... 101

**13  Cells and batteries** ........................................................................... 102
Objectives ...................................................................................................... 102
Introduction ................................................................................................... 102
Primary and secondary cells ........................................................................... 103
Terminology ................................................................................................... 104
Primary cells .................................................................................................. 105
Secondary cells .............................................................................................. 110
Fuel cells ....................................................................................................... 115
Responsible disposal of cells and batteries ..................................................... 117
Batteries ........................................................................................................ 117
Review your learning ....................................................................................... 119

**14  Internal resistance** .......................................................................... 120
Objectives ...................................................................................................... 120
Internal resistance explained ........................................................................... 120
Representing internal resistance in a schematic diagram ................................... 121
Determining the internal resistance of a cell or battery ..................................... 123
Summary ........................................................................................................ 125
Review your learning ....................................................................................... 126

**15  Energy, work, heat and power** ........................................................ 127
Objectives ...................................................................................................... 127
Introduction ................................................................................................... 127
Energy, work and heat ..................................................................................... 127
Joule heating .................................................................................................. 130
Power ............................................................................................................. 130
Units of measurement ..................................................................................... 130
Energy, work, heat and power in electric circuits .............................................. 131
Misconceptions about energy transfer ............................................................. 132
Power ratings .................................................................................................. 133
Power in series, parallel and series-parallel circuits .......................................... 134
Connecting lamps in series: oh, what a surprise! .............................................. 135
Efficiency ....................................................................................................... 135
Maximum Power-Transfer Theorem .................................................................. 136
Residential electricity utility bills ..................................................................... 138

'Smart' energy meters ................................................................................ 140
Electricity and heat.................................................................................. 141
Summary of important electrical equations ............................................ 143
Self-test exercise  ..................................................................................... 143
Self-test exercise – answers...................................................................... 143
Misconceptions......................................................................................... 143
Review your learning ............................................................................... 145

**16  Magnetism** ....................................................................................... 146
Objectives................................................................................................. 146
Introduction ............................................................................................. 146
Magnetic fields ........................................................................................ 149
The Domain Theory of Magnetism ......................................................... 152
Ferromagnetic materials........................................................................... 155
Misconceptions......................................................................................... 156
Review your learning ............................................................................... 156

**17  Electromagnetism**............................................................................. 157
Objectives................................................................................................. 157
Introduction ............................................................................................. 157
Electric currents and magnetic fields ..................................................... 157
Electromagnetic fields.............................................................................. 158
Force on conductors ................................................................................ 164
Basic motor action................................................................................... 165
Review your learning ............................................................................... 166

**18  Magnetic circuits**.............................................................................. 167
Objectives................................................................................................. 167
Introduction ............................................................................................. 167
Electric circuit analogy............................................................................. 167
'Hopkinson's Law' for magnetic circuits ............................................... 170
'Kirchhoff's Laws' for magnetic circuits ............................................... 171
Understanding magnetisation (B-H) curves ........................................... 172
Solving problems on magnetic circuits ................................................... 176
Magnetic materials ................................................................................... 179
Hysteresis loops........................................................................................ 179
Practical applications of magnetic circuits.............................................. 183
Magnetic leakage and fringing................................................................. 184
Limitations of the 'electric circuit' analogy ........................................... 184
Summary ................................................................................................... 185
Review your learning ............................................................................... 187

**19  Electromagnetic induction** .......................................................................... 188

Objectives......................................................................................................... 188

Introduction ..................................................................................................... 188

Generator action ............................................................................................. 190

Action and reaction: Lenz's Law ................................................................... 192

Self-inductance................................................................................................ 194

Behaviour of d.c. inductive circuits ............................................................... 197

Energy stored in a magnetic field ................................................................... 200

Mutual inductance ........................................................................................... 201

Practical examples of mutually inductive circuits.......................................... 203

Inductors........................................................................................................... 203

Review your learning ...................................................................................... 205

**20  Motor principle** ............................................................................................. 206

Objectives......................................................................................................... 206

First, a reminder .............................................................................................. 206

D.C. motor fundamentals ................................................................................ 207

Motor action .................................................................................................... 208

Electrodynamic loudspeakers ......................................................................... 209

D.C. electric motor .......................................................................................... 210

Practical d.c. motors ........................................................................................ 215

Review your learning ...................................................................................... 216

**21  Transformers** ................................................................................................. 217

Objectives......................................................................................................... 217

Introduction ..................................................................................................... 217

Principle of operation ...................................................................................... 220

Advantages of a.c. in electricity transmission/distribution systems.............. 223

Review your learning ...................................................................................... 225

**22  Capacitance** ................................................................................................... 226

Objectives......................................................................................................... 226

Introduction ..................................................................................................... 226

Capacitors......................................................................................................... 227

Capacitance ...................................................................................................... 232

Energy stored by a capacitor ........................................................................... 237

Construction of practical capacitors ............................................................... 237

Behaviour of capacitors in d.c. circuits ......................................................... 239

Supercapacitors ............................................................................................... 240

Natural capacitance ......................................................................................... 241

Capacitors in series and parallel...................................................................... 241

Review your learning ...................................................................................... 244

**23 Capacitors** ............................................................................................ 245

Objectives .................................................................................................. 245

Introduction .............................................................................................. 245

Types of capacitor .................................................................................... 245

Ratings ...................................................................................................... 247

Identifying the capacitance of capacitors ............................................... 248

Review your learning ............................................................................... 250

**24 Geometry and trigonometry revision for a.c** .................................... 251

Objectives .................................................................................................. 251

Introduction .............................................................................................. 251

Terminology .............................................................................................. 251

Pythagoras's Theorem .............................................................................. 252

Trigonometry ............................................................................................ 253

Review your learning ............................................................................... 254

**25 Introduction to alternating current** .................................................. 255

Objectives .................................................................................................. 255

Introduction .............................................................................................. 255

England, Wales and Scotland electricity-supply system ........................ 256

Northern Ireland and Channel Islands electricity-supply system .......... 258

Generation of a sine wave ....................................................................... 259

Alternators: basic constructional features .............................................. 259

Electrical v mechanical degrees .............................................................. 261

Generation of a sinusoidal voltage ......................................................... 262

Terminology .............................................................................................. 266

Measuring sinusoidal values .................................................................... 266

Representing sinusoidal waveforms with phasors .................................. 270

All you need to know to understand alternating current ...................... 271

Review your learning ............................................................................... 272

**26 Series alternating-current circuits** .................................................... 273

Objectives .................................................................................................. 273

Introduction .............................................................................................. 273

Purely resistive circuit .............................................................................. 274

Purely inductive circuit ............................................................................ 275

Purely capacitive circuit ........................................................................... 276

CIVIL ......................................................................................................... 277

Series R-L circuits ..................................................................................... 278

Series R-C circuits ..................................................................................... 282

Series R-L-C circuits ................................................................................. 286

Series resonance ....................................................................................... 291

Review your learning ............................................................................... 294

**27 Power in alternating-current circuits** ........................................................... 295
  Objectives ................................................................................................... 295
  Introduction ................................................................................................ 295
  Behaviour of energy in a.c. circuits ........................................................... 296
  A.C. power waveforms ................................................................................ 299
  Power in resistive-reactive circuits ............................................................ 301
  Apparent power, true power and reactive power ....................................... 302
  Power in a series R-L circuit ...................................................................... 302
  Power in a series R-C circuit ...................................................................... 305
  Review your learning .................................................................................. 308

**28 Parallel alternating-current circuits** ............................................................ 309
  Objectives ................................................................................................... 309
  Introduction ................................................................................................ 309
  Parallel R-L circuits .................................................................................... 310
  Parallel R-C circuits .................................................................................... 316
  Parallel R-L-C circuits ................................................................................ 320
  Review your learning .................................................................................. 325

**29 Summary of equations derived from phasor diagrams** ............................... 326
  Objectives ................................................................................................... 326
  Generating equations for series a.c. circuits .............................................. 326
  Generating equations for parallel a.c. circuits ........................................... 329
  Admittance diagrams .................................................................................. 330
  Review your learning .................................................................................. 333

**30 Power factor improvement** ........................................................................... 334
  Objectives ................................................................................................... 334
  Need for power factor improvement ........................................................... 334
  Power factor improvement in practice ........................................................ 337
  Review your learning .................................................................................. 338

**31 Balanced three-phase a.c. systems** .............................................................. 339
  Objectives ................................................................................................... 339
  Introduction ................................................................................................ 339
  Generating three-phase voltages ............................................................... 340
  Three-phase connections and terminology ................................................ 340
  Three-phase phasor diagrams .................................................................... 343
  Behaviour of currents in balanced three-phase systems ........................... 349
  Power in three-phase systems .................................................................... 350
  Summary of important three-phase relationships ...................................... 351
  Review your learning .................................................................................. 353

# Foreword

I *love* science. I'm an avid reader of books about science and technology, and would rather watch television programs about them than anything else. Physics, astronomy, cosmology, paleontology, earth sciences, wildlife, how things work, and the world we live in – nothing interests me more. And so much of it, delivered through channels like the BBC, Smithsonian, and National Geographic, is very well done. Today the media has the technology to produce educational programs that are not only informative and educational, but also captivating and easy to follow. I expect many young people will develop an interest, and even choose a career, in science and technology because of this exposure. Watching these programs, I think how lucky science teachers are today to have these tools to awaken interest in, and love of, science in their students.

As wonderful as video media is at delivering an overview of science topics, it does not lend itself well to the in-depth study necessary for anyone who chooses to make science and technology his or her career. For that, printed materials are still indispensable. And learners can often find written text less appealing than what they have been exposed to through video and animation. Writers and publishers of science textbooks must produce materials that cater to students who have been conditioned to a sophisticated, interesting, and captivating learning experience.

I am honoured to have been asked to provide a short foreword to Adrian Waygood's newly revised book on electrical science. I know him well and worked closely with him in the Electrical Programs at the Northern Alberta Institute of Technology – NAIT. Adrian was a very effective instructor, popular with his students and highly respected by his colleagues. Through his career he has accumulated a broad experience in technical training with the Royal Navy, with the Royal Navy of Oman, in Canadian colleges, and in several countries in the Middle East.

I expect students will find this textbook on electrical science to be an excellent experience. The presentation is interesting, and the conversational nature of the writing is engaging and encourages reading on. Topics are presented and developed in a logical and easy-to-follow sequence. The author strives to ensure no 'holes' are left in the concepts, and no assumptions made about prior knowledge. The basics of each topic are presented rather than assumed – no lazy presentation here.

Concepts and conventions are made much more interesting and understandable through explanations of their history, development of thought, and evolution of the science. The author is meticulous about the correct use of terminology and scientific accuracy and, where custom or convention tend to create confusion, he provides explanation.

Adrian believes, as do I, that it is not sufficient to skim over the definitions and equations with a simple goal of passing an exam. What is important is a fundamental and thorough understanding of the basics of the science. With that, the learner is armed with the ability to build on his or her knowledge and understanding, and to exercise critical thinking in problem solving.

This is a thorough and comprehensive, yet engaging and easy to understand, introduction to electrical science and circuit fundamentals. It is an excellent resource for anyone preparing for a career in electrical trades, technology, or engineering as well as being an appropriate choice for anyone simply interested in learning for learning's sake.

**Greg Collins**
*Chair (Retired)*
*Electrical Engineering Technology*
*Northern Alberta Institute of Technology – NAIT*
*Edmonton, Alberta, Canada*

# Introduction

The first edition of this book was published in 2013 followed, two years later, by its companion book, *Electrical Science for Technicians*.

Since then, I have revised and expanded *An Introduction to Electrical Science* primarily in order to incorporate topics that, for reasons of space, had to be omitted from the first edition.

Another change that has had to be accommodated in this new edition is the proposal by the **International Committee for Weights and Measures (CIPM)** to change the formal definitions of the SI Base Units, in particular the **ampere**, whose present definition dates back to 1946. The proposed changes are the result of the ability we now have to apply *exact* values to the speed of light together with a further four scientific constants. One of these constants is the 'elementary charge', i.e. the amount of electric charge on an individual electron. This, coupled with the fact that scientists are now able to count the passage of individual electrons, means that the definition of the ampere will be completely changed from its present dependence on the electromagnetic attraction/repulsion between a pair of current-carrying conductors, to the rate of flow of elementary particles.

Unfortunately, at the time of completing the manuscript for this book (January 2018), these changes have yet to be implemented or even finalised. In fact, the changes have already been postponed once, and are now unlikely to be introduced until 2019. However, it is felt that, as the new definitions are imminent, the changes should be brought to my readers' attention, so I have incorporated what we know about the proposals before this book goes to print.

Now, a note on how you should use this book. You will find that each chapter starts with a list of desired learning outcomes, or '**objective statements**'. These guide you as to *what you should endeavour to learn from each chapter*. These statements specify what you *must* know when you have read the chapter, as opposed to what is *nice* to know. By adding a question mark at the end of each objective statement, they become **test items**. For example, an objective statement might state, '*... list the three effects of an electric current*'. After completing the chapter, you should then ask yourself, '***Can I** ... list the three effects of an electric current?*' If you can, then you have *achieved that particular objective, and you can move on to the next.*

I would like to thank my publishers for their support during the preparation of this book, and to my good friend and former colleague, Greg Collins, for writing the Foreword to this revised edition.

Finally, thank you for purchasing this book; I hope you enjoy reading it and find it useful. You are invited to visit my blog site at www.professorelectron.com.

**Adrian Waygood**, MEd (Tech)
(January 2018)

---

**Online resources**

There is also a companion website for this book featuring multiple choice questions, further written questions and an extra chapter on electrical measuring instruments. The website can be accessed via the following link: www.routledge.com/cw/waygood

# Chapter 1

# SI system of measurements

## Objectives

On completion of this chapter, you should be able to

1. explain the term 'SI base units'.
2. list the seven SI base units.
3. explain the term 'SI derived units'.
4. explain the relationship between SI 'base' and 'derived' units.
5. recognise SI prefixes.
6. apply correct SI symbols.
7. use correct SI prefixes.
8. apply SI conventions when writing SI units.
9. be aware of changes to SI base units scheduled for 2019.

## Introduction

In 1948, the **General Conference of Weights and Measures (CGPM)** charged an international committee, the CIPM*, to *'study the establishment of a complete set of rules for (metric) units of measurement'*.

> **\*CIPM (Comité international des poid et mesures)** is an International Committee for Weights and Measurements comprising eighteen individuals, each from a different member state, whose principal task is to promote world-wide uniformity in units of measurement. The Committee achieves this either by direct action, or by submitting proposals to the General Conference on Weights and Measures.

The outcome of this study was a rational system of metric units termed '**SI**'.

The abbreviation **SI** stands for *Système Internationale d'Unités* and this system of measurements has been adopted internationally by the scientific and engineering communities, as well as by businesses for the purpose of international trade. Whereas *most* countries now use SI exclusively, *some* countries – most notably the United States and, to a lesser extent, the United Kingdom – still make wide use of non-metric units, especially for day-to-day use.

Earlier versions of the metric system include the '**cgsA**' ('centimetre, gram, second, ampere') and the '**mksA**' ('metre, kilogram, second, ampere') systems. SI is largely based on the 'mksA' system.

SI comprises *two* classes of units:

- base units
- derived units.

## Base units

There are *seven* **base units** from which *all* other SI units are derived. These are shown in Table 1.1.

Table 1.1

| Quantity | SI unit | SI symbol |
|---|---|---|
| length | metre | m |
| mass* | kilogram | kg |
| time | second | s |
| electric current | ampere | A |
| temperature | kelvin | K |
| luminous intensity | candela | cd |
| amount of substance | mole | mol |

\* The kilogram is a little confusing because it is the only base unit with a prefix (kilo). It has been suggested that the name 'kilogram' should be replaced (to remove the prefix 'kilo') as the unit for mass, but this is likely to cause more confusion than necessary.

An Introduction to Electrical Science, Waygood, ISBN 9780815391821, 2019. © Taylor & Francis

## Derived units

Those SI units which are *not* base units are called **derived units**.

Derived units are formed by combining base units – for example, the '**volt**' is defined as *'the potential difference between two points such that the energy used in conveying a charge of one coulomb from one point to the other is one joule'*.

So the **volt** is defined in terms of the **coulomb** and the **joule**. The **coulomb** (see page 5), in turn, is defined in terms of the **ampere** and the **second** (both base units). The **joule** is defined in terms of the **newton** and the **metre** (a base unit). Finally, the **newton** is defined in terms of the **kilogram**, the **metre** and the **second** (all base units).

So, by 'deconstructing' the **volt**, we find that it is ultimately derived from a combination of each of the base units underlined in Figure 1.1 – i.e. the **ampere**, the **second**, the **kilogram** and the **metre**.

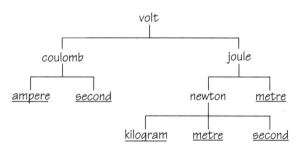

Figure 1.1

Most derived units have been given **special names** in honour of famous physicists whose research has contributed to our knowledge of the quantity concerned – for example, as we have learnt, the derived unit for potential difference is the '**volt**', which is simply a special name given to a '**joule per coulomb**', and is named after the Italian nobleman and professor of physics, Count Alessandro Volta (1745–1827).

Table 1.2 lists **SI derived units** with special names that you will meet in this text.

Table 1.2

| Quantity | Symbol | SI unit | SI symbol |
|---|---|---|---|
| capacitance | $C$ | farad | F |
| capacitive reactance | $X_C$ | ohm | Ω |
| conductance | $G$ | siemens | S |
| electric charge | $Q$ | coulomb | C |
| force | $F$ | newton | N |
| frequency | $F$ | hertz | Hz |
| impedance | $Z$ | ohm | Ω |
| inductance | $L$ | henry | H |
| inductive reactance | $X_L$ | ohm | Ω |
| potential difference | $E, U, V$ | volt | V |
| power | $P$ | watt | W |
| pressure | $P$ | pascal | Pa |
| resistance | $R$ | ohm | Ω |
| magnetic flux | $\Phi$ | weber* | Wb |
| magnetic flux density | $B$ | tesla | T |
| magnetomotive force | $H$ | ampere** | A |
| mutual inductance | $M$ | henry*** | H |
| self inductance | $L$ | henry | H |
| work, energy | $W$ | joule | J |

**Notes:**
*The weber is pronounced 'vay-ber'.
**Often spoken as 'ampere turn'.
***The plural of 'henry' is 'henrys', not 'henries'.

Students often ask *why* some symbols for electrical (and other) quantities sometimes appear rather strange: '*I*' for current, '*U*' for potential difference, for example.

The answer is that much of the research into electricity was conducted by German, French, Italian, and other European physicists so, naturally, some of the symbols are based on those languages – rather than English.

The symbol for current, *I*, for example, is based on the French word *intensité*, from the expression '*l'intensité du courant électrique*'.

The origin of the symbol *U*, for potential difference, is less clear, but is thought to be based on the German word '*unterschied*', meaning 'difference'.

## Non-SI metric units

Not all *metric* units are SI units, although many may be '*used alongside*' SI units. These include the commonly used units shown in Table 1.3.

Table 1.3

| Quantity | Unit | Symbol |
|---|---|---|
| energy | watt hour | W·h |
| mass | tonne | t* |
| volume | litre | L or λ ** |
| rotation | revolutions per second | r/s |
| temperature | degree Celsius*** | °C |
| time | minute (60 s); hour; day; year | min, h, d, a |

**Notes:**
*The unit of mass, the *tonne*, is pronounced, or spoken, as 'metric ton'.
**Since a lower-case 'ell' (l) can be confused with the number 1, we shall use a capital 'ell' (L), in common with North American SI practice.
***The division intervals are identical for the both the *Celsius* and *kelvin* scales. However, 0°C corresponds to 273.15 K, and 100°C corresponds to 373.15 K.

## Multiples and sub-multiples

Frequently, we have to deal with very large, or very small, quantities. For example, the resistance of insulation is measured in millions of ohms, while the resistance of a conductor is measured in thousandths of an ohm.

To avoid having to express very large or very small values in this way, we use, instead, **multiples** and **sub-multiples**. These are indicated by assigning a *prefix* to the SI unit. The more common are listed in Table 1.4.

Table 1.4

| Multiplication factor | Power of ten | Prefix | Symbol |
|---|---|---|---|
| 1 000 000 000 000 × | $10^{12}$ | tera | T |
| 1 000 000 000 × | $10^{9}$ | giga | G |
| 1 000 000 × | $10^{6}$ | mega | M |
| 1 000 × | $10^{3}$ | kilo | k |
| 0.001 × | $10^{-3}$ | milli | m |
| 0.000 001 × | $10^{-6}$ | micro | m |
| 0.000 000 000 001 × | $10^{-12}$ | pico | p |

**Examples**
• 10 000 000 watts can be written as $10 \times 10^{6}$ W, or as 10 MW
• 33 000 volts can be written as $33 \times 10^{3}$ V, or as 33 kV
• 0.025 amperes can be written as $25 \times 10^{-3}$ A, or as 25 mA
Note: the correct spelling for one-millionth of an ohm, is 'microhm', *not* 'microohm' or 'micro-ohm'.

Note that SI recommends using prefixes employed by the '**Engineering System**' – i.e. powers of ten which increase or decrease by a factor of *three*. Accordingly, units such as the 'centimetre', etc., should *not* be used when working in SI.

We *cannot* insert multiples or sub-multiples into equations. For example, we must *always* convert micro-watts, milliwatts, kilowatts, megawatts, etc., into **watts** whenever we insert that quantity into an equation.

To do this:

Table 1.5

| to convert. . . | into. . . | multiply by. . . |
|---|---|---|
| picowatts | watts | $\times 10^{-12}$ |
| microwatts | watts | $\times 10^{-6}$ |
| milliwatts | watts | $\times 10^{-3}$ |
| kilowatts | watts | $\times 10^{3}$ |
| megawatts | watts | $\times 10^{6}$ |
| gigawatts | watts | $\times 10^{9}$ |
| terawatts | watts | $\times 10^{12}$ |

Although, in the above examples, we have used watts, this applies of course to *any* SI unit.

## Multiplying expressions containing Indices

When *multiplying* expressions containing indices, the following rule applies:

$$(a \times 10^x) \times (b \times 10^y) = ab \times 10^{(x+y)}$$

For example: Multiply 2 kilounits by 4 megaunits.

$$(2 \times 10^3) \times (4 \times 10^6) = (2 \times 4) \times 10^{(3+6)} = 8 \times 10^9 \text{ (Answer)}$$

## Dividing expressions containing Indices

When *dividing* expressions containing indices, the following rule applies:

$$\frac{a \times 10^x}{b \times 10^y} = \frac{a}{b} \times 10^{(x-y)}$$

For example: Divide 2 kilounits by 4 megaunits.

$$\frac{2 \times 10^3}{4 \times 10^6} = \frac{2}{4} \times 10^{(3-6)} = 0.5 \times 10^{-3} \text{ (Answer)}$$

## SI conventions

SI specifies *how* its units of measurement should be written. These rules, or **conventions**, apply to the units themselves, to their symbols and to their associated numerals.

You should be aware of the following conventions.

## Rules for writing SI units

1  SI units and their symbols are never italicised:
   e.g.  ampere, not *ampere*
         mV, not *mV*
2  When written in full, units are *never* capitalised:
   e.g.  watt, not Watt
         ampere, not Ampere
3  SI symbols are written in lower-case, *unless they are named after someone*, in which case they are capitalised:
   e.g.  symbol for metre: m
         symbol for ampere: A (after André-Marie Ampère)
4  SI symbols are symbols, not abbreviations, so are *not* punctuated with full stops (periods):
   e.g.  230 V, not 230 V.
         13 A, not 13 A.
5  There is no plural form of an SI symbol:
   e.g.  500 kg, not 500 kgs
         40 W, not 40 Ws
6  Numerals are always followed by the *symbol* for a unit:
   e.g.  400 V, not 400 volts
         10 kW, not 10 kilowatts
7  Written numbers are always followed by a written unit:
   e.g.  Twelve volts, not twelve V
8  A *space* is always placed between a number and the unit symbol:
   e.g.  5000 W, not 5000W
         275 kV, not 275kV
9  A hyphen *may* be used (optionally) between a number and the unit symbol, when the combination is used as an adjective:
   e.g.  'A 66-kV power line' or 'A 66 kV power line'
         'A 13-A socket' or 'A 13 A socket'
10 Compound derived unit symbols are separated by a point placed above the line:
   e.g.  SI unit for apparent power: V·A (volt ampere)
         SI unit for resistivity: Ω·m (ohm metre)
11 No space or hyphen is placed between an SI unit or symbol and its multiplier:
   e.g.  kilowatt, not kilo-watt or kilo watt
         kW, not k-W or k W
   Special case for ohms:
         microhm, not microohm or micro-ohm
         kilohm, not kiloohm or kilo-ohm
12 *Spaces*, not commas, are used as thousand separators with large numbers:
   e.g.  11 000 V, not 11,000 V
         15 000.000 075, not 15,000.000075
   The space is *optional* for four digits: 1500 mW *or* 1 500 mW
13 Square and cubic measurements are written as exponents:
   e.g.  $m^2$ (square metres) not sq m.
   e.g.  $m^3$ (cubic metres) not cu m.

## Proposed new definitions for SI base units

As this book was being prepared for publication, the International Committee of Weights and Measures (CIPM) was examining proposed changes to the formal definitions of these SI base units. These revisions are expected to come into force on 20 May 2019, assuming they are accepted following a final vote at the 26th General Conference on Weights and Measures scheduled to take place on 16 November 2018.

Under the proposed revisions, SI will retain the same seven base units.

Of these, the kilogram, ampere, kelvin, and mole will be redefined by choosing exact numerical values for the physical constants on which they will be based.

However, the metre, second, and candela are already defined by physical constants and it will only be necessary to rephrase their present definitions.

It is argued that the new definitions will improve the SI base units without changing the size of any SI units, thus ensuring continuity with present measurements.

Of particular interest to us is the proposed redefinition of the ampere, which will be discussed later in this chapter.

A draft of the Ninth SI Units Brochure is available online (www.bipm.org/utils/common/pdf/si-brochure-draft-2016.pdf), and may be downloaded as a pdf document.

## Definitions of electrical SI units

### ampere (symbol: A)

The **ampere** is defined as '*the constant current that, if maintained in two straight parallel conductors of infinite length and negligible cross-sectional area and placed one metre apart in a vacuum, would produce between them a force equal to $2 \times 10^{-7}$ newtons per unit length*'.

> However… thanks to the development of new technology, allowing scientists to count the movement of *individual* elementary charges (electrons), from May 2019 this definition is likely to be redefined along the following lines:
>
> The **ampere** (symbol: **A**) is defined as '*a current in the direction of flow of $1/(1.602\ 176\ 620\ 8 \times 10^{-19})$ elementary charges per second*'.
>
> The approved wording of this definition was not available when this book was published. Visit my blog site (www.professorelectron.com) to read about the progress of this proposal.

### coulomb (symbol: C)

The **coulomb** is defined as '*the charge transported through any cross-section of a conductor in one second by a constant current of one ampere*'.

### volt (symbol: V)

The **volt** is defined as '*the potential difference between two points such that the energy used in conveying a charge of one coulomb from one point to the other is one joule*'.

### joule (symbol: J)

The **joule** is defined as '*the work done when the point of application of a force of one newton is displaced one metre in the direction of that force*'.

### ohm (symbol: Ω)

The **ohm** is defined as '*the electrical resistance between two points of a conductor, such that when a constant potential difference of one volt is applied between those points, a current of one ampere results*'.

### newton (symbol: N)

The **newton** is defined as '*the force which, when applied to a mass of one kilogram, will give it an acceleration of one metre per second per second*'.

### watt (symbol: W)

The **watt** is defined as '*the power resulting when one joule of energy is dissipated in one second*'.

### farad (symbol: F)

The **farad** is defined as '*the capacitance of a capacitor, between the plates of which there appears a difference in potential of one volt, when it is charged to 1 coulomb.*

### weber* (symbol: Wb)

The weber is defined as '*the magnetic flux that, linking a circuit of one turn, produces a potential difference of one volt when it is reduced to zero at a uniform rate in one second*'.
(*pronounced 'vay-ber')

### tesla (symbol: T)

The **tesla** is defined as '*one weber of magnetic flux per square metre of circuit area*'.

### henry (symbol: H)

The **henry** is defined as '*the self- or mutual-inductance of a closed loop if a current of one ampere gives rise to a magnetic flux of one weber*'.

## Misconceptions

**The metric system and the SI system are the same thing**

The metric system existed long before the introduction of SI. Former versions of the metric system include the 'cgsA system' (whose base units included the centimetre, gram, second, and ampere), and the 'mksA system', which shared the same base units as SI.

### Celsius and litre are SI units

Celsius and litre are certainly 'metric' units and in common use, but they are *not* SI units. They are classified as units that 'may be used alongside' the SI system. The SI unit for temperature is the kelvin, and the SI unit for volume is the cubic metre.

### The centimetre is an SI unit

Yes, but for engineering purposes, the SI system recommends prefixes based on powers of ten raised to multiples of 3. These include: micro ($10^{-6}$), milli ($10^{-3}$), kilo ($10^{3}$), and mega ($10^{6}$). In SI, other prefixes are considered to be 'non-preferred'. So, for example, we should avoid using centimetres, and use millimetres and metres instead.

### The metre, kilogram, second, and ampere are 'fundamental' SI units

SI doesn't use the expression 'fundamental units' to describe these units. The correct expression is 'base units'.

## Review your learning

Now that we've completed this chapter, we need to examine the **objectives** listed at its start. Placing *'Can I...'* at the beginning, and a question mark at the end, of each objective turns that objective into a **test item**. If we can answer each of those test items, then we've met the objectives of this chapter.

# Chapter 2

# Electrical schematic and wiring diagrams

On completion of this chapter, you should be able to

1 explain the main differences between an electric **schematic diagram** and a corresponding **wiring diagram**.
2 recognise a schematic diagram.
3 identify common IEC circuit symbols.
4 recognise a wiring diagram.
5 recognise a block diagram.
6 recognise a single-line diagram.

## Introduction

Those of us who work in the electrotechnology industry, or who are keen electronics hobbyists, need to be able *to draw* and *to interpret* **electrical diagrams**.

So, in this chapter, we'll look at two different types of electrical diagrams, and learn about their functions.

At one time, all technical drawings, including electrical diagrams, were widely known as '**blueprints**', and this term is still widely used – particularly in North America – although the reproduction process to which it refers has long been replaced. The term 'blueprint' derives from a process used in the past for contact printing an original line drawing, created on tracing paper, on to ultra-violet light-sensitive paper. When chemically processed, the result is a reproduction of the original image, as white lines, on a dark-blue background.

If we ignore architectural electrical plans, which are beyond the scope of this book, most electrical diagrams fall into *two* broad categories: 'schematic diagrams' and 'wiring diagrams', as summarised on the next page.

An Introduction to Electrical Science, Waygood, ISBN 9780815391821, 2019. © Taylor & Francis

## Electrical Diagrams

### Schematic Diagrams

These drawings
- emphasise the *operation* of a circuit.
- use horizontal and vertical lines to show system flow.
- use symbols that indicate the function of components, but which do not necessarily look like those components.
- use a simple layout which does not necessarily represent the physical location of each component.

Figure 2.1 shows a simple **schematic diagram** for a lighting circuit.

Figure 2.1

### Wiring Diagrams

These drawings
- emphasise the *physical connections* between circuit components.
- use horizontal and vertical lines to represent the conductors.
- use simple pictorials that resemble the components.
- show the approximate physical placement of circuit components.

Figure 2.2 shows a simple **wiring diagram** for the same lighting circuit

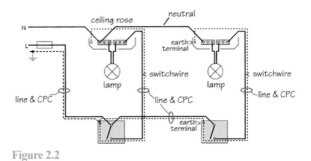

Figure 2.2

So, **schematic diagrams** are made as simple as possible and are used whenever we want to ***design** a circuit* or whenever we want to ***understand**, or to **explain** to someone how a circuit works*.

Whenever we sketch a circuit on a handy scrap of paper in order to explain to someone how a circuit works, we are likely drawing a schematic diagram.

**Wiring diagrams**, on the other hand, tend to be rather more complicated and are used *not* to show how a circuit works, but *how a circuit's components are physically **connected** to each other*.

## Schematic diagrams

So, let's start by describing why we use **schematic diagrams**.

We use a **schematic diagram** to help us *design a circuit* or whenever we want to *explain* to someone *how a circuit works*. By studying the schematic diagram of a circuit, anyone with the appropriate training should be able to *understand how that circuit works*.

Figure 2.3 is an example of a schematic diagram. It illustrates a basic rectifier (a.c. to d.c.) circuit for the current-measuring range of an analogue multimeter.

Figure 2.3

## Circuit symbols

Whenever we create a schematic diagram we should always use standard **circuit symbols** to represent either the *quantities* (e.g. resist*ance*, induct*ance*, capacit*ance*, etc.) or the circuit *components* (e.g. resist*ors*, induct*ors*, capacit*ors*, switches, power supplies, etc.) we are using in that diagram. You could say that circuit symbols are the *vocabulary* of a schematic diagram, so they must be standardised so that they can be readily recognised by any trained person studying that diagram.

So circuit symbols should comply with the relevant international standards. For example, in Europe, we use symbols which comply with standards approved by the **IEC (International Electrotechnical Commission)**. In North America, circuit symbols must comply with standards approved by **ANSI/NEMA (American National Standards Institute/National Electrical Manufacturers Association)**. Other countries may have their own standards but, in many cases, they are based on IEC or ANSI/NEMA standards.

In any event, it is very important that, whenever we construct a schematic diagram, we are *consistent* by using circuit symbols from *either* one standard *or* another, and that *we do not mix them*.

Figure 2.4 shows a selection of some of the most commonly used circuit symbols that accord with the relevant **IEC** standard.

| EXAMPLES OF IEC STANDARD CIRCUIT SYMBOLS | | | | |
|---|---|---|---|---|
| cell | battery | dc supply | ac supply | transformer |
| wire | wires crossing | wires connected | wires connected | wires joining |
| fuse | earth | chassis earth | make contact | break contact |
| resistor | resistor (non-preferred) | variable resistor | potenti-ometer | trimmer resistor |
| inductor | inductor (alternative) | variable inductor | trimmer inductor | d.c. motor |
| capacitor | variable capacitor | trimmer capacitor | electrolytic capacitor | a.c. motor |
| voltmeter | ammeter | ohmmeter | multi-meter | impedance |

Figure 2.4

Many of the more-complicated circuit symbols are actually created by simply *combining* related individual symbols. So, if we understand the functions of the *individual* symbols, then the function of any *combination* of those symbols can easily be worked out.

For example, Figure 2.5 shows the IEC symbols for a set of **normally open (NO) contacts** and a set of **normally closed (NC) contacts**.

Figure 2.5

Now, in Figure 2.6, let's look at the IEC symbols for various **operating devices**.

Figure 2.6

Finally, in Figure 2.7, let's see how these individual symbols can be *combined* to create symbols for various types of contacts according to how they are operated.

Figure 2.7

Incidentally, whenever we construct a schematic diagram, switch contacts are *always* shown in their 'normal' position. For example, **normally open contacts** are *always* drawn in their open position, and **normally closed contacts** are *always* drawn in their closed position. Similarly, relays (remote-controlled contacts) are *always* shown with their contacts in their de-energised positions.

## Basic rules for drawing schematic diagrams

So, whenever we draw a schematic diagram, we must always use approved symbols to represent *quantities* (e.g. resistance) or *components* (e.g. resistors) so that other people can understand *what* we are representing.

We should also

- never mix symbols from different standards (e.g. IEC or ANSI/NEMA).
- label each symbol, if required.
- place each symbol either horizontally or vertically, and connect them using horizontal or vertical lines

which represent conductors. Whenever possible, we should avoid using diagonal lines.

- use dots to show where the lines (conductors) are connected to each other. If wires cross, and there are no dots at the intersections, then the wires are *not* connected to each other.

Figure 2.8 shows a schematic diagram representing a **three-phase induction motor starter**. Anyone with the appropriate training will recognise that this circuit is designed to start and stop the motor by operating a set of heavy-duty, load-breaking, main contacts. He or she will also recognise that the circuit incorporates a protective feature (thermally operated overcurrent sensors) which will automatically disconnect the motor should it become too hot – if, for example, the motor should stall under prolonged excessive load.

In this particular example, to help clarify the schematic, (a) the *heavy lines* represent the main three-phase a.c. supply conductors to the motor, (b) the *thin lines* represent the control circuit conductors and (c) the *broken lines* link all those components that are *mechanically* connected together so that they all operate at the same time. The operating coil, labelled 'X', when energised, mechanically closes *all* those contacts which are also labelled 'X'.

Figure 2.8

The important thing to realise about this schematic diagram is that it is laid out in such a way as to make its *operation* clear to anyone 'reading' that diagram.

The *physical* locations of the various contacts, etc., in a 'real' motor starter bear absolutely no relationship whatsoever to where they are located in the schematic diagram. In fact, *everything* other than the **start** and **stop push button switches**, are contained within a sealed housing – completely out of sight and *only* accessible via the starter's *main terminals,* labelled '**A-B-C**', and the three *auxiliary terminals*, labelled '**1-2-3**'.

As we shall shortly see, the wiring diagram will show the installer how to connect the stop/start pushbuttons

to the motor starter; it provides no information on what components are inside the motor starter, how they are internally connected or how it works.

Don't worry if, at this stage, you are unable to understand how the schematic diagrams shown in this chapter work. As you gain experience, this ability will come to you.

Remember, the purpose of this chapter is *not* to teach us *how* these circuits work, but simply to help us *recognise the type of electrical diagram when we see one*. The majority of diagrams used throughout this book are schematic diagrams.

## Ladder diagrams

With the introduction of the **programmable logic controller** (a type of industrial digital computer designed for controlling manufacturing processes) in the 1960s, a type of visual computer programming, termed '**ladder logic**', was developed which has led to the increasing popularity of a type of simplified schematic diagram called a 'ladder diagram'.

A **ladder diagram** is simply a type of schematic diagram in which the voltage supply is represented by two vertical lines, between which the circuit components are connected rather like the 'rungs' of a ladder – hence the name. We can think of the vertical lines as being the positive and negative (or, in the case of an a.c. circuit, 'line' and 'neutral') conductors, termed 'busbars', which supply current to the 'rungs' of the circuit. Figure 2.9 shows exactly the same circuit represented as a traditional schematic diagram (left) and re-drawn as a ladder diagram (right).

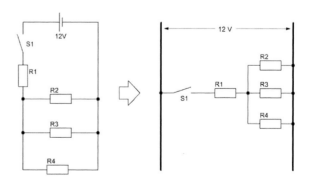

Figure 2.9

Normally, for a d.c. ladder diagram, the positive vertical line is to the left, and the negative vertical line is to the right. For an a.c. ladder diagram, the energised ('hot') line is to the left and the neutral is to the right.

Ladder diagrams are normally 'read' from left-to-right (so fuses or switches should appear to the left of any particular 'rung'), and the various components are simply placed according to neatness and how they are electrically connected to each other, *not* how they are physically located relative to each other.

## Wiring diagrams

Let's move on now, to examine wiring diagrams.

A **wiring diagram** is a somewhat-simplified representation of *how a circuit actually looks*. All the various components are generally represented as *simplified pictorials* of those components, located on the diagram roughly where the *real* components (such as terminal blocks, switches, lamps, etc.) would actually be relative to each other, but usually much closer together, and the diagram shows all terminal markings, together with the conductor colours, etc. Indeed, it shows *any* information that might be helpful to whoever is going to *connect* the circuit together.

Wiring diagrams can be relatively simple or very complicated indeed but, in either case, they are *not intended to help us to understand how a circuit actually works*; that is *not* their primary purpose. In fact, many of the individual components shown in a *schematic* diagram do not necessarily appear in the corresponding *wiring* diagram – as can be seen by comparing Figure 2.7 with Figure 2.4. Wiring diagrams are simply intended to *show which wire goes where* and, so, are intended to help a tradesman or technician *install* or *'wire'* the circuit *not* tell him how it works!

Let's now look at the wiring diagram for the same **three-phase motor starter** whose schematic diagram is shown in Figure 2.8.

First of all, let's look at what an *actual* **three-phase motor starter** looks like.

Figure 2.10

As can be seen in Figure 2.10, the relay coil, together with its main and auxiliary contacts, etc., are *all* enclosed inside a sealed plastic housing and completely

inaccessible to whoever has to install the device. These components have been pre-wired in the factory, so all the installer has to do is to connect the **stop/start station** (i.e. the 'stop' and 'start' pushbutton contacts) to the **auxiliary**, or **control**, **terminals** marked 1, 2 and 3. He also has to connect the (top) main **supply terminals** to the three-phase a.c. supply, and the (bottom) main **motor terminals** to the motor itself.

Now, let's look at the wiring diagram for the motor-starter (Figure 2.11).

Figure 2.11

In this particular example, the wiring diagram happens to be *far* simpler than the corresponding schematic diagram shown in Figure 2.8, but it tells us absolutely *nothing* whatsoever about how the circuit works – indeed, it *can't* because most of the components aren't shown because they're out of sight *inside* the motor starter's housing! The installer only has access to the main terminals marked 'A-B-C' and the auxiliary (control) terminals, marked '1-2-3', and the wiring diagram then tells the installer which wire goes to which terminal. What's *inside* the motor controller is irrelevant, because it's pre-wired and inaccessible.

Although the stop/start pushbutton station is drawn adjacent to the motor starter, in reality it could be several metres away or in a completely different room from the motor starter itself.

In the second example, shown in Figure 2.12 (a central-heating boiler control circuit), the pictorials of the various components (terminal blocks, switches, etc.) are located roughly where the *actual* components are located, relative to each other.

Figure 2.12

When placed next to an actual circuit, the similarity between its wiring diagram and the actual circuit should be immediately apparent to anyone comparing the two.

The terminal block pictorials shown in the examples are labelled in *exactly* the same way and sequence as those engraved on the *actual* terminal blocks, and the various lines are labelled (if it is important) to specify the colours of the actual conductors.

So, with the aid of a wiring diagram, the installer knows *exactly how to connect the various components together*, and *which colour conductors to use*.

## Other types of electrical diagram

Other types of electrical diagrams include **block diagrams** and **single-wire diagrams**.

**Block diagrams**, as the name suggests, use labelled **rectangles** to represent the principal parts of a complex circuit. Each block could represent a straightforward electrical device or it could represent a complex electrical or electronic circuit. Lines join the rectangles together to show the relationship between the individual blocks. Where relevant, signal flow is represented by arrows superimposed on the lines.

Figure 2.13 is an example of a block diagram for an automatic voltage regulator for an alternator (a.c. generator) set. Each block represents a separate circuit and the arrowed lines represent the signal path directions between each of the circuits. In this particular diagram, the black and white line represents a rotating *mechanical shaft* which couples the main and pilot exciters (small generators) to the main a.c. generator.

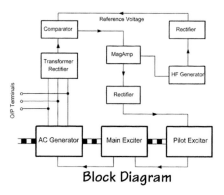

Block Diagram

Figure 2.13

Electricity transmission and distribution systems are 'three-phase' systems (we will learn more about these later in this book), meaning that they transfer energy along **three** *separate* conductors or lines. This

makes any attempt to represent such a system (and in particular its substations) as a drawing particularly complicated and confusing. To simplify things, we often *combine* the three lines into a single line, and the result is a 'single-line' diagram.

So, **single-line diagrams**, or **one-line diagrams** as they are also known, are widely used in electrical power engineering for representing simplified three-phase a.c. transmission and distribution systems. Components, such as alternators, transformers, circuit breakers, etc., are shown using standard symbols, but they are connected together using a single line, rather than drawing each of the three line conductors which are actually used.

None of the elements of a single-line diagram are drawn to scale, or are necessarily located where the actual elements are physically located.

Figure 2.14 shows a single-line diagram showing a 33-kV primary substation supplying an 11-kV ring system which, in turn, supplies a number of separate 400/230-V distribution systems. This diagram would be far more cluttered if all three line-conductors were shown, instead of the single line which represents them.

Single-Line Diagram

Figure 2.14

Once again, remember, the purpose of this chapter is for you to be able to *recognise* and *understand the function of* these types of electrical diagram, and *not* to be able to interpret them. This ability will come later.

## Creating electrical drawings with CAD applications

Although electrical drawings can be produced freehand, or with the aid of plastic stencils, for those of us who are expected to create schematic or wiring diagrams on a regular basis, it is always worth considering purchasing an **electrical CAD (Computer-Aided Draughting) software application**. There are many such applications available, *any* of which will help us create professional-looking electrical drawings quickly and easily. Some are very expensive and aimed at the professional user whereas others are quite inexpensive and aimed at the casual user or hobbyist.

Professional applications tend to have a rather steep learning curve and are primarily aimed at trained electrical draughtsmen or designers, whereas those aimed at the casual user can be exceptionally easy and intuitive to use, yet can produce a final drawing that can be virtually indistinguishable from that produced by a professional application.

For casual users who need to print diagrams up to A3 size, regular **ink-jet printers** are more than adequate, but for professional users who may want to produce high-quality diagrams larger than A3, specialist plotters are necessary. A **plotter** is, essentially, a very large printer which usually uses a pen, pencil or other writing tool to draw line diagrams on to paper. For casual users who occasionally need to produce diagrams beyond A3 size, most high-street print shops offer the use of professional plotters.

Before purchasing an electrical CAD package, it's important to confirm whether the built-in **symbol sets** which it provides meet our needs. For example, some electrical CAD packages may provide *either* IEC *or* ANSI/NEMA symbol sets, whereas others provide *both* symbol sets, which the user can choose from. In some cases, additional symbol sets may have to be purchased separately from that provided with the initial purchase.

The schematic diagrams used throughout this book, together with its companion book *Electrical Science for Technicians*, were created using an inexpensive and easy-to-use application which is aimed at European users, and provides only an IEC symbol set, but provides the facility to create and store one's own symbols.

As can be seen in Figure 2.15, numerous electrical symbols are provided in a floating 'symbol pallet'. The symbols are grouped under various categories (the 'resistors' group of symbols is shown) and they can be simply 'drag-and-dropped' from this pallet into the main working area, or 'layout window', where they can then be rearranged, snap-aligned, duplicated, labelled and connected together using the line-drawing tool. Features such as automatic numbering of symbols can be applied or switched off as convenient, and resistance, capacitance or inductance values can be added.

symbol pallet
(drag and drop)

Electrical CAD Screen

Figure 2.15

Many electrical CAD applications will also automatically compile a list of components, which is constantly updated as the drawing progresses. This feature is particularly useful as it provides us with a ready-made 'shopping list' of components, should we wish to go ahead and construct a 'real' circuit based on our design.

## Review your learning

Now that we've completed this chapter, we need to examine the **objectives** listed at its start. Placing *'Can I...'* at the beginning, and a question mark at the end, of each objective turns that objective into a **test item**. If we can answer each of those test items, then we've met the objectives of this chapter.

# Chapter 3

# The 'electron theory' of electricity

## Objectives

On completion of this chapter, you should be able to

1 explain the difference between
   a elements
   b compounds.
2 describe the structure of Bohr's model of an atom.
3 list the electric charges associated with
   a protons
   b electrons.
4 state the laws of attraction and repulsion between electric charges.
5 explain the term 'free electron'.
6 explain the significance of an atom's valence shell.
7 explain what is meant by 'ionisation', and 'positive' and 'negative ions'.
8 explain the difference between conductors and insulators.
9 describe an electric current in metals, in terms of the flow of free electrons.
10 describe how current flow in non-metallic conductors differs from that in metallic conductors.
11 specify the direction of
    a electron flow current
    b conventional current.
12 explain the terms 'higher potential' and 'lower potential' as they relate to conventional current.
13 list examples of, and describe applications for
    a practical conductors
    b practical insulators.
14 explain what is meant by the term 'displacement current'.

## Introduction

Mankind has known about the existence of **electricity** for thousands of years but, for most of that time, electricity was regarded as either something to be feared (e.g. lightning storms), or simply as a curiosity – with its various phenomena regarded as little more use than to provide entertainment for the curious.

However, the term, 'electricity' has been, and continues to be, misused by the layman and in press reports, leading to a great deal of confusion.

How often, for example, have we read a newspaper report claiming that someone died when '*electricity surged through the victim's body...*'? The term 'electricity', however, has no real meaning: it is not a measurable quantity, such as 'charge', or 'current', or 'energy' – or anything else for that matter!

So, what exactly is it that 'surged through the victim's body'? Is it 'charge'? Or is it 'current'? Or is it 'energy'? Or is it something else again?

Well, as we shall learn in this and the following chapters, an electric current is the quantity of electric charge transported per unit time through a conductor, and the rate at which this charge actually moves is measured in just millimetres per hour. In the case of alternating current, the charge doesn't flow at all; it merely oscillates back and forth within the conductor. Yet the transfer of energy along a conductor is close to the speed of light! So, clearly, 'electricity' cannot be *all* of these things!

So what exactly *is* 'electricity'?

Well, the answer is simple. 'Electricity' is simply the name we give to *a branch of science*. Just like

An Introduction to Electrical Science, Waygood, ISBN 9780815391821, 2019. © Taylor & Francis

'chemistry', or 'thermodynamics', or 'biology', or 'astronomy'.

> We should understand that, these days, the term **'electricity'** is used to describe a **branch of science** in the same way as, for example, we use the word 'chemistry', and it is wrong to use it as though it were a measurable quantity, such as charge or current or energy. It is quite meaningless, these days, to ask, for example, *'How much "electricity" does a residence consume?'* Having said that, until well into the 1960s, it was common for textbooks to describe a *'quantity of electricity'* to mean what we now call a *'quantity of electric charge'*.

By the nineteenth century, despite not really understanding the true nature of 'electricity', scientists such as **Michael Faraday** in Britain, **Joseph Henry** in the United States, **Georg Ohm** in Germany, and many others, were establishing 'rules' regarding its *behaviour*, based on the results of their practical experiments.

However, it wasn't until the gradual accumulation of knowledge on the **structure of the atom**, from the late nineteenth century onwards, that the secrets of 'electricity' finally began to be unravelled.

The problem with **atoms**, of course, is that we can't 'see' them! And without being able to see them, it's difficult to visualise what they are and how they behave. For something to be visible, it must be capable of reflecting light, and atoms are very much smaller than the wavelength of light and, so, are quite incapable of reflection.

So if we can't *see* atoms, how, then, can we possibly know what they 'look' like and understand how they behave?

The answer, of course, is that we have no idea of what they look like and we still don't fully understand how they behave!

In the much-acclaimed BBC television documentary series *The Ascent of Man*, the presenter, physicist Dr **Jacob Bronowski**, summed up this situation by saying,

> *When it comes to atoms, language is not for describing facts, but for creating images. What lies below the visible world is always imaginary; there is no other way to talk about the invisible . . .*

He then continued,

> *When we step through the gateway of the atom, we are in a world which our senses cannot experience. Things are put together in a way we cannot know; we (can) only try to picture it by analogy.*

What Dr Bronowski was saying is that it is *impossible* to describe the structure and behaviour of an atom *as it actually is*, because it is simply *beyond our understanding*; all we can do is try to create an *anology* (a 'likeness'), by comparing it with something that we *can* describe and understand. We call this analogy a '**model**'.

> A '**model**' provides us with a way of imagining what an atom *might* look like, based upon its apparent behaviour and, from that behaviour, predicting how it would behave under different circumstances. A model doesn't necessarily represent the way an atom *really* is, because it is invisible and its actual behaviour is probably beyond our understanding anyway.

One of the earliest models of the atom resembled a mass, within which its various sub-atomic particles were randomly dispersed – rather like the ingredients of a plum pudding. But, over time, this particular model gradually changed as scientists from all over the world sought to refine that model in light of the results of their continuing experiments.

For example, in 1911, the New Zealander **Sir Ernest Rutherford** suggested that an atom consisted mainly of 'empty space', with a heavy 'nucleus', around which electrons moved in different orbits – similar to the way in which the planets circle the Sun (Figure 3.1).

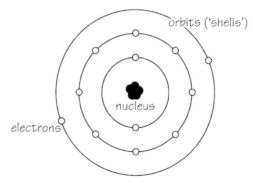

Figure 3.1

However, this 'planetary' model didn't quite account for the way in which the electrons appeared to behave. The planets in our solar system are continuously losing the energy they need to maintain themselves in their orbits.

This causes those orbits to get a little smaller over time – eventually, trillions of years into the future (hopefully!), they will spiral into the Sun and become destroyed! But this *didn't* seem to be the case with electrons; their orbits weren't getting any smaller, which suggested that electrons simply couldn't be 'gradually' losing energy!

Scientists were intrigued about what prevented an electron from gradually losing its energy and eventually spiralling into its nucleus. The eventual answer to this riddle resulted in a completely new way at looking at physics: something called 'Quantum Mechanics', the brainchild of the German physicist, Max Planck (1858–1947).

Amongst other things, Quantum Mechanics explains that an electron's energy can only change in *discrete* (*distinct*) *amounts*. These discrete 'packets' of energy are called '**quanta**'.

The radius of an electron's orbit (called a 'shell') depends on the energy level of that electron; the higher its energy level, the greater the radius of its orbit. If its level of energy cannot change gradually, then neither can the radius of its orbit. So, if the energy levels of electrons can only exist in discrete amounts ('quanta'), then their orbits can only exist at discrete distances from an atom's nucleus. So Quantum Mechanics explains that electrons cannot exist *between* these orbits any more than cars can park between the floors of a multi-storey car park!

To move from an inner orbit or shell, to an outer one, an electron must gain a quantum (discrete amount) of energy, as illustrated in Figure 3.2. Similarly, to move from an outer shell to an inner shell, an electron *must* lose that quantum of energy. So electrons *cannot* 'gradually' gain or lose energy in order to change their orbits.

**Figure 3.2**

In 1913 the Danish physicist **Niels Bohr** (who was to go on to play a major part in developing the atomic bombs which brought about the surrender of Japan in 1945) combined Rutherford's model of the atom with Planck's theory of 'quanta', and published his scientific paper 'On the Constitution of Atoms and Molecules', in which he proposed what is considered to be the basis of the model of the atom that we will now examine.

## Bohr's model of the atom

Bohr's model of the atom consists of a **nucleus**, surrounded by up to seven orbits called **electron shells** (labelled *k*, *l*, *m*, *n*, *o*, *p* and *q*). The electron capacity of each shell being determined by the expression, **2$n^2$** – where *n* represents the sequence of the shell, counting outwards from the nucleus. So, the innermost ('*k*') shell has a capacity of 2, the next shell has a capacity of 8, the next a capacity of 18, and so on.

> An atom's electron shells begin with a '*k*' because the originator of this system of labelling, Charles Barkla, believed that there were undiscovered shells closer to the nucleus, and wanted to be able to maintain the sequence once they had been discovered. It seems that he was wrong!

Bohr's model of the atom, of course, had its limitations, which prevented it from explaining later discovered behaviours, and other physicists have continued to develop that model further. In 1932, for example, James Chadwick proposed that the atom's nucleus was *not* a single heavy mass but, instead, made up of *two* types of 'elementary particle': positively charged **protons**, and **neutrons** which carry no charge at all. And it was the number of *protons* within the nucleus that determined all of nature's elements: hydrogen with a single proton, helium with two, and so on, up to uranium with 92 protons. Since those days, a number of man-made elements have also been created, extending the number of different types of atom to well over 100.

But there remained many questions that still needed answering. And, today, physicists working for organisations such as **CERN** (the **European Organisation for Nuclear Research**) are continuing to unravel the true nature of the atom, and have either discovered, or believe in the existence of, new sub-atomic particles that contribute to explaining its mysteries; these include many with strange names such as '*up quarks*', '*down quarks*', '*neutrinos*', '*pions*', '*muons*', '*kaons*', etc. It's hard to keep up!

In July 2012, CERN scientists claimed what is considered to be the biggest-ever breakthrough in research into the atom: apparent confirmation of the existence of what had been dubbed the 'God particle' (more accurately, the 'Higgs boson'). This sub-atomic particle, the existence of which had been long predicted by a Scottish physicist, Peter Higgs, is believed to be responsible for giving matter its mass – it ultimately being what holds the universe together!

Today's scientists no longer think of the atom in terms of Bohr's model – the present model is still evolving, but bears very little resemblance to his model of electrons travelling around a nucleus!

However, *we* are not physicists, and Bohr's model of the atom is perfectly adequate to help us understand the nature and behaviour of electricity. So, for the remainder of this chapter, we will concentrate on learning a little more about Bohr's model, and how some atoms provide the charge carriers necessary for the phenomenon we call an **electric current**.

## Electrons and electricity

*Anything* that has mass and occupies volume, we call 'matter'. And matter is made up of tiny particles, called **atoms**.

Matter that consists *entirely of identical atoms* is termed an **element**, and there are 92 naturally occurring elements, together with others that are man-made. Examples of elements include hydrogen, helium, oxygen, carbon, copper, uranium, etc.

Matter that consists of *a combination of different atoms* is termed a **compound**, and these combinations of atoms we call **molecules**.

Water, for example, is a compound made up of molecules, each of which comprises two atoms of hydrogen (symbol: H), together with one atom of oxygen (symbol: O) – hence its chemical symbol: '$H_2O$'.

The structure of a molecule determines whether an element or a compound is a solid, a liquid or a gas. With *solids*, molecules form rigid, crystal-like, structures. With *liquids*, the molecules are not bound together as rigidly, resulting in the fluid nature of a liquid. And with *gases*, the molecules drift apart from each other,

and disperse to fill their container. Some elements and compounds can exist as solids, liquids *and* gases (e.g. ice/water/steam), according to their temperature.

## The atom

As we have learnt, Bohr's model of the atom resembles a miniature solar system, with a **nucleus** (corresponding to our sun), surrounded by tiny particles called **electrons** (corresponding to the planets) travelling in different orbits called 'shells'.

Unlike the solar system, however, which is governed by *gravitational* forces, the behaviour of an atom is governed by *electric* forces: that is, by the attraction and repulsion between **electric charges**.

These electric forces are *enormously* greater than the gravitational forces which exist within an atom. For example, in a hydrogen atom the *electric* force between a proton and electron is $2.3 \times 10^{39}$ times as great as the corresponding *gravitational* force! Unfortunately, we cannot even begin to imagine such huge figures!

Electric charges have been arbitrarily assigned as being either **positive** or **negative**, and behave according to a universal law which states that

- *like* charges *repel* each other, while
- *unlike* charges *attract* each other.

Protons are *positively charged*, whereas electrons are *negatively charged*. Accordingly, electrons are held in their orbits by their attraction towards the protons within the nucleus, which is then exactly balanced by a centrifugal reaction which acts in the opposite direction.

Protons are fixed within their nucleus, and all elements are defined by their number of protons.

The simplest element, hydrogen, has just one proton in its nucleus; oxygen has eight; copper has twenty-nine; etc., as listed in the *Periodic Table of Elements*. The naturally occurring element with the greatest number of protons is *uranium* which has 92 protons in its nucleus.

So **electrons** are *negatively charged* particles which whizz around the nucleus within several fixed three-dimensional orbits, called 'shells'. As already explained, the energy level of each electron determines which shell it occupies. Electrons with the lowest energy level occupy the shell nearest to the nucleus, while electrons with the greater energy level occupy the shell further away from the nucleus.

Ernest Rutherford was absolutely right when he believed that most of the volume occupied by an atom was simply 'empty space'. To put this into perspective, the diameter of Bohr's atom is *at least* 100 000 times the diameter of its nucleus! So the atom in Figure 3.1 is *nothing* like its true scale; in fact, if we were to represent the nucleus by printing it as a 1-mm dot in the middle of this page, then the page itself would really need to be 100-m wide to represent the diameter of the atom!

To emphasise the amount of 'empty space' there is within an atom, if it was possible to remove all the empty space in the atoms that comprise, say, the Empire State Building, in New York, then its volume would probably be reduced to something the size of an orange pip – yet its mass would remain unchanged, at thousands of tonnes!

The amount of negative charge on *one* electron is identical to the amount of positive charge on *one* proton. Under normal circumstances, atoms are *neutral* and, so, for every *positively* charged proton, there must be a corresponding *negatively* charged electron (in other words, overall, *the two charges must cancel, or neutralise, each other*).

In Bohr's model of the atom, the most complex atoms have as many as *seven* shells. The outermost shell in any atom is called its **valence shell**, and it is *this shell that determines an atom's electrical (and chemical) properties*. So, for our purposes, we can now ignore the complex structure of the inner shells.

Regardless of its electron *capacity*, the valence shell can actually only *support* up to eight electrons. So, for example, if the valence shell of a particular atom has a 'capacity' of, say, 18, once it is *actually* occupied by eight electrons, no further electrons can be added. Instead, a *new* valence shell is formed and, once this new valence shell has acquired eight electrons, the previous valence shell will then be able to continue to build up its capacity to 18.

So the valence shell has a maximum capacity of just *eight* electrons. If the valence shell has less than four electrons, then that shell is considered to be *unstable* – by which we mean its electrons are loosely secured within that shell and can easily break away from the atom to become what are then termed 'free electrons'.

In Table 3.1, we list a number of **metallic elements**. For each element, if you examine the column for the outermost shell, you will find that it contains *less than four electrons*. Each of these metallic elements, therefore, can release electrons from their valence shells which then drift haphazardly from atom to atom within the element.

We call these haphazardly drifting electrons '**free electrons**'. Elements with large numbers of free electrons are mainly metallic elements, like those listed in Table 3.1, and are called **conductors**. As we shall learn, conductors provide the free electrons necessary to support electric current.

It would be a mistake to assume that, from Table 3.1, because aluminium has three valence electrons, then it must be a better conductor than, say, copper with just one valence electron. What matters is the *overall number of free electrons*, and this depends on the density of free electrons within the conductor. And copper has a greater density of free electrons than a corresponding volume of aluminium, making copper the better conductor. In fact, the best conductor is silver, followed very closely by copper.

Table 3.1

| Element | Symbol | Atomic number | Actual occupancies of shells (capacities shown in brackets) | | | | | | |
|---|---|---|---|---|---|---|---|---|---|
| | | | $k$ (2) | $l$ (8) | $m$ (18) | $n$ (32) | $o$ (50) | $p$ (72) | $q$ (98) |
| aluminium | Al | 13 | 2 | 8 | 3 | | | | |
| copper | Cu | 29 | 2 | 8 | 18 | 1 | | | |
| silver | Ag | 47 | 2 | 8 | 18 | 18 | 1 | | |
| mercury | Hg | 80 | 2 | 8 | 18 | 32 | 18 | 2 | |

If the valence shell contains *more than four electrons,* then that shell is said to be *stable* and its electrons are held tightly within that shell. The resulting scarcity (in relative terms!) of free electrons causes this type of element to behave as an **insulator**.

Although many *elements* behave as insulators, most *practical* insulators are actually manufactured from *compounds* such as plastics, rubber, glass, ceramics, etc.

## Ionisation and ions

When an atom's valence shell temporarily loses or gains an electron, it acquires an electric charge due to the imbalance between its number of electrons and the number of protons contained within its nucleus. We call a charged atom an **ion**, and the process of losing or gaining an electron is called **ionisation**.

If an otherwise neutral atom *loses* an electron (so there are now more protons than electrons), then it acquires an overall positive charge and, so, is called a **positive ion** and tends to attract a nearby free electron – thus becoming neutral again. So within *conductors,* both positive ions and free electrons have a very short lifespan! Immediately an electron breaks away from an atom to become a free electron, that atom becomes a positive ion and attracts a nearby free electron, thus becoming neutral once more.

If an otherwise neutral atom *gains* an electron (so there are more electrons than protons), then it acquires an overall negative charge, and is called a **negative ion** and tends to repel nearby free electrons.

## Electric current

**Practical conductors**, then, are mainly metallic elements with an abundance of free electrons which move extremely rapidly from one atom to another in *haphazard* and *random* directions. You can, if you like, imagine the free electrons forming a sort of negatively charged, rapidly vibrating 'cloud' that fills in the voids between fixed atoms.

In Figure 3.3, we see a length of metal conductor within which free electrons (represented by arrowed dots) move chaotically from atom to atom in random directions.

chaotic electron movement

Figure 3.3

Suppose we now apply an **external positive charge** to one end of this length of conductor, and an **external**

**negative charge** at the other end (don't worry, at this stage, *where* these external charges might come from or *how* they are produced; we'll deal with that later).

negative charge ——drift of electrons——▶ positive charge

Figure 3.4

Free electrons, being negatively charged, will be *repelled* by the external negative charge, and *attracted* towards the positive external charge. So, while the free electrons *still* continue to move frantically and haphazardly from atom to atom, *there will now be a general* **tendency** *for the free electrons to drift from the negative end of the conductor towards the positive end.*

The quantity of free electrons that drift through a metallic conductor, per unit time, is termed an **electric current**, and is expressed in **amperes** (symbol: A).

> In a metallic conductor, an electric current is the quantity of free electrons that drift through a metallic conductor, per unit time, from its negative end towards its positive end.

We can compare an *electric* current with a *river's* current. When we describe a river's 'current', we are describing *the rate of flow of the water* in that river. It's not the '*current*' that's flowing; it's the *water* that's flowing. In the same way, it's *not* the **electric current** that's flowing through a conductor, it's the *electric charges* that are flowing. We simply use 'current' to describe that flow.

So, it's technically incorrect, then, to say that 'a current *is flowing* through a conductor' — although, it has to be said, it is very common to hear it described in this way and, sometimes, it's difficult to avoid!

Figure 3.4 is actually rather misleading because it suggests that the electrons all move in a relatively orderly 'flow' along the entire length of the conductor when under the influence of the external charges. This *isn't* really the case! What *actually* happens is that the electrons continue with their frantic and chaotic movement, but there is a *very gradual tendency* for them to *drift* towards the positive end of the conductor.

This is illustrated in Figure 3.5. The solid line represents the typical chaotic movement of an individual

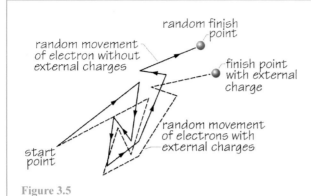

Figure 3.5

electron when there are *no* external charges applied to the conductor. The broken line, on the other hand, shows the same electron when external charges *are* applied. As you can see, it *still* follows a frantic and chaotic path, but *its finish point will be a little further towards the positive end of the conductor*. The new finish point is a mere fraction of the diameter of a single atom.

So, in *both* cases then, *electrons continue to move chaotically and at great speed **within** the conductor*. But, under the influence of external charges, there is a very *gradual drift along* the conductor.

So, while the *effect* of this drift of free electrons, due to their mutual repulsion (like charges repel), is practically *instantaneous* along the entire length of the conductor, the progress of *individual* free electrons along the conductor is *very, v-e-r-y*, slow. So slow, in fact, that *an individual free electron is unlikely to complete its journey through the length of a torch's (flashlight) lamp filament during the lifetime of that torch's battery!*

The average velocity at which free electrons drift past any given point within a conductor is known as their '***drift velocity***', and we will discuss this in more detail in the following chapter.

One of the questions that teachers often hear is, 'If the drift of charge carriers really is so very slow, then why does a lamp illuminate immediately its switch is operated? Why isn't there a delay between operating the switch and the lamp coming on?' Well, the answer is that free electrons are, of course, not waiting behind the switch, ready to pour into the circuit. Rather, they are permanently distributed throughout the entire length of the conductor (indeed, they were there before the raw copper was made into a conductor!) and, while their individual progress may be very slow, all those free electrons are moving at the same time. So the effect of the current is felt immediately around the entire circuit.

The easiest way of visualising this, is to compare this behaviour with that of a row of coupled railway goods wagons, as shown in Figure 3.6: a very small force applied to the wagon at one end of the row will cause a near-instantaneous movement of the wagon at the far end of the row, even though the individual wagons will have moved only a very short distance.

## Why conductors are neutral

It's very important to understand that, despite having enormous quantities of free electrons, *a conductor normally remains electrically* **neutral**. This is because, for *every* single electron (whether 'free' or orbiting within a shell) within the conductor, there is *always* a corresponding proton within the fixed atoms.

So what happens when a free electron exits the conductor at the end connected to an external positive charge? Well, in order to maintain the conductor's neutrality, another free electron is instantaneously drawn into the conductor from the end connected to the external negative charge.

a small movement here...          ...is felt immediately...          ...here.

Figure 3.6

## Non-metallic conductors

So far, we have only discussed conduction within *solid metallic conductors*.

*But not all conductors are metallic and solid!* Some conductors can be *liquids* or *gases*.

For example, some *liquids* are excellent conductors (conducting liquids are called 'electrolytes'). Unlike solids, the atoms in electrolytes are loosely bound together, and are themselves free to move around. So, current flow through electrolytes is *not* due to a drift of electrons, but due to a drift of *ions* (charged atoms) which takes place when the otherwise neutral atoms release free electrons into the external (metallic) circuit.

In electrolytes, a current is not only a drift of ions but, often, a drift of both positive *and* negative ions which move *in opposite directions to each other at the same time!* Despite this, it's conventional to assume the direction of 'current' within an electrolyte is the direction in which the positive ions move.

> In liquid conductors (electrolytes), electric current is a drift of ions (charged atoms). So, it is conventional to assume that the direction of this current is that in which positive ions would flow.

In most cases, of course, we are mainly concerned with current flow through *metal* conductors, so we will not spend any further time studying current flow in electrolytes in this particular chapter. Instead, we'll wait until we examine *cells and batteries*, in a later chapter.

However, *it is important to understand that current flow is* **not** *confined just to a flow of free electrons in metals, but can also be a flow of ions within electrolytes or within ionised gases.*

As electrons and ions are generically described as '**charge carriers**', it would be more accurate, therefore, to describe an electric current as *a drift of* **charge carriers**. This definition of current will then apply to solid, liquid or gaseous conductors.

> *An electric current is defined as 'the quantity of electric charge, transported per unit time'.*

Strictly speaking, this drift of charges is termed a '**conduction current**', because there is *another* type of current, termed a '**displacement current**', which we shall examine next.

The unit of measurement of electric current is the **ampere** (symbol: A), which is one of the seven SI base units, and one that we shall discuss in greater detail in the following chapter.

## Displacement current in insulators or dielectrics

As we have learnt, there are relatively few free electrons available as charge carriers in insulators so, for most practical purposes, within these materials the drift of free electrons is insignificantly small. We call this tiny drift of electrons a '**leakage current**'.

However, whenever we apply external charges across a sample of insulator, something very interesting happens to the fixed atoms. The atoms themselves cannot move but, under the influence of external charges, *the shape of the orbits of their electrons can!* So, as illustrated in Figure 3.7, the orbits of the electrons within a sample of insulation become distorted, or 'stretched', towards the positive external charge. We say that the atoms have become 'polarised'. The greater the difference between the external charges, the greater the amount of distortion.

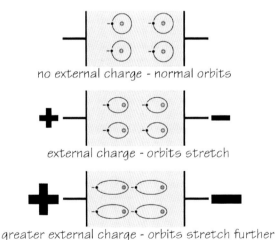

no external charge - normal orbits

external charge - orbits stretch

greater external charge - orbits stretch further

Figure 3.7

This 'stretching' of the electron orbits represents a *momentary* current, which we call a 'displacement current'. So, a displacement current *only* takes place *during* any change to the magnitude of the external charges. If an alternating voltage were applied across an insulating material, then it would be accompanied by a continuous displacement current that would vary in both magnitude and direction, just like that voltage.

A '**displacement current**', then, is associated with insulators, and is a 'momentary' current due to the distortion of the fixed atoms' electron orbits under the influence of external electric charges, whereas a '**conduction current**' is a current due to the movement of free charges (electrons, in the case of metal conductors) through a conductor, or as a tiny 'leakage current' through an insulator.

As we shall learn in a later chapter on *capacitors and capacitance*, displacement currents are very important when dealing with the insulators (or 'dielectrics') used, for example, in the manufacture of circuit devices called capacitors.

## 'Electron flow' versus 'conventional current' flow

By 'direction of current' in metallic conductors, we mean the direction in which current *passes through the load*. In other words, it is its direction through the external circuit, *never* within the source of potential difference.

During the eighteenth century, the great American scientist and statesman **Benjamin Franklin** (1706–1790), along with others, believed that an electric current was some sort of mysterious 'fluid' that flowed inside a conductor from a high-pressure area to low-pressure area. He naturally labelled high pressure as being 'positive' pressure, and low pressure as being 'negative' pressure and so he believed that the direction of an electric current was from '*positive to negative*' – i.e. in a direction *opposite* to that of the drift of free electrons!

Franklin's mistaken theory on current direction was, unfortunately, reinforced during the following century, as a result of experiments in electrolysis conducted by the English scientist **Michael Faraday** (1791–1867). Electrolysis is a method of depositing ('plating') metal on an electrode immersed in an electrolyte. Faraday noticed during his experiments that metal was removed from the positive electrode and deposited on the negative electrode, from which he, too, concluded that the current direction was from positive to negative, although he rejected Franklin's idea that it was a 'fluid', in favour of it being a 'field'.

Over the following years, various rules (e.g. to determine the direction of magnetic fields) were devised, based on the mistaken belief that current in metallic conductors flowed from positive to negative.

So, as strange as it might seem and despite today's knowledge about current in metal conductors being a flow of free electrons, Franklin's current direction *is still widely used as a convention in a great many textbooks*, and is known as '*Franklinian*' or, more commonly, **conventional flow** – as illustrated in Figure 3.8.

---

**Electron flow:**
A drift of free electrons, from *negative to positive*.
**Conventional flow:**
'Current' direction from *positive to negative*.

---

Figure 3.8

In accordance with Franklin's belief that his 'electric fluid' moved from a *higher* pressure to a *lower* pressure, whenever we use **conventional flow**, we say that the direction of current is *from* a **higher potential** *to* a **lower potential**. In other words, a *more-positive potential is considered to be a higher potential than a more-negative potential*.

As you will learn later in this book, knowing the direction of a current is *essential* to understanding the laws of electromagnetism and electromagnetic induction.

So, for consistency with most other textbooks, we have reluctantly adopted **conventional flow**, and *this is the direction that will be used generally throughout this book*.

Having said this, however, there will be occasions when using conventional flow will complicate our understanding of a topic so, on those occasions, we will revert to using electron flow and make it clear that we are doing so.

## Practical conductors and insulators

There is no such thing as a 'perfect' conductor or a 'perfect' insulator ('dielectric'). In fact, if we were to make a list of conductors and insulators, we could arrange the list in such a way that the best conductors (the 'worst insulators') appear towards the *top* of the list and the best insulators (the 'worst conductors') appear towards the *bottom* of that list.

> Even the best insulators contain impurities, and these impurities contribute a relatively small number of free electrons, which enable tiny currents to flow. Furthermore, thermal activity causes free electrons to be generated within most insulators.
>
> So, in practice, there is simply *no such thing as a 'perfect' insulator*, and the worst enemy of any insulation is high temperature. At high temperatures, even the best insulator will break down and conduct.

In practical terms, the very *best* metallic conductor is **silver** which, because of its cost, is only used for special applications such as relay contacts and printed circuit boards. The most *commonly used* conductors are **copper** and **aluminium**; copper comes a close second to silver and is a better conductor than aluminium but far more expensive – which is why (together with its lower mass) aluminium is preferred for high-voltage transmission and distribution line conductors.

Most practical insulators are compounds rather than elements and include toughened **glass** and **ceramics** (used to insulate overhead power lines), certain **gases** (used to insulate high-voltage busbars), **mineral oils** (used to insulate high-voltage transformers and circuit breakers), or **oil-impregnated paper** and **plastics** (used to insulate electric cables).

> Probably the most common insulator is dry air. Thousands of kilometres of overhead lines use bare conductors insulated simply by the surrounding air.

It's important to emphasise that an insulator doesn't 'oppose' an electric current. Rather, it simply *doesn't have sufficient charge carriers available to support an electric current.*

So, what do we mean by 'insufficient' charge carriers? There is a misconception by many students that insulators have 'very few' free electrons. In fact, *it's all relative* – for example, polystyrene is a very good insulator indeed, yet it contains around $60 \times 10^6$ electrons per cubic millimetre! Now, that might sound like an absolutely enormous

figure but, compared to copper, which contains around $85 \times 10^{18}$ electrons per cubic millimetre, that figure is actually very small indeed! To put it another way, copper has around $1.42 \times 10^{12}$ times as many free electrons, per cubic millimetre, as does polystyrene!

## Summary

In this chapter, we have examined Bohr's model of the atom and the part electrons play as charge carriers in metal conductors. In fact, we have gone into a little more detail than is really necessary to understand electric current. So, in this section, we are going to summarise the essential requirement for understanding this topic.

- The term **'electricity'** describes a branch of science, in just the same way as we use the terms 'chemistry', 'thermodynamics', etc.
- **Atoms** consist of a nucleus, containing positively charged **protons**, and **neutrons** that carry no charge; the nucleus is surrounded by negatively charged **electrons**, which travel around the nucleus in orbits called '**shells**'.
- The amount of positive charge on each proton is identical to the amount of negative charge on each electron. As atoms are normally neutral, it follows that **the number of electrons normally equals the number of protons**.
- If a normally neutral atom temporarily loses an electron, it acquires a positive charge and is called a **positive ion**. If an atom temporarily gains an electron, then it acquires a negative charge and is called a **negative ion**. The process of losing or gaining electrons is called **ionisation**.
- **Like charges repel** and **unlike charges attract**. So, electrons are held within their shells by their attraction to the positively charged protons within the nucleus.
- An **element** is made up of identical atoms determined by the number of protons that their nucleii contain. The simplest element is hydrogen, whose atoms contain just one proton. The most complex 'natural' element is uranium, whose atoms contain 92 protons.
- A **compound** is made up of two or more different elements, whose atoms combine to form molecules. An example of a compound is water, each molecule of which contains two atoms of hydrogen and one of oxygen ($H_2O$).
- Electron **shells** exist at fixed distances from the nucleus. An electron's energy level determines

which shell it will occupy. The further a shell is from the nucleus, the greater the energy level of those electrons that occupy that shell.

- As the energy level of each electron exists in discrete amounts, called 'quanta', electrons cannot exist between shells.
- Shells can only contain specific numbers of electrons – the innermost shell can only contain two electrons, the next eight, etc. (determined by the equation $2n^2$ – where $n$ represents the number of the shell, working from the inner to the outer).
- The outermost shell, called the **valence shell**, determines the electrical characteristic of the atom and *cannot contain more than eight electrons* (*regardless* of its $2n^2$ capacity).
- If the valence shell has less than four electrons, then those electrons are weakly held by their atom and can leave their shell to become **free electrons**.
- Materials with a large number of free electrons are called **conductors**.
- If the valence shell has more than four electrons, then those electrons are strongly held by their atom and *cannot* leave their shell to become **free electrons**.
- Materials with relatively few free electrons (compared to conductors) cannot support conduction and are called **insulators**.
- If an external negative charge is applied to one end of a metallic conductor, and an external positive charge to the other end, the free electrons within the conductor will be repelled by the external negative charge and attracted by the external positive charge and **drift** towards the positive end of the conductor.
- The velocity of electron drift within an insulator is **v-e-r-y** slow! But the *effect* of that current within the circuit is immediate.
- The quantity of electrons transported, per second, in a metallic conductor is termed an **electric current** or, more accurately, a 'conduction current'.
- Strictly speaking, **current** *doesn't* drift through a conductor, it's the charge carriers that drift! 'Current' simply describes the *rate* at which those charge carriers drift.
- In conducting liquids (electrolytes), an electric current is usually due to a drift of **charged atoms**, or **ions**.
- In general, then, an electric current is best defined as 'the quantity of electric charge transported, per unit time'.
- 'Displacement currents', as opposed to 'conduction currents', are associated with insulators or

dielectrics. These are momentary currents resulting from the temporary distortion of the electron orbits, under the influence of changes to external charges.

- Before the discovery of atoms, an electric current in metal conductors was thought to be a fluid that 'flowed' from a high-pressure (positive) area to a low-pressure (negative) area. This direction is still used in a great many textbooks and is called '**conventional flow**' to distinguish it from 'electron flow'.
- 'Conventional current' direction is from a 'higher (more *positive*) potential' to a 'lower (more *negative*) potential'.
- In common with other textbooks, this book will assume conventional flow, *except where indicated.*

## Misconceptions

### Atoms look like miniature solar systems

No one has ever seen an atom and no one ever will. Atoms are so complex that they cannot be described in terms that laymen can understand. Scientists are constantly learning new things about the way atoms behave and are discovering more and more new particles within the atom. Our concept of an atom resembling a tiny solar system is nothing more than a 'model' – in other words, we are trying to describe something we *can't* fully understand in terms of something we *can* understand.

### Electrons are tiny particles that orbit the atom's nucleus

Again, this is only a model to help us visualise what 'might' be going on inside the atom. In reality, this is unlikely to be the case. Electrons behave both as charged particles *and* as waves. Sometimes, scientists find it convenient to think of them as charged particles; at other times, they find it convenient to think of them as waves. In reality, they could be neither but something else completely!

### An electric current always describes the quantity of electrons transported per unit time

This is only true in the case of metallic conductors, such as copper and aluminium. This is not necessarily the case in semiconductors, liquids and gases! A far better definition of current is that it is the 'quantity of electric charge transported per unit time'.

### Charge carriers move at the speed of light

While the *effect* of a current may be detected more or less instantaneously, individual electrons drift along *very* slowly. Research suggests that an individual electron will not travel the length of a flashlight's filament within the lifetime of that flashlight's battery!

### Conductors have lots of free electrons, therefore they must be negatively charged

Although conductors do have large numbers of free electrons, for every free electron, there is a corresponding proton within the atoms or positive ions. So conductors don't have an overall charge; they are neutral.

### Insulators 'block' current

Insulators don't 'block' current; they simply don't have sufficient charge carriers to *support* a current.

### Insulators contain few free electrons

Insulators actually contain billions of free electrons per cubic millimetre but, compared to conductors, this figure is relatively small and certainly insufficient to support current flow.

### 'Conventional flow' is a flow of positive charges in the opposite direction to electrons

No. 'Conventional flow' isn't a flow of anything. It's simply a 'direction', mistakenly chosen, for current, from positive to negative.

## Review your learning

Now that we've completed this chapter, we need to examine the **objectives** listed at its start. Placing '*Can I...*' at the beginning, and a question mark at the end, of each objective turns that objective into a **test item**. If we can answer each of those test items, then we've met the objectives of this chapter.

# Chapter 4

# Electric current

## Objectives

On completion of this chapter, you should be able to

1  list the three effects of an electric current.
2  specify the SI unit of measurement of electric current.
3  specify which of the three effects is used to define the SI unit of electric current.
4  state the relationship between electric current and electric charge.
5  state the SI unit of measurement of electric charge.
6  solve simple problems on the relationship between electric current and electric charge.
7  describe the effect on the charge drift velocity of varying the current through, or varying the cross-sectional area of, a conductor.

## Measuring electric current

Earlier, we learnt that the general definition for **electric current** is '*the quantity of electric charge, transported per unit time*' (free electrons, in the case of a metal conductor). This current is driven by external negative and positive charges or, more accurately, by the **electric field** that exists between those external charges. We will learn more about electric fields, later, but, for now, we can imagine them as imaginary lines of force stretching between the external charges and along which the mobile charge carriers slowly drift.

Without the electric field, the movement of the electric charges would be completely random. But, under the influence of an electric field, there is a net drift of electric charge in a particular direction and this constitutes an electric current. Remember, the drift velocity of these charges is *very* low, in the order of millimetres per hour, but the effect is immediate throughout the circuit.

In the case of solid metal conductors, it's a drift of **free electrons**. In other materials, such a conducting liquids ('electrolytes') or ionised gas, it's a drift of **ions** (charged atoms): it can even be a drift of positive ions in one direction, and negative ions in the opposite direction *at the same time!*

There is nothing obvious to indicate the presence of an electric current in a metal conductor or in any other material; after all, we most certainly won't be able to *see* any charges! Even if it were possible to see these charges, then all we would see would be their chaotic movement; we certainly wouldn't be able to perceive any drift in a particular direction because, as we have learnt, this drift is far too slow and would be masked by the charges' random movements.

So the presence of a current can *only* be detected by observing one or more of the *three* **effects** produced by that current. These are the current's

- **heating** effect
- **chemical** effect
- **magnetic** effect.

**Heating effect** – an electric current causes a conductor's temperature to rise. We make practical use of this effect with incandescent lamps, electric heaters, etc. It is also responsible for the operation of fuses, which melt in response to rises in temperature due to excessively high currents. This effect is also responsible for wasteful energy *losses*, due to heat transfer away from a conductor into its surroundings.

**Chemical effect** – an electric current can be responsible for chemical reactions. This can be useful, and we make use of this effect in electrolysis (electro-plating) – but it may also be harmful, as it is also responsible for some types of corrosion.

**Magnetic effect** – an electric current produces a magnetic field, which surrounds the conductor through

An Introduction to Electrical Science, Waygood, ISBN 9780815391821, 2019. © Taylor & Francis

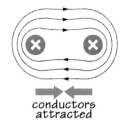

conductors
repelled

conductors
attracted

Figure 4.1

which that current passes. We make use of the forces resulting from the interactions between magnetic fields to drive motors, operate relays, etc.

## The unit of current: the ampere

The SI unit of measurement of electric current is the **ampere** (symbol: **A**), which, as we have learnt, is one of the seven 'base units' of the SI system.

The ampere could be defined in terms of *any* of the three effects described above. And, indeed, since 1947, the definition of the ampere is based on the **force** resulting from the *magnetic effect* of an electric current, as follows:

> The **ampere** (symbol: **A**) is defined as *'that constant current which, when maintained in two straight parallel conductors of infinite length and of negligible circular cross-sectional area, and placed one metre apart in a vacuum, would produce between them a force equal to $2 \times 10^{-7}$ newton per metre of length'.*

The 'force' referred to in this definition is due to the attraction or repulsion (depending in the relative current directions, as seen in Figure 4.1) between two parallel conductors, due to the interaction of the magnetic fields set up around the currents drifting through those conductors.

In Figure 4.1, the dot indicates conventional current drifting *towards* us and the cross represents conventional current drifting *away* from us, and the lines represent the shape of the resulting magnetic fields.

The reason *why* magnetic fields should produce such forces will be discussed in detail in the later chapter on *electromagnetism*. For now, however, we simply need to accept that a force is indeed set up by the interaction between magnetic fields, and it is the resulting strength of this force that allows us to define the ampere.

Unfortunately, the 70-year-old definition of the ampere, as described above, has always been hypothetical and *impossible to physically achieve in practice*. So, from May 2019, the ampere is scheduled to be redefined in terms of *the rate of flow of elementary charges* (i.e. electrons).

This new definition is achievable because, thanks to recent advances in technology, scientists are now able to *accurately count the passage of individual electrons*.

So, the ampere will be defined by taking the fixed numerical value of the elementary charge, *e*, to be 1.602 176 620 8 $\times$ 10$^{-19}$ when expressed in the unit, coulomb, which is equal to an ampere second, where the ampere is defined in terms of the new definition of the second.

At the time of writing this book (December 2017) the new definition of the ampere hasn't been finalised. However, it is likely to be along the following lines:

> The **ampere** (symbol: **A**) is defined as *'a current in the direction of flow of 1/(1.602 176 620 8 $\times$ 10-19) elementary charges per second'.*

To follow the progress of the proposed redefinition of the ampere, visit my blog site at www.professorelectron. com.

## Electric charge

So far, we have described charges as being either 'positive' or 'negative', without making any attempt to assign any values to them. So how *do* we measure electric charge?

Well, the smallest quantity of charge must be the amount of negative charge possessed by an *individual* electron which, of course, is *exactly* the same as the amount of positive charge possessed by an *individual* proton.

As the amount of charge on an individual electron is so incredibly tiny then, in much the same way as we measure sugar by its mass rather than by individual granules, we need to use a much larger and more practical unit for measuring electric charge.

As we have *already* defined the ampere, we have chosen to define electric charge in terms of *the amount of charge transported, per second, by a current of one ampere*, and the name we have given to this amount of charge is the **coulomb** (symbol: **C**), named in honour of the French academic, Charles de Coulomb (1736–1806).

> The **coulomb** (symbol: **C**), then, is defined as *'the quantity of electric charge transported, per second, by a current of one ampere'*.

This definition will remain the same after the redefinition of the ampere after May 2019. We mentioned earlier that the amount of charge on an individual electron is incredibly small; well, to put this in perspective, a coulomb currently equates to the amount of charge possessed by **6.241 509 129×10¹⁸ electrons!**

For most normal purposes, we generally round this off to **6.24×10¹⁸** electrons.

So, for a current of one ampere, that is the number of electrons which are being transported past a given point in that conductor every second!

If there are $6.24×10^{18}$ electrons per coulomb, then an individual electron must possess a charge of $160×10^{-21}$ C.

It's *not* necessary to memorise the definition of a coulomb but, simply, to appreciate that there *is* a relationship between electric charge and electric current. This relationship is written as follows:

$$Q = I\,t$$

where:

$Q$ = quantity of charge, in coulombs (symbol: C)

$I$ = electric current, in amperes (symbol: A)

$t$ = time, in seconds (symbol: s)

If we rearrange this equation, making current the subject, then:

$$I = \frac{Q}{t}$$

This equation confirms that an electric current is, indeed, the quantity of electric charge, transported per unit time.

> An **electric current** is defined as *the quantity of electric charge, transported per unit time'*.

Unfortunately, this has led to some publications to incorrectly 'defining' the *ampere* in terms of a 'coulomb per second'.

While it is certainly true that an ampere *corresponds* to a 'coulomb per second', this is *not* the present SI 'definition' of the ampere, as we *cannot* define a base unit (the ampere) in terms of a derived unit (the coulomb)!

The proposed *new* definition of the ampere will not change this situation, as it will be defined in terms of the rate of flow of a specified number of *electrons* (not 'coulombs'), while the coulomb itself will continue to be equivalent to an ampere second.

> It must be clearly understood that the **ampere** *cannot* be defined as a 'coulomb per second', because you *cannot* define an SI base unit (the ampere) in terms of a derived unit (the coulomb).

Unfortunately, this is precisely what some learned institutions have done in some of their learning materials! It is also very common for North American textbooks to 'define' the ampere as a 'coulomb per second'!

**Worked example 1**   What is the value of current if a charge of 25 mC drifts past a point in a circuit every 5 min?

**Solution**   **Important!** Don't forget, we must first convert millicoulombs (mC) to coulombs (C), and minutes (min) to seconds (s):

$$I = \frac{Q}{t}$$
$$= \frac{(25×10^{-3})}{(5×60)}$$
$$= 83.33×10^{-6}\,\text{A (Answer)}$$

**Worked example 2**   How many electrons will be transported past a given point in a circuit, when a current of 3 A drifts for 10 s?

**Solution**
$$Q = It$$
$$= 3×10$$
$$= 30\,\text{C}$$

Since 1 C = 6.24 ×10$^{18}$ electrons, then:

$$Q = 30 \times (6.24 \times 10^{18})$$
$$= 187.20 \times 10^{18} \text{ electrons (Answer)}$$

## Drift velocity of charge carriers

As we learnt in the previous chapter, the average velocity at which electrons move through a metal conductor is called their '**drift velocity**', and it is surprisingly slow.

Although we are *not* going to attempt to derive the equation here (and, indeed, we don't even need to memorise it), it can be shown that the drift velocity along a conductor is given by the following equation:

$$v = \frac{I}{n\,A\,e}$$

where:   $v$ = drift velocity (metres/second)
   $I$ = current (amperes)
   $n$ = number of electrons per cubic metre
   $A$ = cross-sectional area (square metres)
   $e$ = charge per electron (coulombs)

The number of electrons per cubic metre depends, of course, on the *type* of conductor and its *purity*. For copper, this value is generally taken as **85×10$^{27}$** and, for aluminium, **76.2×10$^{27}$**. The amount of charge on a single electron is generally taken as **16×10$^{-18}$ C** (coulombs).

So, let's calculate the drift velocity for a common **copper** conductor, say a 2.5-mm$^2$ conductor used for wiring ring mains in UK residences. Let's assume it's carrying a direct current of, say, 10 A.

$$v = \frac{I}{n\,A\,e} = \frac{10}{(85 \times 10^{27})\times(2.5 \times 10^{-6})\times(16 \times 10^{-18})}$$
$$= \frac{10}{3.4 \times 10^{6}} = 2.9 \times 10^{-6}\,\text{m/s}$$

That's an incredibly low 2.9 micrometres per second! Which means that it will take 344 828 seconds, or nearly **96 hours**, to travel just one metre! That's *slow!*

From the above equation, we can see that for any given conductor, the drift velocity is *directly proportional to the current, and inversely proportional to the conductor's cross-sectional area*. So, for any given current, the *lower* the cross-sectional area of a conductor, the *higher* the drift velocity. This shouldn't really come as any surprise, as it is also true for the velocity of water as it flows from a larger-diameter tube through a smaller-diameter tube ('venturi effect').

Since the kinetic energy of any object is proportional to the *square* of its velocity, it follows that the energy released when free electrons collide with the fixed atoms in a conductor will be greater when the drift velocity is greater. So, for a given current, the energy expended by a *thinner* wire will be greater than for a thicker wire. This expended energy manifests itself as *heat* and accounts for why the *temperature* of a thinner wire tends to be higher than for a thicker wire carrying the same amount of current.

An electrical **fuse** makes use of this principle. A fuse wire, or 'element', is (a) thinner than the wire which it is designed to protect, and (b) is made from a conductor having a lower melting point. So, when a sustained overcurrent occurs (due to a continuous overload or an electrical fault), the fuse element's temperature rises and reaches its (lower melting) point *faster* than the conductor which it protects.

## Summary

- An electric current is the quantity of electric charge, transported per unit time.
- The unit of electric current is the **ampere** (symbol: **A**).
- The ampere is currently defined in terms of the *force* between two, straight, parallel, current-carrying conductors due to the magnetic fields that surround those currents.
- Thanks to our ability to now count the movement of *individual* elementary charges (electrons), the definition of the ampere is scheduled to change from May 2019. The new definition will likely describe the ampere in terms of the movement of a *specific number of individual electrons* per unit time past a given point in a conductor.
- The amount of charge transported, per second, by a current of one ampere is one **coulomb** (symbol: **C**).
- The **coulomb** is equivalent to the amount of charge generally rounded off to 6.24×10$^{18}$ electrons. This figure will be defined to a greater level of accuracy once the ampere is redefined in 2019.

## Misconceptions

### The ampere is defined in terms of the rate of drift of electric charge

**Current** is defined as *the quantity of charge transported per unit time*. But, at present, its unit, the **ampere** is defined in terms of the force between current-carrying conductors. Proposed changes will likely mean that, from May 2019, the ampere will be redefined in terms of the rate of flow of elementary charges (individual electrons), *not* coulombs.

### The ampere describes the speed of an electric current

The 'speed' of an electric current has nothing to do with its unit of measurement. Electric charge drifts v-e-r-y slowly, regardless of the value of current.

## Review your learning

Now that we've completed this chapter, we need to examine the **objectives** listed at its start. Placing '*Can I...*' at the beginning, and a question mark at the end, of each objective turns that objective into a **test item**. If we can answer each of those test items, then we've met the objectives of this chapter.

### Online resources

The companion website to this book contains further resources relating to this chapter. The website can be accessed via the following link:
www.routledge.com/cw/waygood

# Chapter 5

# Potential and potential difference

An Introduction to Electrical Science, Waygood, ISBN 9780815391821, 2019. © Taylor & Francis

## Objectives

On completion of this chapter, you should be able to

1  explain the need for external negative and positive potentials to cause current in a conductor.
2  describe simple electric field patterns.
3  explain the terms, potential and potential difference in terms of charge movement within electric fields.
4  explain what is meant by 'charge separation'.
5  state the SI unit of measurement of potential difference.
6  briefly explain the differences between each of the following terms:
   a  potential difference
   b  voltage
   c  electromotive force
   d  potential.

## Introduction

In Chapter 3, we learnt that if we applied an external *negative* charge to one end of a *metal* conductor, and an external *positive* charge to the other end of that conductor, *a drift of free electrons* (an electric current) will take place through that conductor, as illustrated in Figure 5.1.

**Figure 5.1**

We describe the external negative charge as having a **negative 'potential'**, and the external positive charge as having a **positive 'potential'**. The difference between these potentials, therefore, is called a **'potential difference'**, and it is this which provides the *'driving force'* for a current.

So, for charge carriers to flow between two points,

1  we need a **conducting path** between those two points.
2  there must be a **potential difference** between the two points.

But what exactly do we mean by **'potential'** and **'potential difference'**, and where do they come from? In order to understand these terms, we need a basic understanding of **electric fields**.

## Electric fields

The area surrounding an electric charge, in which the effects of that charge may be observed, is termed an **'electric field'**.

An electric field is graphically represented using lines of force, called **'electric flux'**. It must be clearly understood that these electric flux lines are *imaginary* – in other words, the lines themselves don't actually exist, but are used simply to provide us with a 'model' (a visual representation) of an electric field in exactly the same way as *magnetic* lines of force are used to represent a magnetic field (as we shall learn later).

These flux lines emanate perpendicularly from an electric charge and spread out in all directions towards infinity, as illustrated in Figure 5.2, with individual flux lines repelling adjacent flux lines. Although they are represented in just two dimensions, it must be understood that they actually extend in all *three* dimensions.

It is conventional to allocate a *direction* to electric flux lines using arrow heads. Although it would *seem* logical to choose a direction in which a mobile *negative* charge would move if placed within the field, by common agreement this direction is determined by the direction in which an isolated, mobile, **positive** charge would move if placed within the field. As a mobile positive charge would be repelled by another positive charge (like charges repel) and attracted towards a negative charge (unlike charges attract), that direction is *along the lines of electric flux, towards a negative charge* – as illustrated by the arrow heads, for a single negative charge, in Figure 5.2.

This is another reason why 'conventional current' (positive to negative) remains in common use, as it is in accordance with the convention described above.

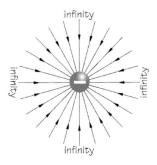

**Figure 5.2**

If point charges of *opposite* polarity are located near to each other, then these electric flux lines will link the

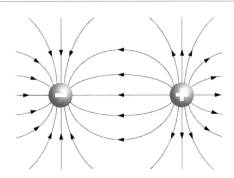

**Figure 5.3**

two charges as well as extending to infinity, as shown in Figure 5.3.

Imagine, now, moving a **free electron** from infinity, along one of these lines of electric flux, *towards a fixed negative charge*, as illustrated in Figure 5.4. As 'like poles repel', **work** must be done to overcome the force of repulsion due to the fixed negative charge, in order to move that electron the distance from infinity to point **B**. The work done in moving the electron results in an increase in that electron's **potential energy** or, simply, its '**potential**'.

The absolute potential at *any* point along a line of electric flux is defined in terms of the work done in moving a negative charge (in Figure 5.4, a single electron but, in practice, a negative charge equal to one coulomb) *from infinity to that particular point (point B)*.

Unfortunately, this is *not* a very practical definition for potential – after all, 'infinity' is hardly 'accessible'!

So, instead, we choose an *accessible*, but arbitrary, point of reference – such as point **A** in Figure 5.4 – and then find the work done in transporting the charge *from* that point *to* point **B**. In other words,

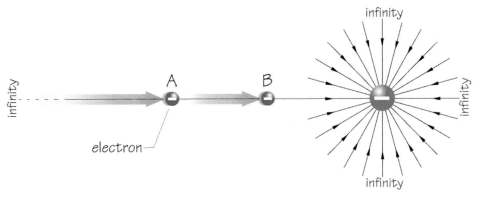

**Figure 5.4**

we determine the potential at point **B** *with respect to point A*.

Or, to put it another way, it allows us to determine the **potential difference** between points **A** and **B**.

> The **potential difference** between two points in an electric field is defined in terms of the *work done* in transporting *electric charge* between those two points.

The term '**voltage**' is synonymous with 'potential difference' (but *not* potential) – i.e. they both mean *exactly* the same thing! The symbol for potential difference or voltage is $E$, $U$ or $V$ – depending on context (more on this later).

We shall be examining the differences between potential and potential difference in more detail later in this chapter.

### The volt

We are now in a position to define the SI unit of both potential and potential difference, which is the **volt**, named in honour of the Italian physicist, Count Alessandro Volta (1745–1827).

> The **volt** (symbol: **V**) is defined as *'the potential difference between two points such that the energy used in conveying a charge of one coulomb from one point to the other is one joule'*.

Again, you do *not* have to memorise this definition, however we *will* need to refer back to it when we discuss energy, work and power, in a later chapter.

In practice, potential differences can vary enormously. For example, a simple AA disposable battery (or, more accurately, 'cell') will provide a potential difference of just 1.5 V, whereas electricity transmission voltages (in the UK) can be as high as 400 kV.

The preceding definition may be expressed in the form of an equation:

$$E = \frac{W}{Q}$$

where:

$E$ = potential difference, in volts (symbol: V)

$W$ = work, in joules (symbol: J)

$Q$ = electric charge, in coulombs (symbol: C)

**Worked example 1**  The work done by a generator in separating a charge of 20 C is 50 kJ. What is the resulting potential difference across its terminals?

**Solution**  **Important!** Don't forget, we must first convert the kilojoules to joules.

$$E = \frac{W}{Q}$$
$$= \frac{(50 \times 10^3)}{20}$$
$$= 2.5 \times 10^3$$
$$= 2500 \text{ V (Answer)}$$

## Creating a potential difference through charge separation

So we now know what we mean by a potential difference and how it is measured. But *how* do we obtain a potential difference in practice?

All materials, including conductors, are usually electrically **neutral** because, under normal circumstances, their atoms contain equal numbers of protons and electrons whose equal, but opposite, charges act to neutralise each other.

In order to acquire a charge, an object must either gain or lose electrons – thereby acquiring an excess or a deficiency of negative charge. For example, if an object has more electrons than protons, it is *negatively charged;* if it has more protons than electrons, then it is *positively charged.*

However, it's *not* necessary for two objects to be *literally* negatively and positively charged for a potential difference to exist between them. For example, if two objects are both negatively charged, but one is *less* negatively charged than the other, then a potential difference will appear between them also.

For example, if object **A** is *less* negative than object **B**, then we can say that object **A** is *'positive with respect to object B'*. Or, if you prefer, *'object B is 'negative with respect to object A'*.

In practice, *this is by far the most common situation we encounter in any circuit,* and is practically *always* the case for electrodes in cells and batteries: with the battery's so-called 'positive' electrode actually being negatively charged, but *less negatively charged* than (or 'positive with respect to') the 'negative' electrode.

The process by which this can be made to happen is called **'charge separation'**.

There are a great many ways in which charge separation can be achieved, so let's look a few of those methods.

## Frictional contact (triboelectricity)

Probably the very earliest-known method of charge separation was through **frictional contact**.

The ancient Greek philosopher, mathematician and astronomer, Thales of Miletus (circa 624–546 BC), recorded that whenever amber was rubbed with wool, the amber would acquire an electric charge (although, of course, he wouldn't have used that expression). It is believed that the ancient Greeks would amuse themselves by charging amber in this way in order to pick up pieces of paper. From this, we might conclude that life must have been particularly boring for an ancient Greek!

In fact, the word 'electricity' is derived from the Greek, '*electra*', meaning 'amber'.

Another Greek word, '*tribo*', meaning to 'rub', has given us the modern term, '**triboelectricity**', which describes the charge separation which occurs whenever one type of material is rubbed by another.

Actually, it's *not* the rubbing that's important; it's bringing the surfaces of two different materials into *contact* with each other. Rubbing the materials together merely brings them into more intimate and repeated contact, thus increasing the amount of charge separation.

Although the reason for this charge separation is not fully understood, it is clear that when different materials come into contact with each other, the surface of one material appears to 'steal' some electrons from the surface of the other. The material that *steals* electrons, therefore, acquires a *negative* charge while the material that *loses* those electrons acquires an equal *positive* charge.

The likely reason for this is that some ions (charged atoms) attract free electrons more strongly than others. So, if we bring two different materials into intimate contact with each other, by rubbing, then it's likely that more electrons will emigrate in one direction than in the opposite, and that the material gathering the greater number of electrons will, therefore, become more negative than the other.

Research into this phenomenon by a Swedish physicist, Johan Carl Wilcke (1732–1796), led him to publish what is known as the '**triboelectric series**': a list of materials in order of the magnitude and relative polarity of the charge they acquire when touched or rubbed by another material.

Materials towards the bottom of the series, when rubbed by materials towards the top of the series acquire a more-negative charge. And, the *further apart* materials are within the series, the *greater* will be the potential difference between their charges when they are brought together by rubbing.

An example of the triboelectric series is reproduced in Figure 5.5. It should be pointed out that different researchers often obtain somewhat different results when determining the position of a particular material within the series. This is due to the numerous factors which appear to affect a material's tendency to transfer charge.

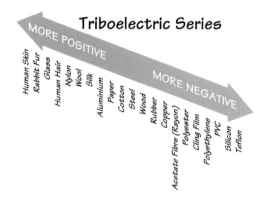

**Figure 5.5**

By examining the series, we see that if, for example, we rub our *hair* with, say, a cotton handkerchief, then our hair will acquire a more-positive charge while the *cotton* handkerchief will acquire a more-negative charge.

We experience a similar effect when we walk across a synthetic carpet when the air is particularly dry. The potential buildup on our body is sufficient to produce a painful shock when we then touch an earthed metallic body such as a radiator!

Although the voltages created through frictional contact can be very large (several thousands of volts), the amount of energy involved is tiny, so triboelectricity is *not* able to sustain currents for more than a few microseconds – certainly not long enough to cause us any harm, but possibly enough to damage sensitive electronic components (which is why we should always 'earth' or 'ground' ourselves before handling electronic circuit boards).

However, current research into triboelectricity is leading to some very interesting developments. For example, 'smart clothing' is a term now being used to describe flexible fabrics which generate voltages as they bend and flex when worn. With clothing that can generate voltages, it might be possible, for example, to recharge mobile telephone batteries on the move. Similarly, triboelectricity materials might, one day, be built into computer-tablet touchscreens, enabling the tablet's battery to be charged whenever the tablet is in use. These are just two examples of the current research into what are known as '**triboelectric nanogenerators**'.

## Thermoelectricity

The term, '**thermoelectricity**', describes the direct conversion of a temperature difference into a potential difference.

When two materials are in intimate contact with each other, there is a tendency for free electrons to diffuse in both directions across the junction from one material into the other. If the materials are *different*, then more free electrons cross the junction in one direction than in the opposite direction. The diffusion of electrons is short-lived but the resulting imbalance causes a small potential difference to appear across the junction.

If the temperature of the junction increases, the resulting higher energy level results in a somewhat greater imbalance, and the resulting potential difference will increase.

This '**thermoelectric effect**' was first observed by the German Physicist, Thomas Seebeck (1770–1831) in the early 1820s, and is known as the '**Seebeck Effect**'.

A simple device that utilises the Seebeck Effect is the '**thermocouple**'. A basic thermocouple consists of two wires, manufactured from different materials, connected together to form *junctions* at opposite ends. For the reasons described in the previous paragraphs, a contact potential difference appears across each of the junctions. If the junctions are at the *same* temperature, then these two potential-differences will be *equal*, but will act in opposition to each other, and there is no overall potential difference within the circuit. If, on the other hand, the temperature of one junction is *higher* or *lower* than the other, then there will be difference between the two junction potential differences which will cause a current around the circuit.

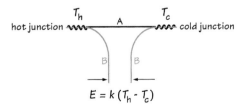

$$E = k\,(T_h - T_c)$$

Figure 5.6 shows a basic thermocouple, consisting of two dissimilar wires, A and B, twisted together at opposite ends. Wire B is broken and, when the junctions are

at different temperatures, a potential difference, $E$ (the difference between the potential differences at the junctions), appears across the break. Research has found the value of this potential *difference* is directly proportional to the temperature *difference* between the 'hot' and 'cold' junctions:

$$E \propto (T_h - T_c)$$

As always, we can change a 'proportional' sign into an 'equals' sign, by introducing a constant, $k$:

$$E = k(T_h - T_c)$$

In this case, the constant is known as the '**Seebeck coefficient**', and is a function of the *combination of wire types used*, expressed in volts per kelvin (V/K) or, in practice, microvolts per kelvin (μV/K).

Seebeck coefficients are typically quoted for a specified metal, *relative to platinum*. For example, a nichrome-platinum thermocouple has a Seebeck coefficient of 25 μV/K.

As we can see, the potential difference appearing across the open circuit is quite small – typically in the millivolt range – so, for this reason, thermocouples are frequently connected in *series* to form what is termed a '**thermopile**', where all the 'hot' junctions are subject to the *higher* temperatures, and all the 'cold' junctions are subject to the *lower* temperatures – as illustrated in Figure 5.7.

While thermopiles are capable of producing sufficient voltage to operate devices such as heating-system gas valves, **thermocouples** are most widely used to measure temperature. For example, large power-system transformers have thermocouples inserted between their windings in order to monitor the temperature of those windings. If that temperature exceeds a safe level (i.e. beyond which the windings' insulation may break down), then the output from the thermocouple will activate a protective relay which, in turn, will disconnect the transformer. Other applications include monitoring high temperatures associated with kilns, gas turbines, etc.

In this particular role, the thermocouple's 'hot' junction is used as the remote **temperature sensor**, while the 'cold' junction is maintained at a controlled temperature (to allow for variations in ambient temperature) and the voltmeter's scale is calibrated to provide a direct temperature readout.

Interestingly, a French watchmaker and part-time physicist, Jean Charles Peltier (1785–1845), who was studying the same phenomenon, discovered that it was *reversible*. In other words, if a potential difference is applied to the circuit shown in Figure 5.6, then *a temperature difference will appear between the two junctions*. This is known as the '**Peltier Effect**', and is made use of in, for example, camping-type 12-V cooler boxes, microprocessor cooling systems, etc.

For this reason, the **thermoelectric effect** is also generally known as the '**Seebeck-Peltier Effect**'.

## Cells and batteries

A detailed description of how a simple electrochemical cell separates charges in order to create a potential difference is covered in a later chapter on *cells and batteries* but, for now, a simplified description will suffice.

There are many different types of cell but, in its simplest form, a cell consists simply of two dissimilar conductors (e.g. zinc and copper), called '**plates**' or '**electrodes**', immersed in a conducting liquid (e.g. dilute sulfuric acid) called an '**electrolyte**' – as illustrated in Figure 5.8.

Figure 5.8

When the zinc electrode is inserted into the electrolyte, it reacts chemically with the electrolyte, and starts to dissolve. As the zinc dissolves, positively

charged zinc ions are released into the electrolyte, leaving electrons behind to accumulate on the zinc electrode – which, therefore, acquires a negative charge.

This action continues until the zinc electrode acquires sufficient negative charge to prevent any further positive ions from escaping.

A similar chemical reaction occurs at the copper electrode, with positive copper ions dissolving into the electrolyte, leaving electrons behind to accumulate on the copper electrode. The reaction of copper, however, is far less vigorous than it is for zinc, and the amount of negative charge acquired by the copper electrode is significantly less than the amount of negative charge acquired by the zinc electrode. We say that the copper electrode, therefore, is '*positive with respect to the zinc electrode*', and is named the 'positive electrode' (or 'positive plate') while the zinc is named the 'negative electrode' (or 'negative plate').

The difference between these two amounts of negative charge results in a potential difference of about 1.1 V appearing between the two electrodes.

This is a gross over-simplification of the chemical process that is *actually* taking place within the cell, but is more than adequate, at this stage, to explain how a chemical cell *separates charges* and *provides a potential difference*. As already mentioned, we will have a more in-depth examination of cells and batteries in a later chapter.

The *open-circuit* potential difference created by the charge separation process is called the **electromotive force** (**e.m.f.**, symbol: $E$) of the source. However, when a load current flows, the open-circuit potential difference will reduce somewhat. We will discuss e.m.f. in greater detail in a later chapter.

There are a great many *other* methods of separating charges, including the use of **light** (photovoltaic cells), **pressure** (piezoelectricity), **heat** (thermocouples) and, of course, **magnetism** (generators). The most important of these various methods is magnetism, and we will learn how a generator uses magnetic fields to separate charges later in the book.

### How charges gain and lose potential

Now, let's put together what we have learnt about charge separation and electric fields, to find out about the **gain and loss of potential** that takes place in a circuit.

This time, we'll use **gravity** as an analogy to describe what is going on.

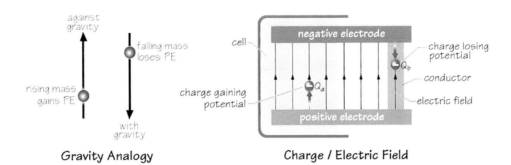

Gravity Analogy          Charge / Electric Field

Figure 5.9

Whenever an object is lifted vertically, *against the force of gravity*, then that object is said to *acquire* potential energy. When that object is allowed to *fall under the influence of gravity*, it *gives up* that potential energy in the form of kinetic energy.

Now, let's turn our simple battery on its side, just so we can continue with the 'gravity' analogy.

So, in Figure 5.9, we see a chemical cell turned onto its side, with its electric field shown between its two electrodes, so that we can compare its electric fields with gravity. Within the cell itself, we see a negative charge moving vertically upwards, through the electrolyte*, along one of the flux lines, and *against* the force of repulsion offered by the negative electrode. The work done in moving this negative charge increases the potential of that charge, in much the same way as raising a mass *against* the force of gravity acquires potential energy. The rise in the potential of this negative charge is acquired at the expense of the energy available through the cell's chemical reaction.

At the same time, another negative charge is being repelled by the negative electrode and is moving through an external conductor that links the two electrodes (for the sake of clarity, no load is shown) – to the right of the cell, in Figure 5.9. As this negative charge moves 'down' through the conductor, along the flux line towards the positive electrode, the potential that this negative charge acquired moving through the electrolyte, is given up, in much the same way as a mass gives up its potential energy as it falls under the influence of gravity.

Again, this is a simplification* of what is *actually* taking place, chemically, within the cell itself (as we have learnt, it's *ions* that move through the electrolyte, not electrons), but the principle applies: the potential *acquired* by *any* charged particle (electrons, ions, whatever) as it moves between electrodes within the cell is then *lost* as it moves through the external conductor.

*As we have learnt, in the case of a cell, electrons don't actually move through the electrolyte but, rather, ions do. But to reduce the complexity of this topic, we have assumed that an electron moving between the positive and negative electrodes is *equivalent* to the actual movement of ions within the electrolyte.

## Electric current – misconceptions

Students sometimes harbour the misconception that a source of potential difference, such as a battery or a generator,'*pumps free electrons into a conductor which, then, flow around that conductor*'. This is a misconception, because the free electrons are already there, whether the conductor is connected to the battery or not!

In Chapter 3, we described an electric current as *a flow of electric charge carriers* which, in the case of a metal conductor, are **free electrons, transported per unit time**. We also learnt that a conductor is normally electrically neutral because the total number of electrons (whether 'free' or orbiting around atoms) is exactly balanced by the number of protons in the atoms' nucleii.

It's important to understand that a source of potential difference, such as a battery, *doesn't* 'pump free electrons into a conductor'; *those free electrons are already in that conductor*. A potential difference simply causes those free electrons to gently drift around the circuit.

So, what happens when an electron leaves the conductor when it reaches its (positive) end and is ejected? Well, in order to maintain the conductor's neutrality, a *new* electron is then drawn into the conductor from the opposite (negative) end. So the number of electrons in the conductor remains the same regardless of whether or not there is any current.

## Voltage drop

Yet another term that we must thoroughly understand is '**voltage drop**'. This is, perhaps, best explained using yet another analogy. Think about a simple central-heating **system** that uses hot water flowing through, say, three radiator panels.

In order for the water to flow we need a **difference in pressure** across the entire heating system, which is provided by the pump shown in Figure 5.10.

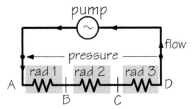

Figure 5.10

The pump produces a difference in pressure across *all three radiators* and causes the hot water to flow through the entire *system*. But at the same time, for water to flow through each *individual* radiator, there must be *a difference in pressure between its inlet and its outlet* – in other words, as well as the difference in pressure across *all three* radiators (i.e. A–D), there must also be individual pressure differences across radiator 1 (i.e. A–B), *and* across radiator 2 (i.e. B–C), *and* across radiator 3 (i.e. C–D). And, obviously, *the sum of these individual pressure differences must equal the pressure difference across the system:*

i.e. **pressure** $_{(A–D)}$ = **pressure** $_{(A–B)}$ + **pressure** $_{(B–C)}$ + **pressure** $_{(C–D)}$

Now, let's compare this heating system with an electric circuit. Instead of a pump, we have a cell or battery and, instead of radiators, we have three lamps, as shown in Figure 5.11.

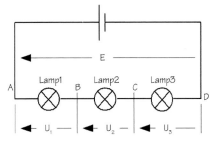

Figure 5.11

The cell produces a potential difference across *all three lamps* which causes charge carriers to flow through the circuit. However, for charge carriers to flow through each individual lamp, there must also be *a difference in potential across each lamp* too – in other words, as well as the potential difference across *all* three lamps ($E$), there must also be individual potential differences across lamp 1 (i.e. $U_1$), *and* across lamp 2 (i.e. $U_2$) *and* across lamp 3 (i.e. $U_3$). And, obviously, *the sum of these individual potential differences must equal the electromotive force across the complete circuit:*

$$\text{i.e. } E = U_1 + U_2 + U_3$$

These individual potential differences that appear across individual circuit components are known as '**voltage drops**'. We will learn more about this, in much greater detail, in a later chapter on *electric circuits*.

## Potential v potential difference (voltage)

Earlier in this chapter, we learnt that the absolute **potential** at any point within an electric field is the energy required to move a charge of one coulomb from *infinity* to that point. But we also learned that this is an impractical means of defining the potential at any particular point. In practice, the potential at any point can be measured from *any* convenient reference point we care to choose. By general consent, for most practical circuits, the 'zero-reference' point for measuring potential is usually **earth**.

This doesn't mean that the earth is *literally* at 'zero potential', it simply means measurements of potential are made *with respect to* (or compared to the potential of) earth.

It's *very* important that you understand the difference between 'potential' and 'potential difference' – as you will learn later, it's particularly important for anyone dealing with earthed (grounded) systems, where it is quite common to measure *potentials* with respect to earth. It's also important for vehicle electricians, who regularly have to measure potentials with respect not to 'true earth', but to 'chassis earth' (where the negative terminal of the battery is connected to the common metal parts or 'chassis' of a vehicle).

To help us reinforce our understanding of the difference between potential and potential difference, we will use yet another analogy which, this time, compares 'potential' and 'potential difference' with the terms 'height' and 'difference in height', where

- 'potential' is equivalent to 'height'
- 'potential difference' is equivalent to 'difference in height'.

Whenever we talk about the 'height' of an object, we have to measure it from some agreed **datum point**. For example, we normally measure the height of a mountain, using *sea level* as its datum **point** (Figure 5.12). So, the heights of two points, *A* and *B*, on a mountain are normally measured from (or 'with respect to') sea level.

Figure 5.12

Suppose the *height* of point *A* is 2000 m from sea level, and the *height* of point *B* is 2250 m from sea level. We can then say that the *difference in height* between points *A* and *B* is 250 m (i.e. 2250 –2000).

If we had chosen a completely different datum point (say, for example, we chose point *A*) then the *heights* of points *A* and *B* would be completely different, with the height of point *A* now being zero metres, and the height of point *B* being *plus* 250 m. On the other hand, if we had chosen point *B* as the datum point, then the height of point *B* would be zero metres, and the height of point *A* would be *minus* 250 m. However, the *difference in heights* between points *A* and *B* would *always* remain the same. In other words, 'height' is *relative* (i.e. it depends from where it is measured), whereas 'difference in height' is *absolute*.

Let's compare this use of the terms 'height' and 'difference in height' with 'potential' and 'potential difference'.

Like 'height', 'potential' exists at a *particular point* in a circuit and is measured from some agreed datum point. The datum point could be *anywhere* (another point in the same circuit, for example), but the most commonly used datum point in electrical-distribution systems is the **earth**.

The general mass of the earth is considered to be a conductor and, by *convention*, is allocated a potential of zero volts. Again, it's important to understand that

the *actual* potential of earth isn't necessarily zero; but this doesn't matter, it's simply zero by convention! For example, in an electrical installation, the neutral conductor is connected to earth and, so, behaves as the point of reference for measurements of potential made at any point along the line conductor.

Figure 5.13 should further clarify the difference between 'potential' and 'potential difference'. If object A has a **potential** of 100 V *with respect to (wrt) earth*, and object B has a **potential** of 80 V *with respect to earth*, then there is a **potential difference** of 20 V between A and B.

Figure 5.13

As another example, let's consider three 1.5-V disposable AA cells, connected end-to-end (in series), as shown in figure 5.14.

Figure 5.14

The 'potential difference' (or 'voltage') across each of the cells is fixed at 1.5 V. But the 'potentials' at points A, B, C, and D, *measured with respect to earth*, vary in both magnitude and polarity, and are *entirely dependent on where the earth connection is made*.

'Potential difference' or 'voltage' means the difference between the potentials at two separate points in a circuit. So, whereas 'potential' exists at a *single* point and its value depends upon where the reference is located, 'potential difference' is measured between *two* points in a circuit, and *is independent of the datum point*. Furthermore, we can describe potential as having 'positive' or 'negative' polarity relative to the chosen datum point. We *cannot* allocate charge polarity to a potential *difference*.

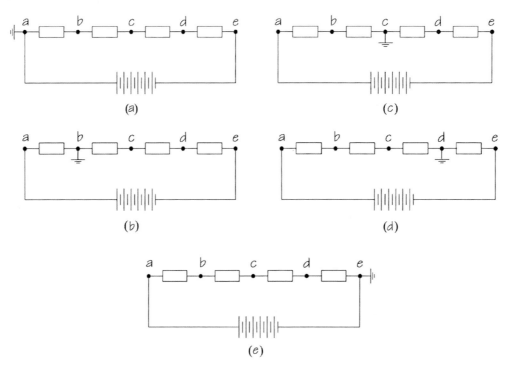

Figure 5.15

So, 'potential' is *relative*, whereas 'potential difference' is *absolute*.

Let's look at the circuits, shown in Figure 5.15. Each represents a circuit supplied by a 200-V supply. Let's assume that each resistor has the same value, in which case (as we shall learn later) the voltage drop across each resistor will be identical: 50 V.

Note how, in this example, the potentials at point *a*, *b*, *c*, *d* and *e*, depend upon where the earth reference is, whereas the potential difference across each resistor remains 50 V regardless of the position of the earth connection.

**Figure 5.15a:**

- the potential at point *a* with respect to earth is **zero** (because its *directly connected* to earth)
- the potential at point *b* with respect to earth is − 50 V
- the potential at point *c* with respect to earth is − 100 V
- the potential at point *d* with respect to earth is − 150 V
- the potential at point *e* with respect to earth is − 200 V

(note that each potential is 'negative', because each point is located between the earthed point and the negative terminal of the battery).

**Figure 5.15b:**

- the potential at point *a* with respect to earth is + 50 V
- the potential at point *b* with respect to earth is **zero**
- the potential at point *c* with respect to earth is − 50 V
- the potential at point *d* with respect to earth is − 100 V
- the potential at point *e* with respect to earth is − 150 V

**Figure 5.15c:**

- the potential at point *a* with respect to earth is +100 V
- the potential at point *b* with respect to earth is + 50 V
- the potential at point *c* with respect to earth is **zero**
- the potential at point *d* with respect to earth is − 50 V
- the potential at point *e* with respect to earth is − 100 V

**Figure 5.15d:**

- the potential at point *a* with respect to earth is + 150 V
- the potential at point *b* with respect to earth is +100 V

- the potential at point *c* with respect to earth is **+ 50 V**
- the potential at point *d* with respect to earth is **zero**
- the potential at point *e* with respect to earth is **– 50 V**

**Figure 5.15e:**

- the potential at point *a* with respect to earth is **+ 200 V**
- the potential at point *b* with respect to earth is **+150 V**
- the potential at point *c* with respect to earth is **+ 100 V**
- the potential at point *d* with respect to earth is **+ 50 V**
- the potential at point *e* with respect to earth is **zero**

You will note that the polarity of the various potentials in the above examples change, *according to where the earth is connected in the circuit*. So it's important to understand that these 'polarities' are not *absolute*, but are *relative to the other points within the circuit*. For example, a point that is labelled '+' isn't necessarily positive in the sense that there are less electrons than protons at that particular point; it simply means that it is 'less negative' than another point.

So, while the potential at any point varies, depending on the position of its point of reference (in the above examples, the earth), the potential difference across each resistor remains at 50 V. You can go ahead and confirm this for yourself if you wish, by simply subtracting two adjacent potentials

in any of the above examples – the answer will *always* be 50 V.

Potential differences or voltages are *never* described as being 'positive' or 'negative' in the sense of electric '**charge**'.

However, you *will* frequently hear the terms applied in the sense of the '**direction**' in which they act. For example, a voltage which acts, say, in a clockwise direction around a circuit might be arbitrarily-described as acting in a '*positive*' *direction*, whereas another voltage which acts in a counter clockwise direction would then be described as acting in a '*negative*' *direction*.

## Summary

'**Voltage**' is simply another word for '**potential difference**'; i.e. they are synonyms. '**Voltage**' is *not* another term for '**potential**'.

- It is *correct* to say, 'The voltage *across* a resistor is so-many volts'.
- It is *incorrect* to say, 'The voltage *at a point* is so-many volts'; instead, we should say, 'The *potential* at a point….'.
- It is *correct* to say, 'The voltage *between* line conductor and earth is 230 V'.
- It is *incorrect* to say, 'The *voltage* of the line conductor with respect to earth is 230 V'; instead, we should say, 'The *potential* of the line conductor with respect to earth is 230 V'.

Table 5.1

| | |
|---|---|
| **potential difference** | The difference in potentials between any two points in a circuit. Symbol: $E$ or $U$ – depending on context. |
| **voltage** | A synonym for 'potential difference'. Symbol: $E$ or $U$ – depending on context. |
| **electromotive force** | The potential difference produced, internally, by a battery, generator, etc., and which appears across its terminals *when it is not supplying a load*. Symbol: $E$. |
| **voltage drop** | The potential difference across an individual circuit component, such as a resistor, responsible for current through that component. Symbol: $U_1$, $U_2$, etc. |
| **potential** | Potential exists at a single point in a circuit, and is measured relative to another randomly chosen point (in practice, often earth). Potential is either negative or positive with respect to the point of reference. Symbol: $U$. |

Traditionally, in the English-speaking world, the symbol for **potential difference**, or **voltage**, has always been '*V*', and the symbol for **electromotive force** has been '*E*' – these are what you will see in most electrical science textbooks.

However, the *IET Wiring Regulations* have adopted the European standard (formerly a German standard) symbol, '*U*', in place of '*V*', and that is what we have done throughout this book.

## Review your learning

Now that we've completed this chapter, we need to examine the **objectives** listed at its start. Placing '*Can I...*' at the beginning, and a question mark at the end, of each objective turns that objective into a **test item**. If we can answer each of those test items, then we've met the objectives of this chapter.

### Online resources

The companion website to this book contains further resources relating to this chapter. The website can be accessed via the following link:

www.routledge.com/cw/waygood

# Chapter 6

# Resistance

## Objectives

On completion of this chapter, you should be able to

1  explain the term 'resistance'.
2  state the SI unit of measurement of resistance.
3  list the four factors that affect the resistance of a material.
4  state the effect upon resistance of varying
   a  length of a material.
   b  cross-sectional area of a material.
5  define the term 'resistivity'.
6  state the SI unit of measurement of resistivity.
7  solve simple problems, based on the equation:
   $$R = \rho \frac{l}{A}.$$
8  explain how the above equation relates to conductor insulation.
9  explain, in general terms, the effect of an increase in temperature upon the resistance of:
   a  pure metal conductors
   b  insulators.

## Resistance

**Resistance** is the natural opposition offered by *any* material to an electric current. In this chapter, we will be examining the resistance of metal conductors and insulators. The resistances of semiconductors, electrolytes (conducting liquids) and ionised gases behave somewhat differently and are beyond the scope of this text.

The origin of the term, 'resistance', in the sense of opposing current, is credited to the German schoolmaster, Georg Simon Ohm (1789–1854). However, the *concept* of electrical resistance predates Ohm, and was the subject of experiments by the English physicist, Henry Cavendish (1731–1810), after whom the famous 'Cavendish Laboratory' at Cambridge University is named. Cavendish was shy and eccentric to such an extent that he would leave notes for his housekeeper, rather than having to talk to a woman face-to-face! Long before the days of electrical measuring instruments, Cavendish studied the effects of the 'opposition to current' by different conductors by subjecting himself to a series of electric shocks – the more intense the shock, the lower the material's opposition to current. Judging from his meticulous and extensive research notes, Cavendish must have subjected himself to *thousands* of such shocks! And, rather surprisingly, his results apparently compare remarkably well with what we know today about the electrical resistance of various materials.

Resistance (symbol: **R**) is, to some extent, dependent upon the quantity of free electrons available as charge carriers within a given volume of material, and the opposition to the drift of those free electrons due to the obstacles presented by fixed atomic structure and forces within that material.

For example, conductors have very large numbers of free electrons available as charge carriers and, therefore, have low values of resistance. On the other hand, insulators have relatively few free electrons in comparison with conductors, and, therefore, have very high values of resistance.

But resistance is also the result of collisions between free electrons drifting through the conductor under the influence of the external electric field, and the stationary atoms. Such collisions represent a considerable reduction in the velocity of these electrons, with the resulting loss of their kinetic energy contributing to the rise of the conductor's temperature. So it can be said that *the consequence of resistance is heat.*

The consequence of resistance is heat.

An Introduction to Electrical Science, Waygood, ISBN 9780815391821, 2019. © Taylor & Francis

Resistance, therefore, can be considered to be a *useful* property as it is responsible for the operation of incandescent lamps, heaters, etc. On the other hand, resistance is also responsible for temperature increases in conductors which result in heat transfer *away* from those conductors into their surroundings – we call these energy *losses*, which, of course, are undesirable.

We can modify the natural resistance of any circuit by adding **resistors**. These are circuit components, which are manufactured to have specific values of resistance.   By changing the resistance of a circuit using resistors, we can, for example, modify or limit the current flowing through that circuit.

---

### Conductance

You may come across the term **conductance** (symbol: *G*). Conductance is the *reciprocal* of resistance, that is:

$$G = \frac{1}{R}$$

Until the adoption of SI, the unit of measurement of conductance was the **mho** – that's 'ohm', spelt backwards (proving that some scientists do, indeed, have a sense of humour)! The SI unit of measurement for conductance, however, is the **siemens** (symbol: **S**).

Conductance is a particularly useful concept to use when we study a.c. theory in a later chapter.

---

## The unit of resistance: the ohm

The SI unit of measurement of resistance is the **ohm** (symbol: Ω), named in honour of Georg Simon Ohm.

---

The **ohm** is defined as *'the electrical resistance between two points along a conductor such that, when a constant potential difference of one volt is applied between those points, a current of one ampere results'*.

---

In other words, an **ohm** is equivalent to a *volt per ampere*, and we should understand the significance of this very important definition, as it will become important when we study *Ohm's Law* in a later chapter.

The **resistance** of any material depends upon the following factors:

- its **length** (symbol: *l*)
- its **cross-sectional area** (symbol: *A*)
- its **resistivity** (symbol: *ρ*, pronounced '*rho*').

### Length

The resistance of a material is *directly proportional to its length*. In other words, *doubling* the length of a conductor will *double* its resistance, while *halving* its length will *halve* its resistance, etc.

### Area

The resistance of a material is also *inversely proportional to its cross-sectional area*. In other words, doubling its cross-sectional area will *halve* its resistance, while halving its cross-sectional area will *double* its resistance, etc.

---

**Important!** The area of a circle is proportional to the *square of its diameter*. So doubling the diameter of a circular-section wire, will actually *quadruple* its cross-sectional area and, therefore, *reduce its resistance to a quarter*. For the same reason, halving the diameter of a wire will *quarter* its cross-sectional area and *quadruple its resistance*.

---

### Resistivity

Knowing how the length and cross-sectional area affects resistance, we can now express these relationships, mathematically, as follows:

$$R \propto \frac{l}{A}$$

As shown on page 35, we can change a proportion sign (∝) to an equals sign (=) by inserting a *constant of proportionality*:

$$R = \text{constant} \times \frac{l}{A}$$

shorter length
= lower resistance

longer length
= higher resistance

Figure 6.1

= higher resistance

= lower resistance

Figure 6.2

This constant is called the **resistivity** (symbol: $\rho$, pronounced 'rho') of the material. So, the final equation for resistance becomes:

$$R = \rho \frac{l}{A}$$

where:

$R$ = resistance, in ohms (symbol: $\Omega$)

$\rho$ = resistivity, in ohm metres (symbol: $\Omega \cdot$ m)

$l$ = length, in metres (symbol: m)

$A$ = cross-sectional area, in square metres (symbol: m$^2$)

If we rearrange this equation to make resistivity the subject of the equation:

$$\rho = \frac{RA}{l}$$

…and insert the units of measurement:

$$\rho = \frac{\Omega \times m^2}{m} = \Omega \cdot m$$

…we see that the unit of measurement for resistivity is the **ohm metre**.

In North America, the cross-sectional area of a circular-section conductor is measured in **circular mils (CM)** and the length of a conductor is measured in **feet**. A circular mil is the diameter of the conductor, expressed in mils (one-thousandth of an inch) squared. If we insert these units into the above equation, we get:

$$\rho = \frac{\Omega \times CM}{ft}$$

…so resistivity is expressed in **ohm circular mil per foot** although, it is more-usually spoken as 'ohms per circular mil foot'!

So what is the real significance of **resistivity**? Well, the problem with *resistance* is that it depends not only

on the material from which a conductor is made, but also on the physical dimensions of that conductor. As we have learnt, if we were to increase the length of a conductor, then the measured resistance would also increase; if we were to increase the cross-sectional area of the conductor, then the measured resistance would decrease.

Resistivity allows us to compare different materials' abilities to pass electrical current that is independent from these geometrical factors.

**Resistivity** is a *physical property* of a material, which varies from material to material, and is affected by **temperature**. *Accordingly, values of resistivity are always quoted at a particular temperature – usually, 293 K (20°C).*

> **Resistivity** is defined as *'the resistance of a unit length of a substance with a uniform cross-section'.*

In SI, the above definition corresponds to *'the resistance between the opposite faces of a one-metre cube of material'.* It's important to understand that a 'metre cube', as shown in Figure 6.3, is *not* the same thing as a 'cubic metre'. A 'metre cube' literally means a cube having sides each measuring a metre, whereas a 'cubic metre' has the same volume, but can be of *any* shape.

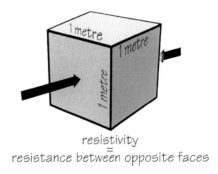

resistivity
=
resistance between opposite faces

Figure 6.3

Resistivity, then, is a fundamental property of a material that describes how easily that material can allow the passage of an electrical current. High values of resistivity imply that the material is very resistant to current; low values of resistivity imply that the material allows current to pass very easily.

In fact, materials are classified as being a 'conductor', an 'insulator' or a 'semiconductor' (semiconductors are materials used in the manufacture of electronic components, such as diodes and transistors) *according to its range of resistivities.*

As you can see from Table 6.1, conductors represent just a *tiny* range of resistivity values, compared to those of semiconductors and insulators.

Table 6.1

| Material | Resistivity / $\Omega \cdot$m |
|---|---|
| conductors | $10^{-8}$–$10^{-6}$ |
| semiconductors | $10^{-6}$–$10^{7}$ |
| insulators | $10^{7}$–$10^{23}$ |

Figure 6.4 puts this into perspective: conductors represent a very small range of resistivity values, compared to semiconductors and insulators.

Figure 6.4

Table 6.2 lists the resistivities (at 20°C) of common conductors.

Table 6.2

| Material | Resistivity / $\Omega \cdot$m |
|---|---|
| silver | $1.64 \times 10^{-8}$ ($16.4 \times 10^{-9}$) |
| copper | $1.75 \times 10^{-8}$ ($17.5 \times 10^{-9}$) |
| aluminium | $2.85 \times 10^{-8}$ ($28.5 \times 10^{-9}$) |
| tungsten | $5.6 \times 10^{-8}$ ($56.0 \times 10^{-9}$) |
| carbon | $3.5 \times 10^{-5}$ ($35.0 \times 10^{-6}$) |

**Worked example 1**   Using Table 6.2, calculate the resistance of 1000 m of copper wire of cross-sectional area 1.5 mm².

**Solution**   **Important!** Don't forget to convert square millimetres into square metres.

$$R = \rho \frac{l}{A}$$
$$= (17.5 \times 10^{-9-3}) \times \frac{1000}{(1.5 \times 10^{-6})}$$
$$= \frac{17.5 \times 10^{-3} \times 1000}{1.5}$$
$$= 11.7 \ \Omega \ (\text{Answer})$$

**Worked example 2**   Using Table 6.2, calculate the resistance of 1000 m of aluminium wire of cross-sectional area 1.5 mm².

**Solution**   **Important!** Don't forget to convert square millimetres into square metres.

$$R = \rho \frac{l}{A}$$
$$= (28.5 \times 10^{-9}) \times \frac{1000}{(1.5 \times 10^{-6})}$$
$$= \frac{28.5 \times 10^{-3} \times 1000}{1.5}$$
$$= 19.0 \ \Omega \ (\text{Answer})$$

**Worked example 3**   *Manganin* is a metal alloy comprising copper, manganese and nickel. Its resistance remains approximately constant over a wide range of temperatures, and it is used in the manufacture of wire-wound resistors. What length of manganin wire, of diameter 1 mm, would be required to make a resistor with a resistance of 0.7 Ω? Its resistivity is 0.4 μΩ·m.

**Solution**   **Important!** Don't forget to convert microhm metres to ohm metres, and the diameter in millimetres into an area in square metres.

$$R = \rho \frac{l}{A}$$
$$l = \frac{RA}{\rho}$$
$$= \frac{R\pi r^2}{\rho}$$
$$= \frac{0.7\pi(0.5 \times 10^{-3})^2}{0.4 \times 10^{-6}}$$
$$= \frac{0.7\pi \times 0.25 \times 10^{-6}}{0.4 \times 10^{-6}}$$
$$= 1.37 \text{ m (Answer)}$$

**Worked example 4** A circular section of copper cable has a resistance of 0.5 Ω. What will be the resistance of a copper cable of the same length but of twice its diameter?

**Solution** We know that the resistance of a conductor is *inversely proportional* to its cross-sectional area. So, what happens to the cross-sectional area of a circular-section cable if its *diameter* is doubled?

Well, doubling its diameter will *quadruple* its cross-sectional area. So, in this example, *the cable's resistance will be reduced by a factor of 4.*

$$resistance = \frac{original\ resistance}{4}$$

$$= \frac{0.5}{4}$$

$$= 0.125\ \Omega\ (Answer)$$

The resistivity of insulators is *not* necessarily constant for a particular temperature, as it is with conductors. Instead, it generally varies considerably according (amongst other things) to the insulator's purity, its surface condition and the duration of the application of a potential difference. Furthermore the resistivity of most insulators *decreases* with an *increase* in temperature.

Table 6.3 lists typical resistivities of common insulators.

Table 6.3

| Material | Resistivity/Ω·m |
|---|---|
| Air | $13 \times 10^{15}$ to $33 \times 10^{15}$ |
| Glass | $100 \times 10^{9}$ to $1 \times 10^{15}$ |
| Porcelain | $1 \times 10^{12}$ |
| Mica | $10 \times 10^{12}$ |
| Polystyrene | $100 \times 10^{12}$ |
| Teflon | $100 \times 10^{21}$ to $10 \times 10^{24}$ |

The enormous differences between the resistivities of insulators and conductors result in *massive* differences in their values of resistance. Comparing the resistance values of conductors and insulators will reveal some truly astonishing figures – as the following worked example will illustrate.

**Worked example 5**

a Determine the resistance of a 25-mm length of mica, of cross-sectional area 2.5 mm², if its resistivity is $10 \times 10^{12}$ Ω·m.

b What length of copper conductor, having the same cross-sectional area, would have the same resistance as this sample of mica (the resistivity of copper is $17.5 \times 10^{-7}$ Ω·m)?

**Solution** For **mica:**

$$R_{mica} = \rho\frac{l}{A} = (10 \times 10^{12}) \times \frac{25 \times 10^{-3}}{2.5 \times 10^{-6}}$$

$$= 100 \times 10^{15}\ \Omega\ (Answer\ a.)$$

For **copper:**

$$R_{copper} = \rho\frac{l}{A}$$

$$l = \frac{R_{copper}A}{\rho} = \frac{(100 \times 10^{15}) \times (2.5 \times 10^{-6})}{17.5 \times 10^{-9}}$$

$$= 14.29 \times 10^{18}\ m\ (Answer\ b.)$$

So, from the above example, sample of mica, just 25 mm long, has the same resistance as a copper conductor, having the same cross-sectional area, but measuring a staggering **14 290 000 000 000 000 kilometres** long!

To put this in some sort of perspective, the average distance between the Earth and the Sun is just 150 000 000 km!

The above worked example dramatically illustrates the scale of difference between the resistance of conductors and the resistance of insulators.

Before we leave the subject of resistivity, we need to know that the reciprocal of resistivity is called '**conductivity**', which is measured in siemens per metre (S/m).

## 'A.C. resistance' due to the 'skin effect'

As we have learned, the resistance of a metal conductor is inversely proportional to the cross-sectional area of that conductor – i.e. the *lower* the cross-sectional area, the *greater* the resulting resistance.

For direct current, the charge carriers are distributed uniformly across the cross-section of the conductor. However, *this isn't the case with alternating current.*

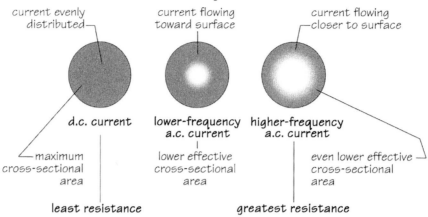

cross-section through a conductor

current evenly distributed

current flowing toward surface

current flowing closer to surface

d.c. current

lower-frequency a.c. current

higher-frequency a.c. current

maximum cross-sectional area

lower effective cross-sectional area

even lower effective cross-sectional area

least resistance

greatest resistance

Figure 6.5

For reasons that are beyond the scope of this particular chapter, for an alternating current, the charge carriers tend to travel closer to the surface of the conductor – we call this the '**skin effect**'.

The 'skin effect' becomes increasingly pronounced at higher frequencies – in other words, at higher frequencies, the charge carriers tend to drift even closer to the surface of the conductor, reducing the effective cross-sectional area even more. This is illustrated in Figure 6.5, where the grey area represents the distribution of charge carriers across a section of conductor.

The effect of the 'skin effect', therefore, is to *reduce the effective cross-sectional area of a conductor, thus increasing the resistance of the conductor.* The resistance of a conductor to alternating current, therefore, is somewhat higher than it would be to direct current, and we call this higher value its '**a.c. resistance**' to distinguish it from its resistance to direct current. 'A.C. resistance' should *not* be confused with another form of opposition to a.c. current, called 'reactance'.

At normal mains frequency (50/60 Hz), the difference between a.c. and d.c. resistance is slight, but it increases significantly at higher frequencies. In fact, at radio frequencies (300 MHz–300 GHz), the skin effect is so pronounced that there is little point in using solid conductors, so tubes (called 'waveguides') are used instead.

## Resistance of conductor insulation

The resistance of the **insulating material** surrounding a conductor conforms to *exactly* the same equation as we have already met, i.e:

$$R \propto \frac{l}{A}$$

As there is no such thing as a perfect insulator, a tiny current (in the order of microamperes), called a **leakage current**, passes through the insulation *from* the conductor to the surroundings (earth, or adjacent conductors).

So, in this context, the 'length' *(l)* of the insulation corresponds to its *thickness* (i.e. the path through which the leakage current passes), and the 'area' *(A)* of the insulation is the product of the *average circumference of the insulation* and the *length of the conductor* – as illustrated in Figure 6.6.

thickness of insulation

conductor

insulation

average circumference of insulation

length of conductor

Figure 6.6

So,

$$R_{insulation} \propto \frac{l}{A} \propto \frac{\text{thickness of insulation}}{\left(\begin{array}{c} \text{average circumference of insulation} \\ \times \text{length of conductor} \end{array}\right)}$$

As the thickness of the insulation and the circumference of any given conductor are fixed values then, for any given cable, its resistance:

$$R \propto \frac{1}{\text{length of conductor}}$$

In other words, the insulation resistance is *inversely proportional to the length of the insulated conductor* – i.e. if we were to, say, double the length of the conductor, then its insulation resistance will be halved!

## General effect of temperature upon resistance

Resistance will vary with **temperature** because, as we have learned, a material's *resistivity* is affected by temperature. It is quite possible to calculate the resistance of a material at *any* temperature but, in this chapter, it is only necessary to learn some *general* rules for pure-metal conductors, insulators, carbon, and certain alloys, as summarised in Table 6.4:

Table 6.4

| **Pure metal conductors** | an increase in temperature of a pure metal conductor will cause its resistance to *increase*. |
| **Insulators and carbon** | an increase in temperature of an insulator, and of carbon (a conductor) will cause their resistance to *decrease*. |
| **Alloys** | certain alloys, such as *constantan* (copper with 10–55% nickel) are manufactured so that an increase in temperature has very little effect upon their resistance. |

The 'hot' (operating) resistance of an incandescent-lamp's tungsten filament is typically around 10–18 times greater than that of its 'cold' resistance. So this type of lamp takes a significantly greater current at the instant it is switched on, compared to its operating current —this is *not* a problem, however, as it only takes less than 0.1 s for the current to fall to, and remain at, its operating value.

Because the resistance of pure metal conductors increases with temperature, special alloys (*such as manganin and constantan*) whose resistance remains approximately constant over a wide range of temperatures have been developed for applications where a constant resistance is important – e.g. precision wire-wound resistors.

Temperature increase, therefore, has an adverse effect upon insulators. As temperature increases, an insulator's resistance will decrease and, eventually, may lead to a catastrophic breakdown of its insulating properties. *Excessive temperature is the major cause of insulation failure*. It's essential, therefore, that the ventilation features of electrical devices should never be obstructed.

## Misconceptions

### Resistance 'blocks' current
No, resistance doesn't 'block' current, it simply limits its value!

### Resistance slows electron drift velocity
For any given value of current, an increase in resistance causes an *increase* in drift velocity!

### Doubling the diameter of a conductor will halve its resistance
No! Resistance is inversely proportional to the cross-sectional area of a conductor. In other words, doubling a conductor's cross-sectional area will halve its resistance. In the case of a circular-section conductor, doubling its diameter will increase its cross-sectional area by a factor of four! So, *doubling* a conductor's diameter will reduce its resistance by a factor of *four*.

### The longer a wire, the greater its insulation resistance
Resistance is directly proportional to the length of a material and inversely proportional to its cross-sectional area. In the case of an insulator surrounding a conductor, 'length' is equivalent to its thickness, and its cross-sectional area is determined by the length of the conductor multiplied by the average circumference of the insulator – in other words, the insulation's 'cross-sectional area' increases with conductor length. So, the longer the conductor, the lower the insulation resistance.

### Temperature affects resistance
Yes, but indirectly. Actually, temperature affects *resistivity* which, in turn, affects resistance.

## Review your learning

Now that we've completed this chapter, we need to examine the **objectives** listed at its start. Placing '*Can I...*' at the beginning, and a question mark at the end, of each objective turns that objective into a **test item**. If we can answer each of those test items, then we've met the objectives of this chapter.

**Online resources**

The companion website to this book contains further resources relating to this chapter. The website can be accessed via the following link:

www.routledge.com/cw/waygood

# Chapter 7

# Resistors

## Objectives

On completion of this chapter, you should be able to:

1  state the function of a resistor.
2  explain the difference between a fixed-value and variable-value resistor.
3  state the relationship between the physical size and power rating of a resistor.
4  briefly outline the differences between a
  • carbon-composition resistor
  • cracked-carbon resistor
  • metal-film resistor
  • wire-wound resistor.
5  briefly explain the difference between a rheostat and a potentiometer.
6  briefly explain why resistors are manufactured in 'preferred values', and explain the significance of their 'E-numbers'.
7  identify the resistance and tolerance of a resistor using a four- and five-band resistance colour code.
8  identify the resistance of a resistor using an alpha-numeric code, in the form '4R7', etc.

## Introduction

We can change the natural resistance of any circuit by using a circuit component called a **resistor**, such as the one illustrated in Figure 7.1.

Figure 7.1

We can also use resistors to create useful circuits, such as a voltage divider (described in a later chapter).

A **resistor** is defined as *'a passive electrical component, whose function is to limit the flow of electric current'.*

Resistors may be **fixed-value** or **variable-value**.

Figure 7.2          Figure 7.3

## Types of resistor

The most common types of resistor are as shown in Figure 7.4.

Figure 7.4

### Fixed-value resistors

**Carbon-composition** resistors have been in use for over a hundred years and, together with **wire-wound** resistors, were the only types of resistor available until well into the 1960s. Since then, carbon-composition resistors have largely been replaced by **cracked-carbon** or **metal-film** types.

An Introduction to Electrical Science, Waygood, ISBN 9780815391821, 2019. © Taylor & Francis

## Carbon-composition resistor

Figure 7.5

**Carbon-composition resistors** (Figure 7.5) are manufactured by compressing a mixture of fine particles of graphite, an insulating material and a binder into short rods. Any value of resistance can be manufactured by adjusting the proportions of the mixture as well as by adjusting the final length of the rod. They were widely used in consumer electronics up until the 1970s but, due to the instability of their resistance values, they are now considered unsuitable for most modern applications.

## Cracked-carbon resistors

Figure 7.6

By heating small ceramic rods up to temperatures around 1000°C in methane vapour, a very thin uniform layer of carbon can be deposited on their surfaces. This process is called 'cracking', so the resistors manufactured in this way are called **'cracked-carbon' resistors** or, simply, **'carbon' resistors** (Figure 7.6). Their resistance depends upon the thickness of the carbon layer, but it can be adjusted to any desired value by cutting a spiral-shaped groove into the carbon. The thickness and length of the resulting 'spiral' of deposited carbon determines the final resistance of the resistor. The resistor is then protected with a coating. This type of resistor can be manufactured to very accurate values of resistance (nominal resistance, ±2% tolerance).

## Metal-film resistors

Cutaway Drawing of a Metal Film Resistor

Figure 7.7

**Metal-film resistors** (Figure 7.7) are now largely replacing cracked-carbon types; they are manufactured in a similar way but using a thin film of vacuum-deposited metal, such as nickel-chromium, on the surface of a ceramic rod. As with cracked-carbon types, the final resistance value is obtained by laser-cutting a helical spiral into the metal film. Metal-film resistors can be manufactured to very accurate values of resistance (nominal resistance, ±0.1% tolerance).

## Wire-wound resistors

Figure 7.8

**Wire-wound resistors** (Figure 7.8) are the oldest type of resistor and are manufactured from 'resistance wire', i.e. a high-resistivity wire such as nickel-chromium or manganin, wound around a ceramic tube and protected by a coating of enamel or plastic.

The accuracy of their resistance values can be very high (better than ±0.01% tolerance) and they are used where high values of precision are needed. They are usually physically larger (for some applications, significantly larger) than the resistors described above, and can carry much higher values of current without overheating.

One drawback of a wire-wound resistor is that it produces a magnetic field which results in an undesirable property, in addition to its resistance, called 'inductive reactance' which opposes the passage of alternating current but not direct current. This drawback can be overcome in various ways, the most simple being the 'bifilar' winding (Figure 7.9), in which the resistance wire is folded back on itself before being wound around the ceramic tube. This results in the current flowing in opposite directions in adjacent turns, which acts to largely eliminate any magnetic field and any resulting inductive reactance.

Figure 7.9

## Surface-mount resistors

Figure 7.10

'Surface-mount' technology describes a range of electronic components designed with small metal tabs, or end caps, that can robotically soldered directly to the surface of a printed-circuit board (PCB). Amongst these components are **surface-mount resistors** (**SMRs**) (Figure 7.10). These are manufactured using the range of technologies already discussed – wire-wound, metal foil, metal film – as well as others outside the scope of this chapter.

## Variable resistors

Variable resistors (variable-value resistors) are used when it is necessary to vary the value of the *current* in a circuit, or to vary the *voltage* across a circuit. Variable resistors are made from carbon-composition or are wire-wound – with wire-wound being used in the manufacture of 'quality' variable resistors, such as high-fidelity amplifier volume controls, sound-board faders, etc.

Figure 7.11

Variable resistors are manufactured in a great many shapes and sizes; Figure 7.11 shows just two examples. What most have in common is a pair of terminals at opposite ends of the resistor, and a third terminal connected to a spring-loaded metal or carbon 'wiper' that slides across the resistive element.

It's the movement and pressure of the wiper that tends to wear and damage carbon-type variable resistors, which is why good-quality variable resistors tend to be wire-wound.

A variation of the variable resistor is the **trimmer resistor**. A trimmer resistor is found in many electronic devices, and is preset to a particular value of resistance

by the manufacturer – in other words, it's not intended to be adjusted by the end user.

Variable resistors are designed to have a characteristic, termed a '**taper**'. When the resistance is proportional to the position of the slider, it is said to have a '**linear taper**'. However, variable resistors intended for use as audio-system volume controls, for example, have a '**logarithmic taper**' which matches the way in which the human ear perceives sound levels.

## Rheostats and potentiometers

Variable resistors are frequently, but incorrectly, called either a 'rheostat' or a 'potentiometer' (Figure 7.12). The terms **rheostat** and **potentiometer**, however, apply to their *applications*, and *not* to the devices themselves, as practically *all* variable resistors can be used *either* as a rheostat *or* as a potentiometer.

A variable resistor connected as a **rheostat** is used to control *current*, while a variable resistor connected as a **potentiometer** is used to control *voltage*.

As already explained, variable resistors have *three* terminals: one at opposite ends of the winding (*a* and *b*), and the third connected to the 'brush' or wiper (*c*). When used as a rheostat, one of the fixed terminals and the wiper terminal are used; when used as a potentiometer, all three terminals are used.

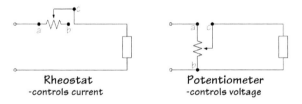

Figure 7.12

## Miscellaneous resistors

Although outside the scope of this chapter, there *are* other types of resistor – including those whose resistance are affected by light, temperature, etc. For example:

- **Photoresistors** are resistors whose resistance varies with the intensity of the light falling on them – these types of resistor, such as cadmium sulphide (CdS) cells, are used, for example, in photographic light meters.
- **Thermistors** are resistors whose resistance varies with temperature – these types of resistor are used, for example, in the manufacture of electronic thermometers. Resistors whose resistance increases with an increase in temperature are termed **positive temperature coefficient** (**PTC**) **resistors**; those

whose resistance falls with an increase in temperature are termed **negative temperature coefficient (NTC)** resistors. Thermistors are used to measure temperature, or to limit temperature change.

## Resistor power ratings

Whenever we select a resistor, not only must we take the value of its **resistance** into account, but we must also consider its **power rating** – i.e. the *rate at which it can dissipate energy through heat transfer into the surroundings by radiation, convection and conduction*. It is its power rating that determines the maximum continuous current a resistor can handle without becoming damaged through overheating.

Since a physically larger resistor has a greater surface area than a physically smaller resistor, it can dissipate energy at a greater rate. The physical size of a resistor, therefore, is indicative of its power rating, but *not* of its resistance.

> The resistance of a resistor *cannot* be 'estimated' from its physical size, but must be determined from its **colour code**.

For comparison, the nominal power-ratings of axial carbon-composition resistors are drawn to scale, *relative to each other*, in Figure 7.13.

**Figure 7.13**

To find the safe maximum continuous operating current *(I)* for a specific resistor, we need to know its **power-rating *(P)*** together with its **resistance *(R)***.

The maximum continuous current that a resistor can carry may be calculated as follows:

$$\text{since } P = I^2 R$$

$$\text{then } I = \sqrt{\frac{P}{R}}$$

For example, the safe operating current for a 0.25-W, 560-Ω, resistor is calculated as follows:

$$I = \sqrt{\frac{P}{R}} = \sqrt{\frac{0.25}{560}} = \sqrt{0.446 \times 10^{-3}} = 0.021\,\text{A or 21 mA}$$

A resistor's nominal power rating is defined for a specified ambient temperature, typically 70°C. For anticipated temperatures above this, we usually have to de-rate its power rating – i.e. the resistor must be operated at a reduced power level.

## Resistance values of resistors

### Tolerance

No resistor is manufactured to an *exact* value of resistance. Instead, the resistance is guaranteed by its manufacturer to fall within a specified range either side of its nominal ('named') value. This range is termed its **tolerance**, and is expressed as a percentage of its nominal value. Inexpensive carbon-composition resistors usually have a tolerance of ±20% while more expensive cracked-carbon or metal-film resistors may have a tolerance as low as ±1% or less. The most commonly encountered resistors, however, typically have a tolerance of ±10% or ±5%.

So, the *actual* resistance of a resistor which has a nominal resistance of, say, 100 Ω ±10% may be of *any* value between 90–110 Ω.

### Resistor preferred values

To manufacture and stock resistors of *every* possible value would, of course, be completely impractical. If, for example, a manufacturer offers a **100 Ω ±10%** resistor, then there would be very little point in also manufacturing resistors having the same tolerance but with resistance values between 90–99 Ω and between 101–110 Ω because these values already fall within the tolerance-value of the 100-Ω resistor.

But it *would* make sense to produce, say, a **120-Ω ±10%** resistor instead, because its value may fall anywhere within 108–132 Ω, giving it a slight overlap with the upper-tolerance value of the 100-Ω resistor.

For this reason, resistors are manufactured using a system of '**Preferred Values**' (termed '**E-series**'), which is based on the principle of **overlapping tolerances**.

By way of example, those resistors manufactured to a tolerance of **±10%**, for example, form what is termed the '**E12 series**' of preferred values. The E12 series is so-called because it comprises *twelve* 'preferred values' of resistor, each having a tolerance range of ±10%. This means the specified resistance of any resistor is $10^{(1/12)}$, or 1.2 times the value of the previous resistor, rounded off. For example, for the range (termed 'decade') 1–10 Ω, resistors are manufactured with the following specified values:

- **1 Ω**
- (1 × 1.2) = 1.2 = **1.2 Ω**
- (1.2 × 1.2) = 1.44 = **1.5 Ω**
- etc.

So, the complete series for decade 1–10 Ω is:

**1, 1.2, 1.5, 1.8, 2.2, 2.7, 3.3, 4.7, 5.6, 6.8, 8.2, and 10 Ω.**

And, for the next decade, 10–100 Ω, these become:

**10, 12, 15, 18, 22, 27, 33, 47, 56, 68, 82, and 100 Ω.**

... and so on.

When someone designs a circuit that requires resistors, they need to determine the resistor values necessary to make that circuit work as well as possible. This may lead to a *very* precise value such as, for example, 14.8 Ω. If the designer has to choose from the E12 preferred value series, then he needs to choose the closest available resistor values to his theoretical values. So, instead of the theoretical design-value of 14.8 Ω, he would need to select a 15-Ω resistor. As the nearest E12 resistor is never in error by more than 10% then, in practice, this kind of error rarely affects the behaviour of a circuit.

The complete range of the **preferred-value series**, comprises: **E6** (for resistors having a 20% tolerance), **E12** (for 10% tolerance), **E24** (for 5% tolerance), **E48** (for 2% tolerance), **E96** (for 1% tolerance) and **E192** (for 0.5% and better tolerance) – where the E-number refers to the number of resistance values within each 'decade'.

So, those resistors manufactured to a tolerance of **±5%** form what is called the '**E24 series**' of preferred values, which comprises *twenty-four* 'preferred values' of resistor, each having a tolerance range of ±5%. For example, for the range (termed '**decade**') 1–10 Ω, resistors are manufactured with the following specified values:

1.0, 1.1, 1.2, 1.3, 1.5, 1.6, 1.8, 2.0, 2.2, 2.4, 2.7, 3.0, 3.3, 3.6, 3.9, 4.3, 4.7, 5.1, 5.6, 6.2, 6.8, 7.5, 8.2, and 9.1 Ω.

## Identifying the resistance of a resistor

On large, wire-wound resistors, the resistance can be printed on the resistor. The small carbon, cracked-carbon and metal-film resistors are too small for this method.

There are *two* systems for identifying the resistance of a resistor:

### Resistance colour code

The most-common system for identifying the resistance of a fixed-value resistor is termed the '**coloured band**' method (Figure 7.14). There are *two* such systems, one which uses four coloured bands, while the other uses five coloured bands.

Some resistors have a *sixth* band, which indicates their **temperature coefficient (of resistance)**, which describes the change in their resistance as a function of the ambient temperature, expressed in parts per million per kelvin (ppm/K). Further discussion on this characteristic is beyond the scope of this chapter.

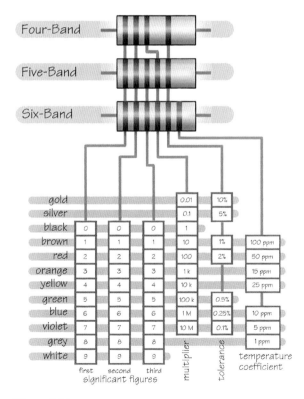

Figure 7.14

Each colour represents a number, starting with black representing zero, then brown which represents one, and so on.

There are various **mnemonics** that will help us remember the sequence of colours, such as:

> ***Billy BROWN Revives On Your Gin But Values Good Whiskey.***

Or, for railway enthusiasts:

> ***Bye, Bye, Rosie, Off You Go (to) Birmingham Via Great Western.***

Remember, too, that the numbers start at '0', *not* at '1'!

The coloured bands are normally (but, unfortunately, *not* always!) offset to one end of the resistor so, to read its value, it's first necessary to orientate the resistor so that the bands are closest to the left-hand end of the resistor (Figure 7.15).

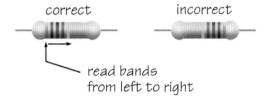

**Figure 7.15**

The bands on some resistors may *not* be closer to one end. This *can* be a problem, but for the majority of resistors (i.e. those with a tolerance of 10% or better) bear in mind that the tolerance band will be a metallic colour (silver or gold) – so the resistor should be orientated so that the tolerance band is to the right.

### Examples of the four-band system

Figure 7.16 shows examples of the four-band system.

**Figure 7.16**

### Examples of the five-band system

Figure 7.17 shows examples of the five-band system.

**Figure 7.17**

## Alpha-numeric resistance code

We *may* come across resistors that use an **alpha-numeric code** printed on them in order to identify their resistance value. We will *most certainly* see the alpha-numeric code used on a circuit diagram to identify the resistance of resistors used in that circuit.

As the name suggests, the alpha-numeric code uses a combination of **letters** and **numerals** to identify resistance values. The letters used are:

- **R** – which represents **ohms**
- **K** – which represents **kilohms**
- **M** – which represents **megohms**

The letter also represents the *position of the decimal point*, as shown in Figure 7.18

**Figure 7.18**

In the examples shown in Figure 7.18,

| | | |
|---|---|---|
| R10 = 0.10 Ω | 1R0 = 1.0 Ω | 10R = 10 Ω |
| K10 = 0.10 kΩ | 1K0 = 1.0 kΩ | 10K = 10 kΩ |
| M10 = 0.10 MΩ | 1M0 = 1.0 MΩ | 10M = 10 MΩ |

In this system, *tolerances* are indicated as follows:

- **C** – which represents ±0.25%
- **D** – which represents ±0.50%
- **F** – which represents ±1.00%
- **G** – which represents ±2.00%
- **J** – which represents ±5.00%

So, for example, **4K7J** would indicate a resistor of **4.7 kΩ ±5%** resistance.

## All-digital code

All surface-mount resistors conform to a **3-digit** or **4-digit code**. Unfortunately, however, there are a *number* of variations of these codes in use, according to the *tolerance* of the resistor – so, it's getting *very* complicated, and the complete system is beyond the scope of this chapter.

However, the basic **three-digit code** works as follows: the *first two digits* represent the *first two significant figures*, while the *third digit* represents the *power of ten* (i.e. the number of zeros).

Figure 7.19

In the example shown in Figure 7.19, the three-digit code is '334', which indicates:

$$33 \times 10^4 = 330\ 000\ \Omega \text{ or } 330\ \text{k}\Omega$$

For resistors having a resistance less than 100 Ω, the third digit is zero (remember that 10, raised to the power of zero is 1), for example:

$$100 = 10 \times 10^0 = 10 \times 1 = 10\ \Omega$$
$$220 = 22 \times 10^0 = 22 \times 1 = 22\ \Omega$$
$$470 = 47 \times 10^0 = 47 \times 1 = 47\ \Omega$$

*However* … to avoid confusion (!) SMRs may, instead, be labelled: 10, 22 or 47.

Finally, SMRs *with resistances less than ten* revert to the **alpha-numeric code**, using the letter **R** to indicate the decimal point. For example:

$$4R7 = 4.7\ \Omega \qquad 0R22 = 0.22\ \Omega \qquad 0R01 = 0.01\ \Omega$$

The **four-digit code** for SMRs is equivalent to the five-band method for conventional resistors, with the first three digits representing the first three significant figures, and the fourth digit representing the multiplier's power of ten.

## Review your learning

Now that we've completed this chapter, we need to examine the **objectives** listed at its start. Placing *'Can I …'* at the beginning, and a question mark at the end, of each objective turns that objective into a **test item**. If we can answer each of those test items, then we've met the objectives of this chapter.

# Chapter 8

# Conductors and cables

## Objectives

On completion of this chapter, you should be able to

1  identify, and state the function of, each of the main components of an electric cable.
2  explain why some conductors are stranded rather than solid.
3  explain how the cross-sectional area of a stranded conductor may be determined.
4  explain why the sectional shape of larger multicore cable conductors are frequently 'wedge' shaped, rather than circular.
5  list the main properties required of a cable's insulation.
6  explain the relationship between a cable's length and the resistance of its insulation.
7  identify the function of a cable conductor by the colour of its insulation.

## Introduction

In the electrotechnology industry, 'wires' are more-properly called '**conductors**'. A conductor may be solid or stranded, insulated or uninsulated.

An electric **cable** normally consists of one or more metal current-carrying **conductors**, each surrounded by a layer of **insulation** which, in turn, is usually covered by a tough outer **protective sheath**. In some cases, the cable may be further protected by metal tape or wire **armour**.

In this chapter, we will confine our descriptions to to low- and medium-voltage cables used for the wiring of residential, commercial and industrial installations.

The complicated construction of *high-voltage* underground cables requires extensive explanations based upon an understanding of electric fields which are beyond the level of this book.

## Conductors

As already explained, in the electrotechnology industry, 'wires' are more-properly called '**conductors**'.

A cable may have one, two, three or more separately insulated conductors. Each insulated conductor is called a **core**, and cables with more than one core are referred to as *multicore cables* (or, in North America, *'multi-conductor'* cables).

The very best metal conductor is **silver** which, because of its cost, is limited to special applications such as relay contacts, printed circuit boards, etc. It's common to compare the conductivity of other conductors to that of silver, as indicated in Table 8.1.

Table 8.1

| Conductor | Relative conductivity |
|-----------|----------------------|
| Silver | 100% |
| Copper | 95% |
| Gold | 67% |
| Aluminium | 58% |
| Tungsten | 30% |
| Iron | 14% |
| Constantan | 3.3% |
| Carbon | 1.5% |

An Introduction to Electrical Science, Waygood, ISBN 9780815391821, 2019. © Taylor & Francis

The most widely used conductor is **copper**, which has a number of advantages over its nearest rival, aluminium. Copper is second only to silver in terms of its relative conductivity (95%), and is a readily available natural resource. Copper used for general conductors is usually annealed, meaning that it is reheated and allowed to slowly cool, rendering it tough, yet soft enough to be easily drawn into single or stranded conductors. Hard-drawn copper, on the other hand, has a higher tensile strength, making it ideal for suspending from overhead electricity distribution poles.

When exposed to air, bare copper forms a conductive oxide which then acts to prevent further corrosion, allowing bare (as opposed to insulated) copper to be used for overhead lines. Copper can also be easily soldered, making it ideal for electronics applications.

The abundance of **aluminium** makes it a more economical alternative to copper, and is the second most widely used conductor. Its conductivity, relative to silver, is 58 per cent, making it a poorer conductor than copper which means it must have a larger cross-sectional area than copper in order to conduct the same amount of current.

Its light weight makes it an ideal conductor for the very long spans between overhead electricity transmission towers, although its relative low tensile strength means it must be reinforced with a steel-wire core. Like copper, an oxide forms on the surface of aluminium, preventing further corrosion but, unfortunately, this oxide is a very poor conductor which can cause problems at terminations. The quickly-forming oxide also makes it very difficult to solder aluminium unless specialist, abrasive fluxes are used.

A major problem with aluminium is termed 'cold flow', which causes the metal to flow away under the pressure from screw terminals and cause the termination to loosen. For this reason, special connectors and techniques are required for terminating aluminium conductors, rendering it unsuitable for residential and commercial wiring. In fact, in some countries, aluminium is banned for residential and commercial wiring following fatal fires.

The central component of any cable, then, is the **conductor** itself, which may be *solid*, or *stranded* – as illustrated in Figure 8.1.

solid conductor          stranded conductor

Figure 8.1

Cables of the type used to wire residential and commercial buildings, having cross-sectional areas below 4 mm$^2$, are usually manufactured using *solid* conductors. Beyond this cross-sectional area, conductors are normally *stranded*. Stranded conductors impart greater flexibility into cables, compared with cables having solid conductors of equivalent size, making them much easier to handle.

A conductor of any given cross-sectional area can be manufactured either from a few strands of thicker wire, or from a large number of strands of thinner wire – in either case, the total area will remain the same. Generally, for any given cross-sectional area, the greater the number of strands, the greater the flexibility of the cable. This is why 'flexible cables' (commonly called 'flex' or 'cords') are used to connect portable appliances to the fixed wiring in buildings. Flexible cables enable such appliances to be regularly moved around without over-stressing and, possibly, snapping the conductors.

The strands of a stranded cable form a mathematical pattern, with the first layer being a single strand, the second layer being formed from 6 strands, the third layer from 12 strands, and so on.

Stranded cables are identified in the following fashion: '7/0.85'. In this example, '7' represents the *number of strands*, and '**0.85**' represents the **diameter** *of one strand*, expressed in millimetres.

The cross-sectional shape, or 'profile', of a conductor is not necessarily circular, whether solid or stranded. For example, in the case of larger multi-core cables (typically over 16 mm$^2$), such as those used for underground services, the conductors' cross-sectional profiles are frequently 'shaped' in order to reduce the overall diameter of the complete cable compared to a cable with circular sections but of the same current rating. These shapes vary depending upon the number of cores, but are typically 'wedge' shaped. In high-voltage cables, these shapes also reduce the need to pack the internal airgaps with fillers, and to help evenly distribute the electrical stress caused by electric fields within the cables.

In the two examples illustrated in Figure 8.2, the circular-section cable has the same size outer diameter as the shaped-section cable to its right, but the cross-sectional area of its conductors is very much smaller than the shaped conductors. It also has larger voids (air gaps) which, even if packed with insulated fillers, make the cable more susceptible to moisture ingress. If these were high-voltage cables, the larger voids in the circular-section cable would also be susceptible to breakdown due to variations in dielectric stress

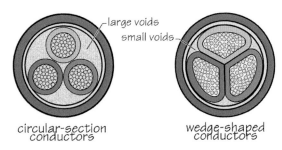

circular-section conductors      wedge-shaped conductors

Figure 8.2

due to the differences in the cable's insulating material and the material used as 'fillers' within the voids (including air).

In order to efficiently carry an electric current, the conductor must have a *low* resistance. If we refer back to the chapter on *resistance*, we will recall that resistance (symbol: *R*) is *inversely proportional to a conductor's cross-sectional area (symbol: A)*, expressed as:

$$R \propto \frac{1}{A}$$

For a given material, then, *doubling* its cross-sectional area will *halve* its resistance, and *vice versa*. This is because the area of a circular-section conductor is proportional to the *square* of its diameter and, so, *doubling the **diameter** will reduce its resistance by a quarter* and so on.

> In the UK and Europe, the cross-sectional areas of conductors are always expressed in **square millimetres** (symbol: **mm²**). In the United States and Canada, however, a completely different method is used, based on the **American Wire Gauge (AWG)** system, which uses US Customary Units of measurement: 'circular mils' for circular-section conductors, or 'square mils' for rectangular-section conductors such as busbars. A 'mil' is one-thousandth of an inch. A 'circular mil' is not really a true unit for area, as it is simply a conductor's diameter, in mils, squared, which results in a number that 'represents' a particular cross-sectional area.

For a solid conductor of circular cross-section, of diameter *d*, the cross-sectional area (*A*) is measured perpendicular to the cable's length, and is given by:

$$A = \frac{\pi d^2}{4}$$

For a circular-section stranded conductor, made up of *n* strands, each of diameter *d*, the total cross-sectional area is given by:

$$A = n\left(\frac{\pi d^2}{4}\right)$$

The cross-sectional area of a **7/0.85** cable, therefore, can be determined as follows (we must first convert millimetres into metres):

$$
\begin{aligned}
A &= n\left(\frac{\pi d^2}{4}\right) \\
&= 7 \times \left(\frac{\pi \times (0.85 \times 10^{-3})^2}{4}\right) \\
&= 4.00 \times 10^{-6}\,\text{m}^2 \\
&= 4.0\,\text{mm}^2 \text{ (Answer)}
\end{aligned}
$$

For cables with a larger number of strand layers, the resistance of the outer strands is actually somewhat higher than that of the inner strands. This is due to the spiralling of the outer strands, which means that the lengths (and, therefore, the resistance) of the outer strands are longer than the lengths of the inner strands. The above worked example, therefore, is only accurate for cables with relatively few strand layers.

> Note that any circuit-protective conductor (often called an 'earth' conductor) within the cable is *not* counted as a 'core'. So, a 'two-core' cable may, or may *not*, include a protective conductor.

## Insulation

A cable's **insulation** is the layer of dielectric material immediately surrounding the conductor(s), whose function it is to electrically isolate the conductors from each other as well as from their surroundings, in order to prevent short circuits or anyone from receiving an electric shock by accidentally touching them.

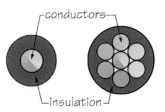

Figure 8.3

The material used as an insulator or dielectric must have each of the following *properties*:

- high resistivity
- high dielectric strength
- impervious to moisture
- strength and toughness
- flexibility
- immune to chemicals
- stability over a wide range of operating temperatures.

An insulator's dielectric strength is defined as *'the maximum electric field that can be sustained by a dielectric before breakdown occurs'*, and is expressed in volts per metre (symbol: V/m) or, in more practical terms: kilovolts per millimetre. In simple terms, it defines the maximum potential difference an insulating material can withstand before breaking down and conducting.

An insulator's dielectric strength must be appropriate for its application – e.g. the insulation of a bellwire operating at 12 V clearly does not require the same degree of dielectric strength as, say, that of a cable operating at 400 V.

As we shall learn in a later chapter on alternating current, **a.c. voltages** are expressed in what is termed **'effective'**, or **'root-mean-square'**, values. We needn't go into this subject right now, but we should be aware that an a.c. voltage with a nominal (named) value of, say, 230 V, actually 'peaks' at 325 V. So, if cables are to operate with alternating currents, they must be insulated to withstand these 'peak' values, and not the 'effective' values.

As we learnt in the chapter on *resistance*, the resistance of insulation is *inversely proportional to the length of the conductor* – i.e. the *longer* the conductor, the *lower* its insulation resistance!

For residential wiring, the minimum values of insulation resistance, measured between the live\* conductors and between the live conductors and the protective ('earthing') conductor are specified in the **IET Wiring Regulations**. For nominal voltages up to and including 500 V, this value must be ≥ 1.0 MΩ, and the insulation must withstand a test voltage of 500 V(d.c.).

\*The term **'live conductor'** refers to any conductor intended to be energised in normal use – in other words, it applies to both line and to the neutral conductors, but *not* a protective (earth) conductor.

## Insulation (dielectrics)

The type of insulating (or 'dielectric') material determines the safe operating temperature of a conductor (for residential wiring, typically 70°C or 90°C), as early insulation failure may result if its actual temperature is allowed to exceed its specified safe operating temperature.

Most residential installation cables are insulated using what, in earlier editions of the IET Wiring Regulations *used* to be called '**PVC**' or '**rubber**'. However, since the introduction of the 2008 edition of these Regulations, these terms are no longer used to describe insulating materials; instead, 'PVC cables' are now called '**thermoplastic-insulated**' cables, and 'rubber cables' are now termed '**thermosetting**' cables.

> **Thermosetting plastics** are synthetic materials which strengthen when heated, but cannot then be remoulded or reheated without burning.
>
> **Thermoplatics**, on the other hand, soften when heated then harden and strengthen after cooling and this can be repeated as often as necessary.

These new terms are related to the anticipated operating temperature of the cable, and provide a simpler method of classifying insulation. This offers a solution to the problem of the ever-increasing variety of materials being used these days in the manufacture of cable insulation, by simply classifying them in terms of their *properties*, instead of their *ingredients*.

**Mineral-insulated (MI) cable** was patented by a Swiss engineer, Arnold Francois Borel, in 1896. His cable (Figure 8.4) was constructed from inorganic dielectric materials, (i.e. compounds which do not contain carbon) completely enclosed within a metallic sheath, and designed so that it would be resistant to severe mechanical stress, as well as being capable of operating at very high temperatures, including the ability to survive a fire — hence the origin of its trade name, 'Pyrotenax' (from the Greek, 'pyro', meaning 'fire').

Originally manufactured in France, from 1936 this cable was produced under licence by the British company, Pyrotenax. Electricians often refer to these cables simply as 'pyro'.

MI cables are manufactured in a range of conductor cross-sectional areas, typically from 1 mm$^2$ to 240 mm$^2$, and containing up to 19 cores. Usually, for cables above 25 mm$^2$, only single-core cables are available.

Special tools are required for preparing terminations, in order to strip back the copper sheath to allow access to the conductors, and special kits are provided to terminate and seal the cables.

Figure 8.4

The inorganic dielectric (magnesium oxide) used in the construction of an MI cable provides it with characteristics that are significantly different from any other type of cable. In particular, MI cables can operate continuously at temperatures up to 250°C and, for short periods, even at temperatures close to 1083°C, the melting point of copper! This property enables the cable to continue operating in a fire, supplying energy to essential services and, in many cases, surviving in working condition.

MI cable can also withstand severe mechanical abuse, such as twisting, bending, and hammering without any appreciable deterioration in its electrical properties. The copper sheath is resistant to most forms of chemical attack but, in situations where the copper sheath *can* corrode, then the cable is available with an protective layer of thermoplastic material which will not propagate flames or emit harmful smoke particles, in the event of a fire. Alternatively, the cable can be supplied with a metal sheath manufactured from a different material, such as stainless steel.

The cable's dielectric, magnesium oxide, is *non-toxic, chemically-stable*, has a *high melting-point*, and *high thermal and electric resistivities*. Unfortunately, though, it is *hygroscopic*, which means it does tend to *attract* and *absorb moisture* which, if unheeded, will lead to a breakdown in its insulating properties. Special termination kits are, therefore, provided for use with MI cable which are designed to seal terminations to prevent ingress of moisture – as illustrated in Figure 8.5.

Figure 8.5

### Colour coding insulation

Insulation is always colour coded so that the function of each conductor can be identified. These colours must conform to national or international codes of practice. For example, the approved colours for identifying cable cores in Europe for a.c. power and lighting circuits are shown in Table 8.2.

Table 8.2

| Function | Alphanumeric | Colour |
| --- | --- | --- |
| protective (earthing) conductor | PE | yellow & green stripe |
| single-phase line conductor | L | brown |
| line 1 conductor of a three-phase system | $L_1$ | brown |
| line 2 conductor of a three-phase system | $L_2$ | black |
| line 3 conductor of a three-phase system | $L_3$ | grey |
| neutral | N | blue |

In North America, the colours are different from those used in Europe. Those specified by the **National Electrical Code** (United States) differ slightly from those specified by the **Canadian National Code**, as listed in Table 8.3.

Table 8.3

| Function | Alphanumeric | Colour |
| --- | --- | --- |
| protective (earthing) conductor | PG (US) / PE (CAN) | green |
| single-phase line conductor | L | black |
| split-phase, line 1 conductor* | | black |
| split-phase, line 2 conductor* | | red |
| line 1 conductor of a three-phase system | $L_1$ | black (US) / red (CAN) |
| line 2 conductor of a three-phase system | $L_2$ | red (US) / black (CAN) |

| Function | Alphanumeric | Colour |
|---|---|---|
| line 3 conductor of a three-phase system | $L_3$ | blue |
| neutral | N | white |

[*North American residential systems use a 'split-phase' system, based on two line (energised) conductors and a neutral which provide nominal voltages of 240V between lines and 120 V between either line and neutral.]

## Protective sheath

Cables used in residential installation work will normally have a **thermoplastic sheath**, as the likelihood of mechanical damage is low because such cables are normally reasonably-protected from their surroundings by being run below floorboards, sunk into plasterwork, etc.

The main functions of this sheath are to provide limited *mechanical protection* to the insulation, to *prevent ingress* of moisture into the cable and, in the case of multicore cables, to *contain the individual cores*. A two-core thermoplastic cable with a protective sheath is illustrated, in Figure 8.6, below:

Figure 8.6

Underground multi-core distribution cables may have an extruded-lead sheath whose main function is to prevent moisture ingress into the cable. A puncture to this sheath due to mechanical damage is one of the main causes of insulation breakdown of underground cables.

## Armouring

Cables, particularly underground distribution cables, intended for use where the possiblity of mechanical damage is high, require additional protection.

Such cables are usually **armoured**. This consists either of overlapping flat metal tapes or of circular-section metal wires (often two layers, wound in opposite directions to increase the cable's flexibility), manufactured from steel or, in some cases, aluminium.

Armouring is intended to protect the cable from the impact of hand tools such as shovels, pick axes, etc., and to prevent the possibility of penetration by rocks, etc., once the cable trench has been backfilled.

## Review your learning

Now that we've completed this chapter, we need to examine the **objectives** listed at its start. Placing '*Can I...*' at the beginning, and a question mark at the end, of each objective turns that objective into a **test item**. If we can answer each of those test items, then we've met the objectives of this chapter.

# Chapter 9

# Effect of temperature change upon resistance

## Objectives

On completion of this chapter, you should be able to

1 state the general effect of changes in temperature upon:
   a  pure metal conductors
   b  insulators
   c  carbon (special case).
2 explain the term *temperature coefficient of resistance*.
3 state the unit of measurement of *temperature coefficient of resistance*.
4 explain the difference between *positive* and *negative* temperature coefficients of resistance.
5 solve problems on the effect of changes in temperature upon the resistance of materials.

## Introduction

Variations in temperature can affect the physical properties of any material – including, amongst others, its optical, electrical and magnetic properties.

One of the physical properties of a material is its **resistivity** and, since resistance is directly proportional to resistivity, any change in resistivity will result in a change in **resistance**.

> A change in temperature will *directly* affect the resistivity of a material, thereby *indirectly* affecting its resistance.

The way in which a material's resistivity and, therefore, resistance changes depends upon the nature of the material itself. Of particular interest to us, in general terms, is the behaviour of **pure metal conductors**, **alloys**, **insulators** and (a special case) **carbon**.

### Pure metal conductors

In the case of **pure metal conductors**, such as copper or aluminium, an *increase* in temperature will cause an increase in its resistance. Conversely, reducing the temperature of a pure metal conductor will cause its resistance to fall. The *amount* of change in resistance for a given change in temperature, varies from metal to metal.

### Alloys

Some alloys, such as **constantan** (a copper-nickel alloy), have been specially developed to maintain a fixed resistance over a wide range of temperature variations. Such alloys are used in the manufacture of measuring instruments, for example, where any change in resistance due to temperature variations can affect the accuracy of such instruments.

### Insulators

In the case of an insulator, an increase in temperature causes a corresponding increase in the internal energy of its atoms which cause free electrons to be liberated to act as additional charge carriers, effectively *reducing its resistance*. If the temperature is allowed to increase, then it will eventually lead to a complete breakdown

An Introduction to Electrical Science, Waygood, ISBN 9780815391821, 2019. © Taylor & Francis

of the insulator's ability to withstand voltage. High temperatures are a major cause of insulation breakdown, which is why it is important never to interfere with the ventilation provided with electrical equipment.

In electrical installations, whenever cables are bundled together, or installed within thermal insulation (e.g. within a thermally insulated wall), they must be *de-rated* – that is, their published current ratings must be reduced in order to prevent them from overheating, which would likely result in insulation failure.

## Carbon

Carbon is widely used in the manufacture of contacts used to connect the stationary and rotating conductors of electrical motors and generators. These contacts are termed '**brushes**'.

Carbon is ideal for this application because it is self-lubricating and is a relatively good conductor. It is a *special case* because, unlike most other conductors, its resistivity decreases when its temperature increases. In other words, it behaves in the *opposite* way to metal conductors.

## Temperature coefficient of resistance

If the resistance of a length of **copper** wire is measured at various values of temperature from around 20°C (which is normally considered to be ambient temperature) up to around 200°C, it will be seen to *increase* in a linear fashion – as represented by the solid graph line in Figure 9.1.

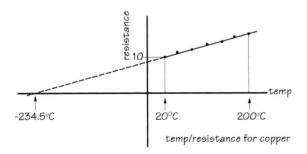

temp/resistance for copper

Figure 9.1

If this graph line is projected (a process called *extrapolation*) to the left, as represented by the broken graph line, it will eventually 'cut' the horizontal (temperature) axis at **–234.5°C**. In other words, at a temperature of –234.5°C, the resistance of the copper wire will be zero ohms.

Now, this should *not* be interpreted as meaning that, for copper, the linear relationship between resistance and temperature continues to be maintained below 20°C. What it means is that, for the range of temperatures *over which the conductor is expected to operate*, its resistance is approximately linear, and *behaves as though it would reach zero ohms at –234.5°C*.

If we now assume that the resistance at 20°C is, say, 1 Ω then, as we follow the extrapolated line back from –234.5°C to 20°C (a total of **254.5** degrees), the resistance of the copper wire will have increased by 1 Ω.

So, if a 254.5-degree rise in temperature causes the resistance of the copper wire to rise by one ohm then, for every one-degree rise in temperature, its resistance will rise by $\frac{1}{254.5}$ Ω.

> *A one-degree rise in temperature has caused a corresponding $\frac{1}{254.5}$ ohm increase in the resistance of the copper wire.*

This fraction is more commonly expressed as a decimal: $3.93 \times 10^{-3}/°C$ or, in SI, $3.93 \times 10^{-3}/K$ **(per kelvin)**.

This figure has special significance, and is called the **temperature coefficient of resistance** (symbol: $\alpha_{20}$, pronounced '*alpha-twenty*').

The subscript '20' simply indicates that the temperature coefficient of resistance is quoted for an ambient temperature of 20°C.

> The term **temperature coefficient of resistance** is defined as '*the incremental change in the resistance of any material as a result of a change in temperature*', and is considered to be a property of any conducting material. Its SI unit of measurement is '**per kelvin**' (/K) or, for everyday use, '**per degree Celsius**' (/°C).

The temperature coefficient of resistance of $3.93 \times 10^{-3}/°C$, quoted above, applies to *copper*. The values for some other common conductors are listed in Table 9.1. Be aware, though, that these values *vary somewhat according to the purity of the various metals* and, therefore, may vary according to which source the figures are obtained from.

Of particular interest in Table 9.1 are the temperature coefficients of resistance for **constantan** and for **carbon**. The figure for constantan indicates that its graph line is practically *horizontal*, whereas the figure for carbon indicates that its graph line slopes in the

Table 9.1 Temperature Coefficients of Resistance at 20°

| Conductors | $\alpha_{20}$ (/ °C) |
|---|---|
| silver | $3.80 \times 10^{-3}$ |
| copper | $3.93 \times 10^{-3}$ |
| aluminium | $3.90 \times 10^{-3}$ |
| tungsten | $4.50 \times 10^{-3}$ |
| constantan | $0.008 \times 10^{-3}$ |
| carbon | $-0.50 \times 10^{-3}$ |

*opposite direction* to that for a pure metal! We say that carbon has a **negative temperature coefficient of resistance**.

Although tables of temperature coefficients of resistance usually assume an ambient temperature of 20°C, you should also be aware that *some* tables assume an ambient temperature of 0°C (or their equivalents expressed in kelvin). Temperature coefficients of resistance at 0°C are shown as $\alpha_0$.

### Finding resistance at various temperatures

Figure 9.2 represents the effect of temperature upon a copper conductor. Two points are considered on the graph: the resistance ($R_1$) of the conductor at temperature ($T_1$) and the resistance ($R_2$) of the conductor at a second temperature ($T_2$).

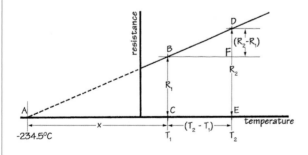

Figure 9.2

Examination of the graph will reveal two right-angled triangles, *ABC* and *BDF*. Because angles $\angle BAC$ and $\angle DBF$ are the same, these two triangles are termed **similar triangles**, and the following ratios apply:

$$\frac{BC}{DF} = \frac{AC}{BF}$$

where:

$$BC = R_1$$
$$DF = (R_2 - R_1)$$
$$AC = x$$
$$BF = (T_2 - T_1)$$

Substituting these values into our original equation:

$$\frac{R_1}{(R_2 - R_1)} = \frac{x}{(T_2 - T_1)}$$

Rearranging, to make $R_2$ the subject of the equation:

$$\frac{R_2 - R_1}{R_1} = \frac{(T_2 - T_1)}{x}$$

$$R_2 - R_1 = R_1 \frac{(T_2 - T_1)}{x}$$

$$R_2 = R_1 + \left( R_1 \frac{(T_2 - T_1)}{x} \right)$$

$$R_2 = R_1 \left( 1 + \frac{T_2 - T_1}{x} \right)$$

Rearranging further:

$$R_2 = R_1 \left( 1 + \frac{1}{x}(T_2 - T_1) \right)$$

The reciprocal of $x$ is, of course, the **temperature coefficient of resistance** at temperature $T_1$ (symbol: $\alpha_{T_1}$). So, the the equation becomes:

$$\boxed{R_2 = R_1 \left( 1 + \alpha_{T_1}(T_2 - T_1) \right)} \qquad \text{—equation (1)}$$

where:

$R_2$ = resistance at temperature $T_2$

$R_1$ = resistance at temperature $T_1$.

$\alpha_{T_1}$ = temp coefficient of resistance, quoted at temperature $T_1$.

$T_2$ = upper temperature

$T_1$ = lower temperature (ambient)

This is a general equation, and applies to a temperature coefficient of resistance quoted at *any* temperature. Most commonly, however, the coefficient is quoted at an ambient temperature of 20°C – in which case, we can replace $T_1$ with 20°C, the equation becomes:

$$R_2 = R_{20}\left(1 + \alpha_{20}(T_2 - 20)\right) \qquad \text{—equation (2)}$$

If, however, the coefficient is quotated (less commonly) at zero-degrees Celsius, then the equation becomes:

$$R_2 = R_0\left(1 + \alpha_0(T_2 - 0)\right)$$

Or simply:

$$R_2 = R_0\left(1 + \alpha_0 T_2\right) \qquad \text{—equation (3)}$$

**Worked example 1**  The resistance of a copper conductor, at 20°C, is found to be 50 Ω. Calculate its resistance at a temperature of 60°C.

**Solution**  From the table of temperature coefficients of resistance, $\alpha_{T_1} = 3.9 \times 10^{-3}$.

**either**

*(using equation 1):*

$R_2 = R_1\left(1 + \alpha_{T_1}(T_2 - T_1)\right)$

$= 50 \times \left[1 + (3.9 \times 10^{-3})(60 - 20)\right]$

$= 50 \times \left[1 + (3.9 \times 10^{-3})(40)\right]$

$= 50 \times \left[1 + (0.156)\right]$

$= 50 \times \left[1.156\right]$

$= 57.8\ \Omega$ (Answer)

**or**

*(using equation 2):*

$R_2 = R_{20}\left(1 + \alpha_{20}(T_2 - 20)\right)$

$= 50 \times \left[1 + (3.9 \times 10^{-3})(60 - 20)\right]$

$= 50 \times \left[1 + (3.9 \times 10^{-3})(40)\right]$

$= 50 \times \left[1 + (0.156)\right]$

$= 50 \times \left[1.156\right]$

$= 57.8\ \Omega$ (Answer)

## Variation on the general equation

An alternative way of approaching this topic results in the following variation of the **general equation**, which can be used to find the resistance at *any* temperature, *without the need to know an initial resistance at a reference temperature of either 0°C or 20°C* – providing you know the *coefficient ($\alpha_0$) quoted at zero degrees Celsius.*

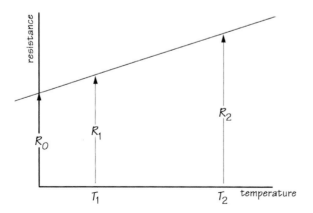

Figure 9.3

In the resistance/temperature graph shown in Figure 9.3, the resistance, $R_1$ is found using equation (3), above, that we have already learnt, i.e:

$$R_1 = R_0(1 + \alpha_0 T_1) \qquad \text{—equation (4)}$$

Similarly, the resistance $R_2$ is found from:

$$R_2 = R_0(1 + \alpha_0 T_2) \qquad \text{—equation (3)}$$

If we now divide equation (3) by equation (4),

$$\frac{R_2}{R_1} = \frac{\cancel{R_0}(1 + \alpha_0 T_2)}{\cancel{R_0}(1 + \alpha_0 T_1)}$$

As you can see, the $R_0$ appearing in the numerator and denominator cancel, giving us the following equation:

$$\boxed{\frac{R_2}{R_1} = \frac{(1 + \alpha_0 T_2)}{(1 + \alpha_0 T_1)}} \qquad \text{—equation (5)}$$

where:

$R_2$ = final resistance

$R_1$ = initial resistance

$T_2$ = final temperature

$T_1$ = initial temperature

$\alpha_0$ = temp coefficient at 0°

**Worked example 2**  The resistance of a coil at the beginning of an experiment is found to be 25 Ω at an ambient temperature of 18°C. What will be its resistance at a temperature of 300°C? Take the coil's temperature coefficient of resistance as 0.000 24/°C at 0°C.

**Solution**

$$\frac{R_2}{R_1} = \frac{(1 + \alpha_0 T_2)}{(1 + \alpha_0 T_1)} \qquad \text{—equation (5)}$$

$$\frac{R_2}{25} = \frac{(1 + 0.000\ 24 \times 300)}{(1 + 0.000\ 24 \times 18)}$$

$$R_2 = 25 \times \frac{(1 + 0.000\ 24 \times 300)}{(1 + 0.000\ 24 \times 18)}$$

$$= 25 \times \frac{1.072}{1.004} = 25 \times 1.068 = 26.7\ \Omega \text{ (Answer)}$$

**Worked example 3**   The resistance of a coil of copper wire at the start of an experiment is measured at 173 $\Omega$, at a temperature of 16°C. At the end of the test, the resistance was found to have risen to 215 $\Omega$. Calculate the final temperature of the coil. Take the coil's temperature coefficient of resistance as 0.004 26/°C at 0°C.

**Solution**

$$\frac{R_2}{R_1} = \frac{(1+\alpha_0 T_2)}{(1+\alpha_0 T_1)} \qquad \text{—equation (5)}$$

$$\frac{215}{173} = \frac{(1+0.004\,26\times T_2)}{(1+0.004\,26\times 16)}$$

$$1+0.004\,26\times T_2 = \frac{215}{173}\times(1+0.004\,26\times 16)$$

$$1+0.004\,26\times T_2 = \frac{215}{173}\times 1.068 = 1.327$$

$$1+0.004\,26\times T_2 = 1.327$$

$$0.004\,26\times T_2 = 1.327-1$$

$$T_2 = \frac{0.327}{0.004\,26} = 76.76°C \ (Answer)$$

## Negative temperature coefficient of resistance

Earlier in this chapter, we learnt that carbon has a *negative* temperature coefficient of resistance ($\alpha_{20}$), quoted as **−0.5×10⁻³/°C** in Table 9.1. This means that as its temperature *rises*, its resistance will *fall*.

This will be demonstrated in the following worked example.

**Worked example 4**   If the resistance of a block of carbon is 0.25 $\Omega$ at 20°C, calculate its resistance at 200°C, where $\alpha_{20}$ is −0.5 × 10⁻³/°C.

**Solution**

$$R_2 = R_{20}\left(1+\alpha_{20}(T_2-20)\right) \qquad \text{—equation (2)}$$

$$= 0.25\times\left[1+(-0.5\times 10^{-3})(200-20)\right]$$

$$= 0.25\times\left[1+(-0.5\times 10^{-3})(180)\right]$$

$$= 0.25\times\left[1+(-90\times 10^{-3})\right]$$

$$= 0.25\times\left[1-0.009\right]$$

$$= 0.25\times 0.91$$

$$= 0.228\ \Omega \ (Answer)$$

This gives carbon a characteristic which proves very useful when it is used as 'brushes' in electric motors. As the brushes heat up due to the friction of the machine's commutator, their resistance falls.

## Constantan and similar alloys

With critical measuring instrument components, it's usually essential to maintain an approximately constant resistance value over a wide range of temperature values if false readings are to be avoided. For this reason, critical circuit resistors are wound using wire manufactured from metal alloys such as 'constantan'.

'**Constantan**', also known as 'Eureka wire', is the trade-name for a copper-nickel alloy (approx. 60:40 ratio) formulated in the late 1800s by Edward Weston, an English-born American who founded the world-famous *Weston Electrical Instrument Company*, manufacturing precision measuring instruments (including their famous photographic exposure meters).

With an $\alpha_{20}$ of just **0.000 02 per degree Celsius**, constantan maintains an *approximately constant resistivity and, therefore, resistance over a wide range of temperatures*.

For example, if the resistance of a coil of constantan resistance wire, at 20°C, is 50$\Omega$, let's find out what its resistance will be at 500°C.

$$R_2 = R_{20}\left(1+\alpha_{20}(T_2-20)\right) \qquad \text{—equation (2)}$$

$$= 50\times\left[1+(0.000\,02)(500-20)\right]$$

$$= 50\times\left[1+(0.000\,02)(480)\right]$$

$$= 50\times\left[1+(0.0096)\right]$$

$$= 50\times\left[1.0096\right]$$

$$= 50.48\ \Omega \ (Answer)$$

As you can see, this is a very small increase in resistance over quite a large temperature variation.

## Summary

The **temperature coefficient of resistance** (symbol: $\alpha$) is a physical property of a material which directly affects the resistivity and, indirectly, the resistance of that material.

**Temperature coefficient of resistance** is defined as '*the incremental change in the resistance of any* *material as a result of a change in its temperature'.* Its SI unit of measurement is '**per kelvin**', although it is more commonly expressed as '**per degree Celsius**'.

The temperature coefficient of resistance of most metal conductors is **positive**, meaning that an *increase* in temperature results in an *increase* in resistance. Most insulating materials (as well as the conductor, carbon), however, have a **negative** temperature coefficient of resistance, meaning that their resistances will *decrease* as the temperature *increases*.

The value of the temperature coefficient of resistance varies according to its **reference temperature**. Whereas some tables list temperature coefficients of resistance at an ambient temperature of 20°C (shown as '$\alpha_{20}$'), other tables list the coefficients at 0°C (shown as '$\alpha_0$'). *So care must be taken when reading these tables.*

To find the resistance of a material at any given temperature, the following equations may be used.

A **general equation**, which can be used for *any* reference temperature ($\alpha_{T_1}$), where $R_2$ is the final resistance, $R_1$ is the initial resistance, $T_2$ is the final temperature, $T_1$ is the initial temperature, and $\alpha_{T_1}$ is the coefficient at $T_1$:

$$R_2 = R_1 \left(1 + \alpha_{T_1}(T_2 - T_1)\right) \qquad \text{—equation (1)}$$

The following variation on the above equation can be used when the reference temperature is 20°C, where

$R_2$ is the final resistance, $R_{20}$ is the initial resistance at 20°C, $T_2$ is the final temperature, and $\alpha_{20}$ is the coefficient at 20°C:

$$R_2 = R_{20}\left(1 + \alpha_{20}(T_2 - 20)\right) \qquad \text{—equation (2)}$$

If the reference temperature is 0°C, then the following variation of the equation applies:

$$R_2 = R_0\left(1 + \alpha_0 T_2\right) \qquad \text{—equation (3)}$$

Finally, if the value of $\alpha_0$ (coefficient at zero degrees Celsius) is known, then the following variation of the equation may be used, where $R_2$ is the final resistance, $R_1$ is the initial resistance, $T_2$ is the final temperature, and $T_1$ is the initial temperature:

$$\frac{R_2}{R_1} = \frac{(1 + \alpha_0 T_2)}{(1 + \alpha_0 T_1)} \qquad \text{—equation (5)}$$

## Review your learning

Now that we've completed this chapter, we need to examine the **objectives** listed at its start. Placing '*Can I…*' at the beginning, and a question mark at the end, of each objective turns that objective into a **test item**. If we can answer each of those test items, then we've met the objectives of this chapter.

**Online resources**

The companion website to this book contains further resources relating to this chapter. The website can be accessed via the following link:
www.routledge.com/cw/waygood

# Chapter 10

# Ohm's Law of constant proportionality

## Objectives

On completion of this chapter, you should be able to

1 given any two of the following quantities, calculate the third:
   - voltage
   - current
   - resistance.
2 state Ohm's Law.
3 explain the conditions under which Ohm's Law applies.
4 explain the difference between 'linear' ('ohmic') and 'non-linear' ('non-ohmic') materials and circuit components.
5 discuss whether or not Ohm's 'Law' should be considered as being a true scientific 'law'.

## Relationship between resistance, voltage and current

In the earlier chapter on **Resistance**, we learnt that

> the **ohm** is defined as electrical resistance between two points along a conductor such that, when a potential difference of one volt is applied between those points, a current of one ampere results.

In other words, an **ohm** is equivalent to a **volt per ampere**.

We can express that as:

$$\text{ohm} = \frac{\text{volt}}{\text{ampere}}$$

Rewriting this in terms of *quantities*, rather than of units, we have:

$$R = \frac{E}{I} \qquad \text{—equation (1)}$$

This equation can be rewritten, making **E** and **I** its subjects, as follows:

$$\boxed{E = IR} \text{—equation (2) and } \boxed{I = \frac{E}{R}} \text{—equation (3)}$$

These three equations are, arguably, the most fundamental and, certainly, amongst the most important equations in electrical engineering showing as they do the relationship between the three fundamental quantities: *potential difference (E), current (I) and resistance (R)*.

It's *very* important to understand that equation (1) simply tells us *what the resistance **happens to be** for any particular ratio of voltage to current*. This ratio does *not* determine the resistance. Resistance, as we learnt earlier, is determined by a conductor's length, cross-sectional area and resistivity (which, in turn, is affected by temperature) – it is not directly affected by either voltage or current.

> **Worked example 1**   What is the resistance of a conductor when a potential difference of 100 V results in a current of 50 mA?
>
> **Solution**
>
> $$R = \frac{E}{I} = \frac{100}{50 \times 10^{-3}} = 2 \times 10^3 \ \Omega$$
>
> or 2 kΩ (Answer)

An Introduction to Electrical Science, Waygood, ISBN 9780815391821, 2019. © Taylor & Francis

**Worked example 2**    What potential difference would produce a current of 3 A in a conductor of resistance 0.002 Ω?

**Solution**

$$E = IR = 3 \times 0.002 = 0.006 \text{ V (Answer)}$$

**Worked example 3**    What value of current would result if a potential difference of 120 V was applied across a conductor of resistance 20 kΩ?

**Solution**

$$I = \frac{E}{R} = \frac{120}{20 \times 10^3} = 6 \times 10^{-3} \text{ A or } 6 \text{ mA (Answer)}$$

Many students (not to mention qualified electricians and technicans!) believe that equation (3) is an expression of '**Ohm's Law**'. This is hardly surprising, as a great many textbooks reinforce this belief, by insisting that:

> **Ohm's Law** states that 'the current in a conductor is directly proportional to the voltage across that conductor, and inversely proportional to its resistance'.

*However, that is <u>not</u> what Ohm's Law says!*

Unfortunately, this is so well ingrained into the teaching of electrical science that equation (3) will likely to continue to be known as the '**formula for Ohm's Law**'!

Equation (3) will *always* tell us what the current *happens to be* for a *particular* combination of potential difference and resistance. And it applies under *all* circumstances, *whether a circuit obeys Ohm's Law or not!*

And most circuits do ***not*** 'obey' Ohm's Law!

So *what*, then, is Ohm's Law?

## Ohm's Law

In 1827, a physics teacher from Cologne in the German Rhineland, by the name of **Georg Simon Ohm** (1789–1854; Figure 10.1), published the results of a series of experiments in which he had established a relationship between *the current in a wire* and *the potential difference applied across the ends of that wire*.

Figure 10.1

Ohm's findings were initially greeted with derision by the scientific community, and were not accepted until Ohm was nearing the end of his life. Surprisingly, recognition came first not from his native Germany but, rather, from Britain when, in 1842, he was granted membership of the Royal Society.

So, let's start by quoting **Ohm's Law**.

> **Ohm's Law** states that '*the current in a wire at constant temperature is directly proportional to the potential difference across its ends*'.

We can express this as:

$$I \propto E$$
$$I = k\,E \text{ — (where } k = \text{a constant)}$$

… or, putting it another way, we can say that the *ratio of voltage to current is a constant*:

$$\frac{E}{I} = k \text{ (constant)}$$

During the course of his experiments, Ohm found that changing the type of wire resulted in a *different* ratio of voltage to current, resulting in different value to the constant, *k*. From this, he concluded that *k* must represent some form of *physical property* for a particular wire.

Furthermore, since, for a given potential difference, the current must *decrease* if the value of *k increases*, then *k* must, he concluded, represent *some form of opposition* to the drift of current.

Ohm is justly credited with naming this opposition to current '**resistance**', and he was later honoured by having his name used as the unit of measurement for resistance.

Up to this point, then, it would seem that equation (3) *does* agree with the results of Ohm's experiments. And if the resistance (i.e. the ratio of voltage to current) of a conductor *always* remained constant *for variations in applied voltage*, then Ohm's Law would *always* be true.

But, as we shall learn, the resistance of a conductor (and other circuit components) does *not* always remain constant for variations in applied voltage.

*And this means that Ohm's Law is **not** always true!*

## Ohm's experiments

Ohm's experiments are very simple by today's standards, and they can be performed with great accuracy in any school or college laboratory using readily available equipment. We'll describe this shortly.

But, in the middle of the nineteenth century, the experiments would have been extremely difficult for Ohm to perform because, of course, there were no electrical measuring instruments as we know them today available for him to use. In fact, Ohm had to design and build his own instruments. Neither were there any standard units of measurements for him to use, with the ampere and the volt not making an appearance until nearly forty years after his death. This indicates the stature of scientists like Ohm.

We have already mentioned that, for Ohm's Law to apply, the *temperature* of the conductor must remain constant but, in fact, *all* physical conditions must remain constant. For example, even bending or stretching the conductor can upset the results, as could placing the conductor within a strong perpendicular magnetic field.

At the time, Ohm believed that his 'law' applied to *all* conductors. In other words, he believed that it was *universal* – rather like Newton's Laws of Motion. But, as we shall learn, he was wrong! And *this* is where, as we shall learn, the common *misinterpretation of Ohm's Law kicks in!*

*Despite what we now know as 'Ohm's Law' being probably the best-known 'law' in electrical engineering, it remains the **most misinterpreted** and **least understood** by most students, electricians and technologists!*

So let's make it clear, right now: Ohm's Law is *not* – despite what many people think – a 'universal' law. That is to say, unlike most other scientific laws, *Ohm's Law does **not** apply to all circuits or electronic components under all circumstances!*

In fact, some scientists have even gone so far as to suggest that Ohm's Law should be 'demoted', and no longer called a 'law'(!) because it is *cannot* be

universally applied – this led one science correspondent to write:

*Ohm's law isn't a very serious law. It's the 'jaywalking' of physics. Sensible materials and devices obey it, but there are plenty of rogues out there that don't!*

Those 'sensible' materials and devices which actually obey Ohm's Law are termed '**linear**' or '**ohmic**'; those 'rogue' materials which do *not* obey Ohm's Law are called '**non-linear**' or '**non-ohmic**'.

*And there are significantly fewer 'ohmic' materials than there are 'non-ohmic'!*

Only 'linear' or 'ohmic' materials obey Ohm's Law, and there are far fewer of these than 'non-ohmic' or 'non-linear' materials.

Most pure metals are 'linear' or 'ohmic' to some extent, and obey Ohm's Law, *providing their temperatures remain approximately constant.*

But conducting liquids (electrolytes) and ionised gases are 'non-linear' or 'non-ohmic' conductors, and do *not* obey Ohm's Law.

In the case of electronic devices and components, we would discover, from examining their characteristic curves, that **diodes**, **transistors**, **thermionic valves (tubes)**, etc., are either completely or partially 'non-linear'.

## Repeating Ohm's experiments in the laboratory

### Linear conductors

To perform the modern-day equivalent of Ohm's experiment, we can use the simple circuit shown in Figure 10.2, comprising a variable-voltage supply, a voltmeter to measure changes in the potential difference across a resistor, and an ammeter to measure the resulting currents.

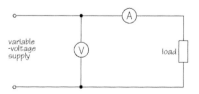

Figure 10.2

By simply increasing the potential difference across the resistor in uniform increments, and noting the corresponding values of current, we can use our results to construct a graph.

As it's important to prevent the temperature of the resistor from increasing during the experiment, we must ensure that the range of voltages we apply to the resistor don't result in currents that are high enough to cause the resistor's temperature to increase.

It is standard practice, when constructing *any* graph, to plot the quantity which we intentionally vary (called the *'independent variable'*) along the horizontal axis, and the quantity which changes as a result of changes in the first quantity (the *'dependent variable'*) along the vertical axis.

So, for our experiment, it is usual to plot the voltage along the *horizontal* axis, and the resulting current along the *vertical* axis.

For our resistor, the resulting graph should look something like the one shown in Figure 10.3.

graph for a 'linear' material that obeys Ohm's Law

Figure 10.3

The *straight-line (i.e. 'linear') graph*, of course, indicates *constant proportionality* between the dependent and independent variables. Which is precisely what led Ohm to conclude that, for each of the wires that he tested, providing the temperature (and *all* other physical conditions) remained constant, *the ratio of voltage to current is a constant*.

So we can describe **Ohm's Law** as being *a law of constant proportionality*. It's important to understand what this means because it is rarely mentioned in most textbooks.

So, how can we 'test' for the validity of Ohm's Law? It's simple! We need only ask **whether the ratio of voltage to current remains constant** *for variations in voltage*. If it does, then Ohm's Law applies.

So a proof of **Ohm's Law** would be:

> If the ratio of potential difference to current for a particular conductor or device remains proportional for variations in that potential difference, then that conductor or device obeys Ohm's Law.

Moving on, we should, hopefully, remember from our geometry lessons, that the **gradient** of a line is defined as the *change in vertical distance*, divided by the *change in horizontal distance* – as illustrated in Figure 10.4.

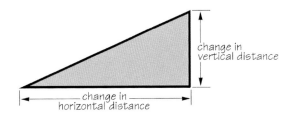

$$\text{gradient} = \frac{\text{change in vertical distance}}{\text{change in horizonal distance}}$$

Figure 10.4

For our results' graphs, this is equivalent to:

$$\text{gradient} = \frac{\text{change in current}}{\text{change in voltage}} = \frac{1}{\text{resistance}}$$

Or to put it another way, resistance must be equal to *the reciprocal of the gradient of the plotted graph line:*

$$R = \frac{1}{\text{gradient}}$$

So, for a straight-line graph, the *steeper* the gradient, the *lower* the resistance it represents. For example, in Figure 10.5, load A has the highest resistance, and load C the lowest.

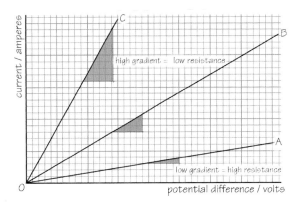

Figure 10.5

Providing their temperatures are kept constant, most metal conductors are 'linear'. Some alloys, such as *constantan* are deliberately manufactured so that their resistance remains approximately constant over a wide range of temperatures.

Fixed-value **resistors**, which are circuit components used to modify the natural resistance of a circuit, *must* be manufactured from 'linear' conductors of course, because it is important that their specified values of resistance remain constant for voltage/current variations within their power ratings.

These are the sort of results Ohm would have obtained during the course of his own experiments. And this, no doubt, would have led him to believe that this linear relationship applied universally – in other words, that it was a universal 'law'. But he was wrong!

## Non-linear conductors

One of the conditions necessary for Ohm's Law to apply is that *the temperature of the conductor must remain constant during the experiment*. So what would happen if that temperature were allowed to vary during the above experiment?

For example, what would happen if we were to repeat the same experiment but using, say, a **tungsten-filament lamp** and incrementally increased the voltage right up to the lamp's rated voltage? Well, instead of producing a straight-line graph, our experiment would produce a *curved- (i.e. a 'non-linear') line* – similar to that shown in Figure 10.6.

graph for a 'non-linear' material that does not obey Ohm's Law

**Figure 10.6**

The reason for this is that, as the current through the lamp increases during the course of the experiment, the *temperature* and, therefore, the *resistance* of its tungsten filament continually increases, and increases significantly!

In fact, the operating or 'hot' resistance of a tungsten filament can be up to around *18–20 times* its 'cold' temperature!

For example, a common 100-W/230-V incandescent lamp filament has an operating, or 'hot', resistance of around 350 Ω; so, at room temperature, its 'cold'

resistance would be around 20 Ω.

The curved-line graph confirms that changes in potential difference *do not* produce proportional changes in the resulting currents. That is, the ratio of voltage to current *varies continuously throughout the experiment* – confirming that the increasing temperature of a tungsten filament *prevents it from obeying Ohm's Law!*

For a curved-line graph, *the gradient continually changes along the graph* – as illustrated in Figure 10.7 – with the *steeper* gradient at point **A** representing the *lower* resistance, while the *lower* gradient at point **B** represents the *higher* resistance. For this particular curve, then, the resistance increases with higher values of potential difference.

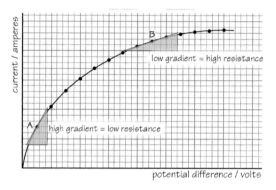

**Figure 10.7**

Although these non-linear curves mean that Ohm's Law doesn't apply, the equation ($R = E/I$) remains valid, and can be applied at *any* point along those curves, and *will tell us what the resistance happens to be only at those particular points.*

What you *cannot* do, for non-linear materials, is to use equation (1) to *predict* what the resistance will be at some other point along the curve. This is because the resistance is *not* dependent on this equation, but upon the temperature at those other points.

## Non-linear electronic components

If we were to study the voltage/current characteristic curves of various electronic components, such as diodes, transistors, thermionic valves (tubes), etc., we would find most are either *entirely* non-linear, or are a *combination* of linear/non-linear!

A particularly interesting example is the '**tunnel diode**', the application of which is well beyond the scope of this book. Instead, what we are interested in is its rather unusual voltage/current characteristic curve.

Characteristic Curve for a Tunnel Diode

Figure 10.8

As can be seen from Figure 10.8, the part of the curve A–B is *linear* and, so, over this part of the characteristic curve, the tunnel diode *obeys* Ohm's Law. However, between points B–E, the curve is *non-linear* so, over this part of the characteristic curve, the tunnel diode does *not* obey Ohm's Law.

But, interestingly, the curve between C–D is not only non-linear, but it has a *negative slope*, which means it displays a **negative-resistance** characteristic – that is, as the voltage *increases*, the current *falls* (non-linearly, in this case)! And this, of course, *completely contradicts Ohm's Law!!*

Again, the equation ($R = E/I$) can be applied at *any* point along this characteristic curve, and *will tell us what the resistance happens to be at that particular point*.

## Further worked examples

**Worked example 4**  If the potential difference across a circuit is 150 V and the circuit's resistance is 50 Ω, calculate the resulting current. Does the circuit obey Ohm's Law?

**Solution**

$$\text{e.g. } I = \frac{E}{R} = \frac{150}{50} = 3\,\text{A (Answer)}$$

*There is insufficient information to determine whether the circuit obeys Ohm's Law —there's no information supplied which tells us whether the ratio of potential difference to current remains constant at other voltages.*

**Worked example 5**  If the current through a conductor is 5 A, and its resistance is 25 Ω, find the potential difference across its ends. Does the circuit obey Ohm's Law?

**Solution**

$$E = IR = 5 \times 25 = 125\,\text{V (Answer)}$$
$$\text{e.g. } I = \frac{E}{R} = \frac{150}{50} = 3\,\text{A (Answer)}$$

There is insufficient information to determine whether the circuit obeys Ohm's Law – there's no information supplied which tells us whether the ratio of potential difference to current remains constant at other voltages.

**Worked example 6**  If a circuit's potential difference is 6 V, and current is 3 A, find the resistance of the circuit. Does the circuit obey Ohm's Law?

**Solution**

$$\text{e.g. } R = \frac{E}{I} = \frac{6}{3} = 2\,\Omega\ \text{(Answer)}$$

*There is insufficient information to determine whether the circuit obeys Ohm's Law – there's no information supplied which tells us whether the ratio of potential difference to current remains constant at other voltages.*

So, do the circuits in each of the above worked examples obey Ohm's Law? The answer is that *we simply don't know* – there's no information supplied which tells us whether the ratio of potential difference to current remains constant at other voltages. If ratio *does* remain constant, then they do; if the ratio *changes*, then they don't! But whether they obey Ohm's Law or not is irrelevant to the solutions, as the equations we have used apply under *any* circumstances.

**Worked example 7**  The voltage applied to a particular circuit is 12 V, and the resulting current is 3 A. If the voltage is then increased to 36 V, the resulting current is found to be 4 A. Does the circuit obey Ohm's Law?

**Solution**

At the beginning of the experiment,

$$R = \frac{E}{I} = \frac{12}{3} = 4\,\Omega$$

At the end of the experiment, $R = \dfrac{E}{I} = \dfrac{36}{4} = 9\,\Omega$

*As the resistance has increased, the ratio of voltage to current has changed, so Ohm's Law does not apply.*

**Worked example 8** The voltage applied to a particular circuit is 12 V, and the resulting current is 6 A. The voltage is then increased to 60 V, and the resulting current is found to be 30 A. Does the circuit obey Ohm's Law?

**Solution**

At the beginning of the experiment,

$$R = \frac{E}{I} = \frac{12}{6} = 2\,\Omega$$

At the end of the experiment, $R = \frac{E}{I} = \frac{60}{30} = 2\,\Omega$

*As the resistance has remained constant, the ratio of voltage to current has changed, so Ohm's Law does apply, within that range of voltages.*

## Summary

If you've made it as far as here, then you now understand *far* more about Ohm's Law and its limitations than most other students, electricians, or technicians!

But we've just spent an entire chapter studying Ohm's Law, only to learn that *'it isn't really* a *very serious law: the "jaywalking" of physics'*!

So it would be quite reasonable to ask, 'If it's really *not* a very "serious law", and it doesn't even apply to most materials and electronic components, then *why are we even bothering to learn about it?'*

That's a *very* good question! So, *why* are we?

Well, the problem is that Ohm's Law is included in probably *every* electrical engineering textbook that's ever been written, and is taught in *every* school and college. So, we certainly cannot ignore it. But, in so many cases, it is taught poorly, which is why so few people seem to understand its limitations. Far better that we learn it properly

and *understand its limitations* – i.e. when it applies and, importantly, when it doesn't.

But, it's the view of this book's author, at least, that *the time has probably come to quietly consign Ohm's Law to the history books*. Whether that will ever happen, of course, is another question! At the moment, it looks very unlikely!

## Misconceptions about Ohm's Law

**Ohm's Law states that 'current is directly proportional to voltage and inversely proportional to resistance'**

No. Ohm's Law simply describes the linear relationship between potential difference and current, for variations of voltage, for certain ('linear' or 'ohmic') materials under certain conditions (e.g. constant temperature).

**Ohm's Law applies to *all* circuits under *all* conditions**

No. Ohm's Law only applies to 'linear' or 'ohmic' circuits. These are limited to a relatively small range of metallic conductors, provided their temperatures and other factors aren't allowed to change. Most other materials and electronic components are 'non-linear' or 'non-ohmic'.

**The 'Ohm's Law formula' is: *I = E/R***

No. Ohm's Law *isn't* a 'formula' and, actually, makes no mention of resistance! The equation referred to is derived from the definition of the ohm!

## Review your learning

Now that we've completed this chapter, we need to examine the **objectives** listed at its start. Placing *'Can I ...'* at the beginning, and a question mark at the end, of each objective turns that objective into a **test item**. If we can answer each of those test items, then we've met the objectives of this chapter.

# Chapter 11

# Series, parallel and series-parallel circuits

## Objectives

## Objectives

On completion of this chapter, you should be able to

1. recognise a series circuit, a parallel circuit and a series-parallel circuit.
2. recognise and interpret voltage and current 'sense' arrows.
3. explain Kirchhoff's Voltage Law.
4. explain Kirchhoff's Current Law.
5. calculate the total resistance of a series resistive circuit.
6. calculate the current flow through a series resistive circuit.
7. calculate the voltage drop appearing across each resistor in a series resistive circuit.
8. explain the potential hazard of an open circuit in a series circuit.
9. calculate the total resistance of a parallel resistive circuit.
10. calculate the current flow through each branch of a parallel resistive circuit.
11. calculate the voltage drop appearing across each resistor in a parallel resistive circuit.
12. explain the major advantages of a parallel circuit.
13. calculate the total resistance of a series-parallel resistive circuit.
14. calculate the current flow through each resistor in a series-parallel resistive circuit.
15. calculate the voltage drop appearing across each resistor in a series-parallel resistive circuit.
16. calculate the voltage drop along conductors supplying a load.

## Introduction

In order for charge carriers to flow, there must be a *continuous* external conducting path, called a **circuit**, between the terminals of a source of **electromotive force** (e.g. a battery, generator, etc.) and a **load** (e.g. a lamp, electric heater, etc.). This *continuous* 'conducting path' is termed a **closed circuit**. If there is a *break* anywhere in this conducting path, then there can be no current and it's termed an **open circuit**.

**Circuits** are categorised according to the way in which they are connected. There are *four* such categories that you should be aware of.

A typical example of each type of circuit is shown in Figures 11.1–11.4.

Figure 11.1

An Introduction to Electrical Science, Waygood, ISBN 9780815391821, 2019. © Taylor & Francis

**Figure 11.2**

**Figure 11.3**

**Figure 11.4**

As we will see, a **series-parallel** circuit combines series and parallel elements which, as we shall learn, can be easily resolved into an equivalent series circuit and, ultimately, to a single component. There's an *infinite* number of such combinations.

The term '**complex circuit**' is actually rather misleading. It *doesn't* necessarily mean that the circuit is *complicated* (although it very often is!), but is simply the term given to a category into which we can lump *any* circuit that isn't either a series circuit, a parallel circuit or a series-parallel circuit.

Again, there is an infinite number of examples of complex circuits, but they are beyond the scope of this book, as they require special techniques (called '**network theorems**') to solve them – but you should be aware of their existence, and the fact that they *cannot* be simplified using the techniques you will learn in this chapter. Network theorems are covered in the companion book, ***Electrical Science for Technicians.***

*Any* electrical components may be connected in the ways described above. Since most electrical components have resistance, we will now consider the effect of connecting *resistances* in series, parallel and in series-parallel.

## Understanding sense arrows

Throughout this chapter, and elsewhere in this book, we will be using arrows in circuit diagrams to represent the 'directions', or 'sense', in which potential differences act and currents flow.

We call these arrows '**sense arrows**', and they help us form a 'mental picture' of the 'direction' or 'sense' in which the potential differences and currents act in any given circuit at any given instant in time.

Even though we have learnt that, in metallic conductors, electric current is a flow of free electrons from a negative potential to a positive potential, it has been traditional to show '**conventional flow**' (positive to negative) in circuit diagrams. A reminder that, with conventional flow, a *positive* potential is considered to be a 'higher' potential than a negative potential.

For direct-current (d.c.) circuits, the rules for sense arrows are straightforward. As we shall learn later, they also apply to alternating-current (a.c.) circuits – but we'll worry about that later.

### Voltage-source sense arrows

For potential differences, a single-headed sense arrow is used. The arrow head *always* represents the higher (positive) potential of a **voltage source *(E)*** – as illustrated in Figure 11.5.

### Current and voltage-drop sense arrows

For **current**, an arrow placed in the circuit always points *in the direction of conventional current flow (positive-to-negative)*.

Figure 11.5

Figure 11.6

Figure 11.7

For current to flow through individual circuit components, such as resistances, there must be a difference in potential across each of those components. As we have learnt, we call this difference in potential, a **voltage drop** $(U_1, U_2, etc.)$. The arrowhead for a sense arrow representing a voltage drop always points towards the *higher* (*positive*) potential – or in the *opposite direction to the current sense arrow* (see Figure 11.6).

Figure 11.7 shows **voltage source (E)**, **current (I)**, and **voltage drop (U)** sense arrows in a simple series resistive circuit.

You will notice that the sense of each voltage drop, $U_1$, $U_2$ and $U_3$, act in the opposite sense to the potential difference, $E$, across the source (battery). This agrees

with Kirchhoff's Voltage Law, as described in the next section.

Once we have established our sense arrows in a circuit, all *actual* voltages and currents acting in the *same* directions as those sense arrows are assumed to be 'positive' (in the sense of the 'directions', not polarity, in which they act). Any *actual* voltages and currents that then act in the *opposite* directions to those established sense arrows are then assumed to be 'negative'. For example, in Figure 11.8, sense arrows are established for a circuit in which a battery is supplying a resistor. If the resistor is removed, and replaced with, say, a battery charger which supplies a charging current back to the battery, then the 'charging' current is considered to be *negative* relative to the established sense arrow.

Sense arrows are particularly important when we come to study a.c. theory, and we will be making a great deal of use of them at that point.

## Kirchhoff's Laws

Before we move on to examine series, parallel and series-parallel circuits, we need to understand *two* very important 'laws' which will help us understand the behaviour of voltages and currents in electric circuits.

These two laws are called, respectively, '**Kirchhoff's Voltage Law**' and '**Kirchhoff's Current Law**', and they are both credited to the Prussian physicist, **Gustav Kirchhoff** (1824–1887) who, astonishingly, established them while he was still a university student!

These laws describe the behaviour of *voltages* and *currents* in *all* electric circuits, and an understanding of these laws is essential to understanding series, parallel, series-parallel *or* complex circuits.

### Kirchhoff's Voltage Law

**Kirchoff's Voltage Law** states that, *'for any closed loop, the sum of the voltage drops around that loop is*

battery supplying resistor

battery charger
supplying battery

Figure 11.8

*equal to the applied voltage'*. A 'voltage drop', you will recall, is the potential difference that appears across a component, such as a resistor, and which drives current through that component ($U = I\,R$).

In many textbooks, **Kirchhoff's Voltage Law** is expressed in the following form: *'In any closed loop, the algebraic sum of the voltages is zero'.* This is simply another way of stating what we have already stated, above.

Let's examine this law for various circuits, starting with a simple series circuit, as shown in Figure 11.9.

Figure 11.9

For a series circuit, there is only one 'closed loop', and this is shown in bold. If we were to individually measure the voltage drop across each resistor, $U_1$, $U_2$ and $U_3$, and compare them with the applied voltage, $E$, we would find:

$$E = U_1 + U_2 + U_3$$

To fully understand what we mean by a 'closed loop' and the relationship between the voltages around that closed loop, let's now look at another example: this time, the series-parallel circuit illustrated in Figure 11.10.

Figure 11.10

For our purposes, a 'closed loop' is *any* **route** around the circuit from the battery's positive terminal to its negative terminal.

For example, let's first examine the 'closed loop' shown in bold in Figure 11.11, which includes resistors $R_1$, $R_2$, $R_4$ and $R_5$:

Figure 11.11

If we were to measure the voltage drop across each resistor in this particular 'closed loop', $U_1$, $U_2$, $U_4$ and $U_5$, and compare them with the applied voltage, $E$, we would find:

$$E = U_1 + U_2 + U_4 + U_5$$

Now, let's look at a different 'closed loop' through the same circuit, which includes resistors $R_1$, $R_3$, $R_4$ and $R_6$ (see Figure 11.12).

Figure 11.12

If we were to measure the voltage drop across each resistor in this particular closed loop, $U_1$, $U_3$, $U_4$ and $U_6$, and compare them with the applied voltage, $E$, then we would find:

$$E = U_1 + U_3 + U_4 + U_6$$

Next, let's look at yet a different 'closed loop' through the same circuit, which includes resistors $R_1$, $R_3$, $R_4$, $R_7$ and $R_8$ (Figure 11.13).

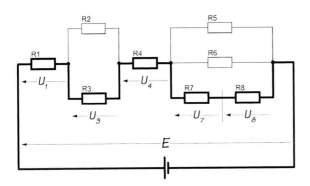

Figure 11.13

Once again, if we were to measure the voltage drop across each resistor in the closed loop, $U_1$, $U_3$, $U_4$, $U_7$ and $U_8$, and compare them with the applied voltage, $E$, we would find:

$$E = U_1 + U_3 + U_4 + U_7 + U_8$$

Finally, let's look at the 'closed loop', formed by $R_1$, $R_3$, $R_4$ and $R_5$ (Figure 11.14).

Figure 11.14

If we were to measure the voltage drop across each resistor in the closed loop, $U_1$, $U_3$, $U_4$ and $U_5$, and compare them with the applied voltage, $E$, we would find:

$$E = U_1 + U_3 + U_4 + U_5$$

> From each of the above examples, it should be clear that it doesn't matter *which* 'closed loop' route you take around *any* electric circuit, *the sum of the voltage drops around that particular loop will* **always** *equal the applied voltage*. **It's essential that you understand this concept.**

## Kirchhoff's Current Law

**Kirchhoff's Current Law** states that '*the sum of the individual currents approaching a junction is equal to the sum of the currents leaving that junction*'.

In many textbooks, **Kirchhoff's Current Law** is expressed in the following form: '*At any junction, the algebraic sum of the currents is zero*'. This is simply another way of stating what we have already stated above.

Let's examine another series-parallel circuit in order to understand this law.

Figure 11.15

In the circuit shown in Figure 11.15, there are three 'junctions', labelled **A**, **B** and **C**.

At junction **A**, the current ($I_1$) *approaching* that junction is equal to the sum of the two currents, $I_2$ and $I_3$, *leaving* that junction. i.e:

$$I_1 = I_2 + I_3$$

At junction **B**, the current ($I_2$) *approaching* that junction is equal to the sum of the two currents, $I_4$ and $I_5$, *leaving* that junction. i.e:

$$I_2 = I_4 + I_5$$

At junction **C**, the sum of the three currents ($I_3 + I_4 + I_5$) *approaching* that junction is equal to the current, $I_6$, *leaving* that junction. i.e:

$$I_3 + I_4 + I_5 = I_6$$

And, of course, current $I_6$ is exactly the same current as $I_1$.

There is a misconception amongst many students that the current approaching a junction 'splits up' at that junction. It's far more accurate to say that the current 'approaching' a junction is 'made up of' (or 'the sum of') the two currents 'leaving' that junction. In other words, the magnitude of the current approaching a junction is decided by the sum of the currents leaving that junction.

## Series circuits

If a number of different circuit components are connected 'end-to-end' (or 'daisy-chained'), so that there is only *one* continuous path for the current to flow along, then these components are said to be connected in **series**. An example of a **series circuit** is a set of old-fashioned Christmas tree lights – the disadvantage with a series circuit (as you will probably have experienced!), is that if there is a break *anywhere* in the circuit, then no current can flow.

*Any number of components can be connected in series.* However, for convenience we will consider a series circuit with just three components (in this case, resistances), each having resistances labelled: $R_1$, $R_2$ and $R_3$ (see Figure 11.16).

**Figure 11.16**

### Current in a series circuit

In a series circuit, then, because there is only *one path* for the current to flow along, the *same* current must flow through each component. We say that *'the same current is common to each component'*. Wherever we place an ammeter (a current-measuring instrument) in the circuit, it will give exactly the same reading because it's measuring exactly the same current – each of the ammeters in the series circuit shown in Figure 11.17 will register *exactly* the same value.

**Figure 11.17**

### Voltage drops in a series circuit

If we placed four voltmeters (instruments for measuring voltage) across each resistance, as shown Figure 11.17, we would find that the sum of the voltage readings across each resistance would equal the supply voltage (i.e. the potential difference across the circuit). As explained in the earlier chapter on *potential and potential difference*, these individual voltage readings are termed **voltage drops**, and are the product of the current through the individual resistance and that resistance (i.e. $U = IR$).

**Figure 11.18**

As we have seen, this relationship is credited to Kirchhoff, who realised that *the sum of the voltage drops around any closed path is equal to the supply voltage*. This may be expressed as follows:

$$E = U_1 + U_2 + U_3$$

We know, from the earlier chapter on Ohm's Law, that $E$ or $U = IR$.

So, substituting $E$ and $U$, we have:

$$IR_T = IR_1 + IR_2 + IR_3$$

Since the current is common to each resistance, we can divide throughout by $I$:

$$\frac{\cancel{I}R_T}{\cancel{I}} = \frac{\cancel{I}R_1}{\cancel{I}} + \frac{\cancel{I}R_2}{\cancel{I}} + \frac{\cancel{I}R_3}{\cancel{I}}$$

Simplifying this equation, we end up with:

$$R_T = R_1 + R_2 + R_3$$

where: $R_T$ = total resistance of the circuit.

So, for a **series circuit**, the total resistance is simply ***the sum of the individual resistances***:

$$\boxed{R_T = R_1 + R_2 + R_3 + etc.}$$

### Summary

In a *series circuit*:

- the same current flows through each component.
- the sum of the individual voltage drops will be equal to the supply voltage applied to the circuit.
- the total resistance is equal to the sum of the individual resistances.

**Worked example 1**  A circuit comprises four components, having resistances of 2 Ω, 4 Ω, 6 Ω and 8 Ω connected in series. If the circuit is connected across a 200-V supply, calculate:

a  the total resistance of the circuit
b  the current flowing
c  the voltage drop across each component.

**Solution**  Always start by sketching a circuit diagram (as shown in Figure 11.19), and inserting all the information that you are given in the question (remember, in a circuit diagram, '2R' represents '2 Ω', as explained in Chapter 7).

**Figure 11.19**

a  Total resistance:

$$R_T = R_1 + R_2 + R_3 + R_4$$
$$= 2 + 4 + 6 + 8$$
$$= 20 \ \Omega \text{ (Answer a.)}$$

b  Circuit current:

$$I = \frac{E}{R} = \frac{200}{20} = 10 \text{ A (Answer b.)}$$

c  Voltage across each component:

$$U_1 = IR_1 = 10 \times 2 = 20 \text{ V (Answer c.1)}$$

$$U_2 = IR_2 = 10 \times 4 = 40 \text{ V (Answer c.2)}$$

$$U_3 = IR_3 = 10 \times 6 = 60 \text{ V (Answer c.3)}$$

$$U_4 = IR_1 = 10 \times 8 = 80 \text{ V (Answer c.4)}$$

*To confirm the answers to part c, check that the sum of the voltage drops is equal to the supply voltage:*

$$E = U_1 + U_2 + U_3 + U_4 = 20 + 40 + 60 + 80 = 200 \text{ V}$$

## Potential hazard of series circuits

Many students believe that if there is a gap in a series circuit (e.g. when a lamp is removed), the voltage across that gap is zero. *Nothing could be further from the truth!*

In fact, the voltage across the gap will be *equal to the circuit's supply voltage!*

At mains level (i.e. 230 V) or higher supply voltages, an **open circuit** occurring in a series circuit, could result in *a potentially hazardous situation*. As no current flows, no voltage drops (i.e. the product of current and resistance) can occur across the healthy resistors – leaving *the full circuit voltage to appear across the break in the circuit*!

Let's look at the worked example above. Assume that $R_2$ fails and creates an open circuit. Since no current can now flow, then: $U_1 = 0$; $U_3 = 0$ and $U_4 = 0$. Therefore:

$$\text{since } E = U_1 + U_2 + U_3 + U_4$$
$$\text{then } U_2 = E - U_1 - U_3 - U_4$$
$$U_2 = 200 - 0 - 0 - 0$$
$$= 200 \text{ V}$$

**Potential hazard when working with series circuits!** In the event of an **open circuit in a series circuit**, *the full supply voltage will appear across the break in the circuit – creating a potentially hazardous situation.*

## Series circuits in practice

A very useful application for a series circuit is as a 'voltage divider'.

**Voltage dividers** are widely-used in electronics' circuits, where a single d.c. voltage source is often required to provide lower voltages to various load circuits fed from the same supply. In an a.c. circuit, this would be easily achieved by using a **transformer** but, as we will learn in a later chapter, transformers are alternating-current machines and they do *not* work with direct current.

A simple **voltage divider** consists of the **load** circuit, which it supplies, connected in series with a '**dropping resistor**', connected across a d.c. **voltage source**, as illustrated in Figure 11.20.

**Voltage Divider**

Figure 11.20

Note that, in Figure 11.20, we have represented the load circuit as a simple **resistance** but, it could be *anything* – including, for example, a lamp, a transistor circuit, or some other electronic device or circuit. But, whatever it is, we shall assume that this load circuit is drawing a load current from the supply.

The principle of operation of a voltage divider is best described through a simple worked example (Figure 11.21). Suppose, then, we have a particular load circuit that requires a constant operating voltage of 15 V at which it will normally draw a continuous load current of, say, 20 mA. Assuming that the supply voltage is 25 V, what value of dropping resistor will be required to achieve the necessary voltage reduction across the load circuit?

Figure 11.21

We know that the load circuit (whatever it is; it really doesn't matter!), when supplied with 15 V, will draw a load current of 25 mA, so we can determine its *equivalent resistance ($R_{load}$)* as follows:

$$R_{load} = \frac{U_{load}}{I_{load}} = \frac{15}{20 \times 10^{-3}} = 750\,\Omega$$

In order to drop the supply voltage of 25 V down to 15 V across the load circuit, the amount of voltage we must 'lose' across the dropping resistor is:

$$U_{dropping\ resistor} = E - U_{load} = 25 - 15 = 10\,\text{V}$$

So, now we can determine the necessary resistance of the dropping resistor:

$$\text{dropping resistance} = \frac{U_{dropping\ resistor}}{I_{load}}$$

$$= \frac{10}{20 \times 10^{-3}} = 500\,\Omega\ \text{(Answer)}$$

So, by placing a 500-$\Omega$ resistor in series with our load circuit, we have achieved our aim: to provide a voltage of 15 V across that load circuit.

This 'voltage divider' method of supplying a lower voltage to a load circuit is simple, effective, and widely-used in electronic circuits —*providing the load current is constant, and **not** variable!*

But, if the load current does change (due, for example, to a change in the resistance of the load circuit), then the voltage drop across the dropping resistor will also change – resulting in *a change in the voltage being applied to the load circuit.*

For example, let's look at what will happen if the load current, described in the above example, should change from 20 mA to, say, 10 mA.

This will cause the voltage drop across the 500-$\Omega$ dropping resistor to *fall* from 10 V down to:

$$U_{dropping\ resistor} = R_{dropping\ resistor} \times I_{load}$$

$$= 500 \times \left(10 \times 10^{-3}\right) = 5\,\text{V}$$

…which, in turn, will cause the voltage across the load circuit to *rise* to:

$$U_{load} = E_{supply} - U_{dropping\ resistor} = 25 - 5 = 20\,\text{V}$$

…and this increase in load voltage is too high, then it *could* cause the load circuit (which, remember, was designed to operate at 15 V) to fail!

In many cases, a load circuit requires a specific voltage at which to operate, but (depending on its function) *its load current may vary significantly* – certainly possible if, for example, the load circuit is a transistor.

In the above example, the worse-case scenario would be for the load current to fall to zero, in which case *no* voltage drop will appear across the dropping resistor *and the full 25-V supply voltage will be applied to the load circuit*, almost inevitably causing to it to break down.

This can be prevented from happening by connecting what is termed a '**bleed resistor**' in parallel with the load.

The resistance-value of the bleed resistor is chosen to draw a continuous current of between 10 – 25% of the total current drawn from the supply. This will guarantee a continous current through the dropping resistor in the event of the load current drawn by the load circuit falling to zero, thus guaranteeing a voltage drop across the dropping resistor, and preventing the load voltage from rising to that of the supply voltage level.

## Parallel circuits

When individual components are connected, as shown in Figure 11.21, where there is *more* than one path for current flow, then the components are said to be connected in **parallel**. Each of these individual paths is termed a **branch**.

One advantage of a **parallel circuit** is that, should a break occur in one branch, it will *not* affect the operation of the components in the other branches, as they are still connected to the supply voltage.

*Any number of components can be connected in parallel.* Again, for the sake of convenience, we will consider a parallel circuit with just three components, each having resistances labelled: $R_1$, $R_2$ and $R_3$.

Figure 11.22

## Voltage in a parallel circuit

Figure 11.23

The supply voltage ($E$) applied across a parallel circuit *is common to each branch* – regardless of the number of branches (Figure 11.23).

$$E = U_1 = U_2 = U_3 = etc.$$

This complies with Kirchhoff's Voltage Law, where *each branch represents an individual closed loop* (i.e. an alternative route through the circuit), so the voltage drop across a component in an individual branch will equal the supply voltage.

This is the *second major advantage* of a parallel circuit, and is the reason why most everyday circuits are connected in parallel – it ensures that *the same voltage is applied across every component*. For example, every circuit in a house is connected in parallel, ensuring that 230 V (or 120 V in North America) will appear across every component (individual lamps, socket outlets, etc.).

## Current in a parallel circuit

If we placed ammeters into a parallel circuit, as shown in Figure 11.24, we would find that the sum of the current readings in each branch would equal the current drawn from the supply.

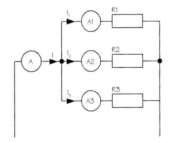

Figure 11.24

As we have learned, this relationship was discovered by Kirchhoff, who realised that *the sum of the currents approaching any junction in a circuit is equal to the sum of the currents leaving the same junction*, and can be expressed as follows:

$$I = I_1 + I_2 + I_3$$

We know, from Ohm's Law, that $I = \dfrac{E}{R}$

So, substituting for *I*, we have:

$$\frac{E}{R_T} = \frac{E}{R_1} + \frac{E}{R_2} + \frac{E}{R_3}$$

Since the supply voltage, $E$, is common to each resistance, we can divide throughout by $E$:

$$\frac{\cancel{E}}{\cancel{E}R_T} = \frac{\cancel{E}}{\cancel{E}R_1} + \frac{\cancel{E}}{\cancel{E}R_2} + \frac{\cancel{E}}{\cancel{E}R_3}$$

Simplifying this equation, we end up with:

$$\frac{1}{R_T} = \frac{1}{R_1} + \frac{1}{R_2} + \frac{1}{R_3}$$

where: $R_T$ = total resistance of the circuit.

So, for a **parallel circuit**, the total resistance is given by:

$$\boxed{\frac{1}{R_T} = \frac{1}{R_1} + \frac{1}{R_2} + \frac{1}{R_3}}$$

We could also express this relationship in terms of **conductance (G)**, expressed in siemens (S), as shown below. Remember, conductance is simply the *reciprocal* of resistance.

$$\boxed{G_T = G_1 + G_2 + G_3}$$

Once we have found the total conductance, we simply invert that figure to determine the total resistance.

There is a *very* useful 'shortcut' variation of the usual equation that we can use whenever we need to determine the equivalent resistance of two resistances. It's called the **'product over sum'** method, and is expressed as:

$$\boxed{R_T = \frac{R_1 R_2}{R_1 + R_2}}$$

## Summary

In a *parallel* circuit:

- the supply voltage appears across each component.
- the sum of the individual branch currents will be equal to the supply current.
- the total resistance is found from the equation:
$$\frac{1}{R_T} = \frac{1}{R_1} + \frac{1}{R_2} + \frac{1}{R_3} + \text{etc.}$$
- the total conductance is found from:
$$G_T = G_1 + G_2 + G_3$$
- for *two* resistances in parallel, we can use:
$$R_T = \frac{R_1 R_2}{R_1 + R_2}$$

**Worked example 2**   A circuit comprises four components, each having a resistance of 2 Ω, 4Ω, 6 Ω and 8 Ω respectively. If the circuit is connected across a 96-V supply, calculate:

a   the current through each branch
b   the supply current
c   the total resistance.

**Solution**   Always start by sketching a circuit diagram (as shown in Figure 11.25), and inserting all the information that you are given in the question (remember, in a circuit diagram, '2R' represents '2 Ω', as explained in Chapter 7).

a   Current through each branch:
Since the supply voltage, $E$ (= 96 V), is common to each branch

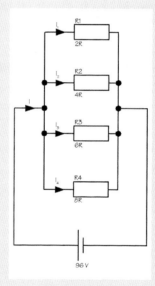

**Figure 11.25**

$$I_1 = \frac{E}{R_1} = \frac{96}{2} = 48 \text{ A (Answer a.1)}$$

$$I_2 = \frac{E}{R_2} = \frac{96}{4} = 24 \text{ A (Answer a.2)}$$

$$I_3 = \frac{E}{R_3} = \frac{96}{6} = 16 \text{ A (Answer a.3)}$$

$$I_4 = \frac{E}{R_4} = \frac{96}{8} = 12 \text{ A (Answer a.4)}$$

b   Supply current:
Applying Kirchhoff's Current Law

$$I = I_1 + I_2 + I_3 + I_4$$
$$= 48 + 24 + 16 + 12$$
$$= 100 \text{ A (Answer b)}$$

c   Total resistance:

This can be determined in either of *two* ways:

*either:*

$$\frac{1}{R_T} = \frac{1}{R_1} + \frac{1}{R_2} + \frac{1}{R_3} + \frac{1}{R_4}$$

$$\frac{1}{R_T} = \frac{1}{2} + \frac{1}{4} + \frac{1}{6} + \frac{1}{8}$$

Using the lowest common factor:

$$\frac{1}{R_T} = \frac{12 + 6 + 4 + 3}{24}$$

$$\frac{1}{R_T} = \frac{25}{24}$$

$$R_T = \frac{24}{25}$$

$$= 0.96 \ \Omega \text{ (Answer c.)}$$

*or you can use your calculator!*

*or:*

$$R_T = \frac{E}{I} = \frac{96}{100} = 0.96 \ \Omega \text{ (Answer c.)}$$

**Note**: In any parallel circuit, the total resistance is *always* less than the lowest-value branch resistance.

**Worked example 3**   A circuit comprises two resistors, of 10 Ω and 5 Ω respectively, connected in parallel. What is the equivalent resistance of this combination?

**Solution**

Using the '*product over sum*' method:

$$R = \frac{R_1 R_2}{R_1 + R_2} = \frac{10 \times 5}{10 + 5} = \frac{50}{15} = 3.33 \ \Omega \text{ (Answer)}$$

## Parallel circuits in practice

Individual electrical components (lights, socket outlets, etc.) are *all* connected in parallel with each other to ensure that they share the same supply voltage. This is essential because, as we shall learn later, for any component to operate at its rated power, it *must* be subjected to its rated voltage (which must correspond to the supply voltage). This is true whether you are looking at the wiring system of a house or the wiring system in a car.

## Series-parallel circuits

A **series-parallel circuit** (some textbooks call these '*combinational circuits*') is a circuit that combines the characteristics of series and parallel circuits. Unfortunately, there is an infinite number of possible combinations but, fortunately, all such circuits can be solved by applying the same logical approach as we will now apply to the following examples.

Essentially, this approach comprises reducing any resistances connected in series to a single equivalent resistance, any resistances connected in parallel to a single equivalent resistance ... and continuing this process until we end up with a single resistance.

### Examples of series-parallel circuits

Figure 11.26

Figure 11.27

Figure 11.28

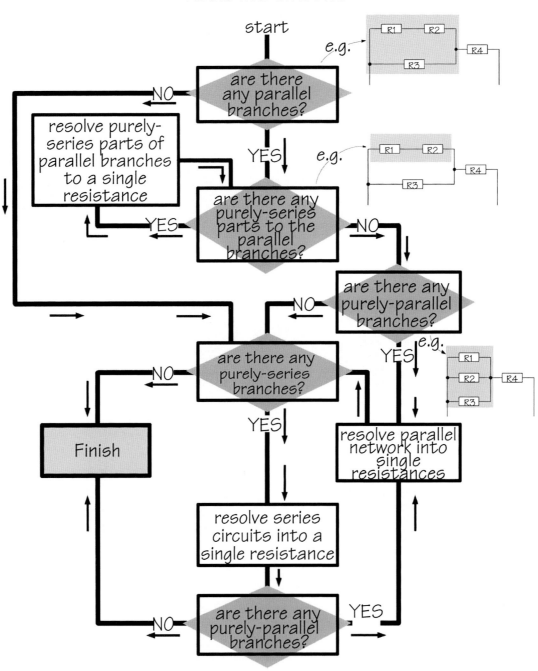

Figure 11.29

The circuits shown in Figures 11.26–11.28 are simply representative examples of series-parallel circuits. In Figure 11.26, there are two resistors in series in the upper branch, which is in parallel with a third resistor in the lower branch. In Figure 11.27, there are two resistors in parallel with each other and this combination is in series with a third resistor. In Figure 11.28 we have two resistors in parallel with each other; this combination is in series with a third (forming the upper branch), and this combination is in parallel with a fourth resistor in the lower branch.

Unfortunately, as we have already learnt, there is an *infinite* number of combinations of circuit that can be classified as series-parallel!

## Solving series-parallel circuits

'Solving' a typical series-parallel circuit means that we must be able to: (a) *determine its total equivalent resistance*, and (b) *determine the voltage drop across, and the current through, each component.*

Determining the **total equivalent resistance** of a series-parallel circuit requires a logical step-by-step approach, which you cannot reasonably be expected to develop instantly! So to help you develop the skills necessary to solve series-parallel circuits, the algorithm in Figure 11.29 may prove a helpful tool.

Work through the following example, using the algorithm. Then apply it to other circuits until you are confident of the logical sequence of steps necessary for solving them. Then, phase out your use of the algorithm as you gain confidence.

Let's use this algorithm to solve the relatively simple series parallel circuit shown in Figure 11.30. We are required to find (a) its total resistance, (b) the current through, and the voltage drop across each component.

Figure 11.30

So, starting at the top of the algorithm, we are asked '**Are there any parallel branches?**' The answer is, of course, 'yes' – we've highlighted this in Figure 11.31.

Figure 11.31

The next question is '**Are there any purely series parts to the parallel branches?**' Again, the answer is 'yes', as highlighted in Figure 11.32.

Figure 11.32

The algorithm now tells us to '**resolve purely series parts of parallel branches to a single resistance**'. So, let's go ahead and do that (we'll call the resulting resistance $R_A$):

$$R_A = R_1 + R_2 = 2 + 4 = 6\ \Omega$$

Having done this, we can redraw the circuit as shown in Figure 11.33.

Figure 11.33

The algorithm then repeats, '**Are there any purely series parts to the parallel branches?**'. The answer is 'no', so we move on to the next question, '**Are there any parallel branches?**' and, of course, the answer is 'yes', so we are instructed to '**resolve parallel network into a single resistance**'. So, let's do that (we'll call the resulting resistance, $R_B$):

$$\frac{1}{R_B} = \frac{1}{R_A} + \frac{1}{R_3}$$

$$\frac{1}{R_B} = \frac{1}{6} + \frac{1}{12}$$

$$\frac{1}{R_B} = \frac{2+1}{12}$$

$$\frac{1}{R_B} = \frac{3}{12}$$

$$R_B = \frac{12}{3}$$

$$= 4\,\Omega$$

The equivalent circuit now looks like that shown in Figure 11.34.

two resistors in series

**Figure 11.34**

The algorithm now asks, '**Are there any purely series branches?**', to which the answer is 'yes', so we are instructed to '**resolve series circuit into a single resistance**'. Let's do that (we'll call the resulting resistance $R_T$):

$$R_T = R_B + R_4 = 4 + 16 = 20\,\Omega \text{ Answer (a)}$$

So, *by following our algorithm*, we have resolved (simpified) the series-parallel circuit into one equivalent resistance of 20 Ω (as shown in Figure 11.35).

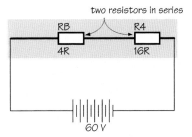

single equivalent resistance

**Figure 11.35**

We have been asked to determine the *currents* flowing through, and the *voltage drops* across, each of the components. So, we'll start by redrawing the original circuit, and labelling all the voltage drops and currents (see Figure 11.36).

**Figure 11.36**

Before beginning, though, it's worth pointing out that *we cannot necessarily work out the currents and voltages in the same sequence in which they have been labelled.*

So, let's examine the circuit. We've labelled the supply voltage *I*. In accordance with Kirchhoff's Current Law, this current divides at point A in the circuit, with $I_1$ flowing through the upper branch of the parallel combination, and $I_2$ flowing through the lower branch. At point B, these two currents recombine and flow through $R_4$ as the supply current, *I*.

As we have already worked out the *total resistance of the circuit*, we can find the value of the supply current very easily:

$$I = \frac{E}{R_T} = \frac{60}{20} = 3 \text{ A (Answer)}$$

So, we can now quite easily find the voltage drop across $R_4$, which we'll label $U_4$:

$$U_4 = IR_4 = 3 \times 16 = 48 \text{ V (Answer)}$$

Now, we know, from Kirchhoff's Voltage Law that, in any closed loop, the sum of the individual voltage drops will equal the supply voltage. So, if we take the loop or path through $R_3$ (i.e. the lower branch of the parallel combination) and $R_4$, then:
since: $E = U_3 + U_4$
then: $U_3 = E - U_4 = 60 - 48 = 12$ V (Answer)

We can now work out the value of the current, $I_2$, through the lower branch of the parallel combination:

$$I_2 = \frac{U_3}{R_3} = \frac{12}{12} = 1\,\text{A (Answer)}$$

If we apply Kirchhoff's Current Law to junction A, we can find the value of current, $I_1$, flowing through the top branch of the parallel combination:

since: $I = I_1 + I_2$

then: $I_2 = I - I_1 = 3 - 1 = 2$ A (Answer)

Finally, we are now able to determine the voltage drops, $U_1$ and $U_2$:

$$U_1 = IR_1 = 2 \times 2 = 4\,\text{V (Answer)}$$

$$U_2 = IR_2 = 2 \times 4 = 8\,\text{V (Answer)}$$

So, solving a series-parallel circuit, regardless of its complexity, involves the logical, step-by-step approach demonstrated in the above example.

Once again, you are urged not to *rely* on using the algorithm described above; instead, use it to help you *understand* the logical process for solving series-parallel circuits. Use it to solve a few such circuits, then try to solve them *without* the algorithm.

## Voltage drops along cables

The lines that represent the connections (the 'wires', if you like) between components in the schematic diagrams, shown in this chapter, are considered to have no resistance. But 'real' wires, of course, *do* have resistance – albeit *low* values of resistance.

This means that, whenever a load current passes through a real wire or conductor, a voltage drop will occur along its length. So the potential difference appearing across the load must equal the supply voltage, *less* the voltage drop along the conductors supplying that load – that is:

*potential difference across load = supply voltage – conductor voltage drops*

Figure 11.37 should make this clear. The two voltmeters will measure the voltage drops along each of the two supply conductors.

This, of course, is really a simple series circuit, in which the resistance of each conductor is in series with the resistance of the load.

So the potential difference *(U)* appearing across the load will be the difference between the supply voltage *(E)* and the sum of the two voltage drops (voltmeter readings $V_a$ and $V_b$) along the two conductors:

$$U = E - (I_{load}R_a + I_{load}R_b)$$
$$= E - I_{load}(R_a + R_b)$$

Normally, both conductors will have the same cross-sectional area and length and, so, will have identical resistances. So the above equation could then be simplified to:

$$U = E - 2(I_{load}R)$$

. . . where $R$ is the resistance of either of the two conductors.

It is very important, of course, that the voltage drops along conductors are not excessive or it may result in a deterioration in the performance of the load being supplied. Lighting, in particular, is very susceptible to reduced voltage, and dim lighting usually indicates a low voltage. This is why the *IET Wiring Regulations* specify that the voltage drop between the supply terminals and any fixed current-using devices must not exceed **3 per cent** in the case of lighting circuits, whereas it must

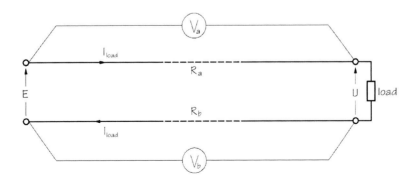

Figure 11.37

not exceed **5 per cent** for other circuits. For a nominal supply voltage of 230 V, this works out at 6.9 V for lighting circuits, and 11.5 V for all other circuits.

For the more common two-core cables used in residential buildings, the resistance (at 20°C) of each core, expressed in milliohms per metre, is shown in Table 11.1.

**Table 11.1**

| Cross-sectional area | Applications | Resistance in mΩ/m |
|---|---|---|
| 1.0 mm² | lighting circuits | 18.10 |
| 1.5 mm² | lighting circuits | 12.10 |
| 2.5 mm² | ring main and radial power circuits | 7.41 |
| 4.0 mm² | radial power circuits | 4.61 |

**Worked example 4**   The resistance of the heating element of an electric iron is 180 Ω. It is connected across a 230-V supply using a cable comprising a pair of conductors, each having a resistance of 1.2 Ω. Calculate (a) the current drawn by the electric iron, (b) the voltage across the iron's heating element, (c) the total voltage drop along the cable and (d) each core.

**Solution**

**Figure 11.38**

a   The load current will be the supply voltage divided by the total resistance of the circuit:

$$I = \frac{E}{\left(R_{load} + R_a + R_b\right)} = \frac{230}{\left(180 + 1.2 + 1.2\right)} = \frac{230}{182.4}$$
$$= 1.26 \text{ A (Answer a.)}$$

b   The voltage *(U)* appearing across the iron's heating element will be the product of the load current and the resistance of the heating element:

$$U = IR_{load} = 1.26 \times 180 = 226.8 \text{ V (Answer b.)}$$

c   The voltage drop along the cable will be the difference between the supply voltage *(E)* and the voltage appearing across the load *(U)*:

$$\text{cable voltage drop} = E - U$$
$$= 230 - 226.8 = 3.2 \text{ V (Answer c.)}$$

d   So, the voltage drop across each core must be:

$$R_a = R_b = \frac{3.2}{2} = 1.6 \text{ V (Answer d.)}$$

## Review your learning

Now that we've completed this chapter, we need to examine the **objectives** listed at its start. Placing '*Can I...*' at the beginning, and a question mark at the end, of each objective turns that objective into a **test item**. If we can answer each of those test items, then we've met the objectives of this chapter.

# Electrical measurements

On completion of this chapter, you must be able to

1 recognise and name the electrical measuring instruments used to measure current, voltage and resistance.
2 explain how an ammeter and a voltmeter must be inserted into a circuit in order to measure electric current and voltage, respectively.
3 explain how an ohmmeter is used to measure resistance.
4 describe the main construction features of a moving-coil instrument, together with the function of each.
5 describe the steps necessary to prepare an analogue multimeter for taking a measurement.
6 compare an analogue multimeter with a digital multimeter, in terms of their advantages/ disadvantages.
7 identify the main features of a typical digital multimeter.

## Introduction

In the earlier chapter on *Ohm's Law*, we learnt how the German schoolmaster discovered the relationship between voltage and current, and recognised that a circuit had resistance. By today's standards, his experiments appear childishly simple and can easily be performed in a school or college laboratory.

'Simple', that is, until we consider that, in Ohm's time, there were *no* instruments for accurately measuring voltage or current. Furthermore, there were *no* 'volts' and *no* 'amperes', as these units of measurement didn't exist until nearly thirty years after

his death! And, of course, there was *no* 'resistance', as this was a word coined by Ohm himself, and its unit of measurement named in his honour after he had died.

Experimental scientists, like Ohm, had to build their own measuring instruments and apparatus. To measure current, for example, Ohm constructed a primitive galvanometer, which made practical use of Ørsted's discovery that a current would cause a nearby compass needle to deflect, and the greater the current, the greater the amount of deflection. In fact, galvanometers were one of the very few electrical measuring instruments available commercially in those days, and consisted of a vertical coil of wire surrounding a horizontally balanced magnetised needle, rather like a compass. But, as Ohm knew, galvanometers built by different people would not produce standard measurements (two different galvanometers, connected to measure the same current, would likely produce completely different deflections!) – which is why he built his own. To add to these difficulties, the instrument had to be carefully aligned so that the earth's magnetic field would provide a restraining torque on the needle. Furthermore, the deflection of the magnetic needle was *not* proportional to the current!

So Ohm had (1) to make his own instruments to measure voltage and current, for which (2) there were no units of measurement, and (3) whose scales were not even proportional to the values being measured!

In light of all these difficulties, we can only admire the ingenuity of researchers, such as Ohm, and be in awe at the results they were able to achieve.

Today, we have a standard and internationally agreed set of units of measurement, and instruments that can measure those values with great accuracy and precision.

In this chapter, then, we are going to examine how we *detect* and *measure* the three basic electrical quantities: current, voltage and resistance.

An Introduction to Electrical Science, Waygood, ISBN 9780815391821, 2019. © Taylor & Francis

To do this, we use an **ammeter**, a **voltmeter** and an **ohmmeter** (Figures 12.1–12.3).

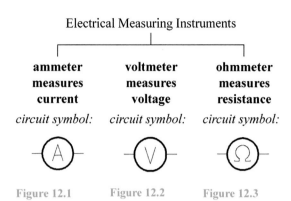

Electrical Measuring Instruments

| ammeter measures current | voltmeter measures voltage | ohmmeter measures resistance |
|---|---|---|
| circuit symbol: | circuit symbol: | circuit symbol: |

Figure 12.1    Figure 12.2    Figure 12.3

In practice, we are very unlikely to use these individual instruments outside a college laboratory. In the field, it is far more likely that we will use a **multimeter**, either analogue or digital (Figure 12.4), which *combines the functions of these three separate instruments into one.*

digital multimeter

analogue multimeter

Figure 12.4

Although analogue instruments have now largely been replaced by digital instruments, we are still likely to meet them *both* in a college laboratory and certainly in the field.

So, in this chapter, we'll first consider how precision *analogue* instruments work and how they measure current, voltage and resistance. And, then, we'll spend some time understanding how to use them, as these are rather more complicated to set up and use compared with *digital* instruments.

*And, if we can handle an analogue instrument, then we will have absolutely no trouble transferring the necessary skills to a digital instrument.*

## Analogue instruments

The '**moving-coil**', or '**d'Arsonval**', movement (named after its inventor Jacques-Arsène d'Arsonval) is the most common type of meter movement employed by *precision* analogue instruments such as ammeters, voltmeters, ohmmeters and multimeters. Its main components are illustrated in Figure 12.5.

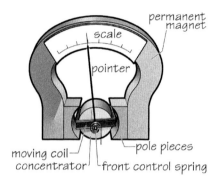

Figure 12.5

As we can see, a horseshoe-style permanent magnet, together with a pair of shaped pole-pieces and a solid iron cylinder, called a 'concentrator', forms what is termed the instrument's **magnetic circuit**. The concentrator does not rotate, but is fixed in place between the pole pieces by means of a pair of non-magnetic brackets (not illustrated). The pole pieces are curve-shaped to minimise the width of the airgap between the poles and the concentrator in order to achieve a uniformly distributed radial magnetic field of maximum intensity within that gap. This uniform field is important, because it results in a linear scale whose graduations are evenly spaced, making it (relatively) easy to read accurately.

The instrument's **coil** is manufactured from fine insulated copper wire, wound around a rectangular-shaped aluminium 'former', which is pivoted between frictionless, jewelled bearings so that it can rotate through a limited arc within the narrow air gap. A pointer, attached to the coil assembly, moves across a scale as the coil rotates through its arc. The moving coil is electrically connected to the external circuit via a pair of coiled hairsprings, located at opposite ends of the coil, which also act to *control* the movement of the coil, and to *restore* the coil and pointer to their 'rest' (or 'zero') position when there is no current flowing around the coil. These **control springs** are wound in opposite directions, to cancel the effects of expansion due to fluctuations in ambient temperature.

Figure 12.6

Figure 12.6 represents a front view of the moving coil movement showing (for the purpose of clarity) just three turns (there are, of course, many more), with a current flowing through the coil in the directions shown – i.e. with current flowing *towards us* (represented by dots) on the left-hand side of the coil, and flowing *away from us* (represented by crosses) on the right-hand side of the coil.

With current passing through the coil in this particular direction, the resulting 'motor action' results in the left-hand side of the coil being subject to an *upward-force (F)*, with the right-hand side of the coil being subject to a *downward force (F)* of equal magnitude. These forces couple together to provide a torque, which then acts to rotate the coil in a clockwise direction.

The resulting torque acting on the coil is directly proportional to the current passing through the coil.

If the moving coil were *only* subjected to this torque, then it would continue to rotate clockwise until it comes up against its stop. However, as the coil rotates clockwise, it *tightens* the two control springs (hairsprings) which together act to provide a counterclockwise '**restraining torque**'. This restraining torque is proportional to the force applied (in accordance with Hooke's Law, which states that *'the extension of a spring is directly proportional to the load applied to it'*).

When the restraining force is *exactly* equal to the deflecting force, the coil will stop rotating, and *the angle through which it has turned will be proportional to the value of the current flowing through it* – so if we were to, say, *double* the current, then we will also *double* the angle of deflection.

The **scale** across which the moving-coil's pointer moves, then, is *linear* – that is, its graduations are evenly spaced. Linear scales, of course, are relatively easy to read, which is one of the major advantages of a moving-coil movement compared with some other types of meter movement, such as the 'moving-iron' types, which have non-linear scales and are difficult to read in comparison.

*All* analogue instruments incorporate a mechanical **zero set** adjustment for precisely locating the pointer over the scale's zero position. With moving-coil instruments, this is normally provided by an eccentric-screw arrangement which will slightly adjust the 'relaxed' position of the hairsprings, enabling the pointer to be adjusted so that it hovers *exactly* over its zero position. *This action must **always** be performed before using the instrument whenever the pointer is seen to be off-zero.*

Because the direction of the torque depends upon the direction of the current flowing through the coil, *all* moving-coil instruments are d.c. instruments. However, to enable moving-coil instruments to measure a.c. values, a **rectifier** (a circuit that change a.c. into d.c.) is incorporated into these meters, and their scales calibrated to indicate **a.c. values**. Some d.c./a.c. instruments have different scales for d.c. and a.c. values, so care must be taken to read from the correct scale.

## Measuring current: ammeters

To measure **current**, the circuit should be temporarily disconnected from its supply, and must then be broken at the point where we want that current to be measured, and the ammeter inserted at that point. In other words, *an ammeter must be connected in series* with the load under test, as shown in Figure 12.7.

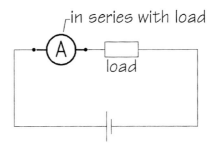

Figure 12.7

It's very important that the insertion of the ammeter into a circuit has very little effect on the circuit's existing resistance which would alter the current normally flowing in the circuit and therefore give an inaccurate reading. So, ammeters are manufactured with *extremely low* values of internal resistance.

Because ammeters have a very low internal resistance, it is vitally important that they are *never* inadvertently connected across, or in parallel with, any circuit component – and *especially* with the supply. Failure to do so will result in a short-circuit current flowing through the instrument which may damage the ammeter (although most ammeters are fused) or even result in personal injury.

> **Ammeters** have a very low internal resistance, and must *always* be connected in **series** in a circuit.

For multi-range ammeters, we should *always* start by selecting the *highest* current range and, then, select whichever range gives the *greatest deflection*, as this will give us the most accurate reading.

## Measuring voltage: voltmeters

To measure **potential-difference**, or **voltage**, a voltmeter must be connected between *two* points at different potentials. In other words, *a voltmeter must always be connected across, or in **parallel** with,* the part of the circuit under test, as shown in Figure 12.8.

**Figure 12.8**

In order to operate, a voltmeter must, of course, draw *some* current from the circuit under test to drive its operating coil, and this can lead to inaccurate results because it interferes with the normal condition of the circuit. We call this the 'loading effect' and, to minimise this 'loading effect' (and, therefore, improve the accuracy of a reading), this operating current drawn from the circuit must be as small as possible and, for this reason, voltmeters are manufactured with a *very high* value of internal resistance – usually many megohms.

> Voltmeters must *always* be connected in **parallel** in a circuit, and have a very high internal resistance.

For multi-range voltmeters, we should *always* start by selecting the highest range and, then, select whichever range gives the greatest deflection, as this will give us the most accurate reading.

## Exercise

Examine the following circuit, and identify which circles represent ammeters, and which circles represent voltmeters, by placing an 'A' or a 'V' within each circle shown in Figure 12.9.

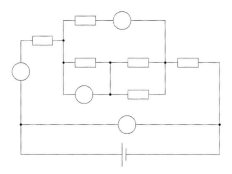

**Figure 12.9**

The answers are shown at the end of this chapter.

## Measuring resistance: ohmmeters

To measure **resistance**, we use an instrument called an **ohmmeter**, or we can use the 'ohms' range of a multimeter. Ohmmeters also provide us with a convenient way in which to check *continuity* – that is, to find out whether there are any breaks in a circuit. For example, we might want to check whether or not a fuse has 'blown'. When checking continuity, we are usually only interested in observing a deflection, and *not* necessarily the value of the resistance reading.

An ohmmeter works by using an internal battery to pass a small test current through the unknown resistance, and measuring the value of that current: the higher the resulting current, of course, the lower the resistance and *vice-versa*. Its scale, of course, is graduated in ohms and kilohms.

The schematic diagram (Figure 12.10) shows the basic internal circuit for a typical ohmmeter.

ohmmeter internal circuit

**Figure 12.10**

The ohmmeter's moving-coil movement is connected in *series* with a **battery**, a fixed-value **limiting resistor** and a pair of **terminals** across which the unknown resistance will be connected.

Connected in *parallel* with the movement is a variable **'zero-adjust' shunt resistor**, which is used to zero the instrument in order to compensate for any changes in the battery's voltage. Using this variable resistor to obtain a full-scale deflection is called '**zero-ohms adjustment**', and *this action **must** be carried out prior to taking any resistance measurement*. In the case of a multiple-range ohmmeter, the zero-ohms adjustment may also be necessary *after* changing the range, but *before* taking a new measurement. This compensates for any variations in the voltage of the instrument's built-in battery; if a zero-ohms adjustment *cannot* be achieved, then the voltage is too low and the battery must be replaced.

The function of the limiting resistor is to protect the movement from burning out, by preventing the current that flows during the zeroing process from significantly exceeding the movement's full-scale deflection current.

The scale of an ohmmeter differs from that of an ammeter or voltmeter in *two* very important ways. Firstly, its scale is *reversed* – i.e. it reads from right to left – with 'zero ohms' corresponding to its full-scale deflection. Secondly, the scale is *non-linear*, with its graduations becoming closer and closer together and, therefore, more difficult to read, at the higher values of resistance (i.e. towards the left-hand end of the scale), as illustrated in Figure 12.11.

Figure 12.11

When using an ohmmeter, we must *always* observe the following rules:

- ***Never EVER connect an ohmmeter to a live circuit*** – doing so will likely result in a burnt-out instrument, and may cause personal injury.
- Beware of measuring resistance of any component that may be connected in parallel with other components (this is not necessarily obvious), as you'll end up measuring the combined resistance of *all* those components! It may be necessary to remove at least one of the component's connections

(e.g. disconnect one end of the resistor from the circuit to which it's connected).

- When using a *multimeter* to measure resistance (or to check continuity), it is important to *switch the instrument away from its resistance setting* (preferably to a current range) when the measurement has been completed. Failure to do so might result in the battery becoming discharged should the test-leads accidentally short circuit during storage.

## Analogue multimeters

Figure 12.12

An analogue **multimeter** is (usually) a moving-coil instrument that can measure both d.c. and a.c. currents and voltages, together with resistance. Because of their versatility, multimeters are far more widely used in the field than separate ammeters, voltmeters and ohmmeters, which are mainly found in laboratories. Some multimeters can measure other quantities, too.

The first multimeter is attributed to a GPO (General Post Office) telephone engineer, Donald Macadie, who, in the late 1920s, became so fed up with having to carry around so many different test instruments, that he decided to design a single instrument which could replace them all. His multimeter design evolved into the famous '**AVO**' (the acronym for the values it was designed to measure: **a**mpere, **v**olts, and **o**hms) meter, and it went on to become a brand in its own right.

Until 1982, the GPO was the United Kingdom's state-owned sole telephone service provider.

Subscribers rented their telephones from the GPO, and the organisation was responsible for the residential telephone service wiring and its connection to the local network.

Various versions of the AVO analogue multimeter (Figure 12.12) were in continuous production, in its various versions, from the late 1920s until it finally went out of production in 2008 because its manufacturer, *Megger Group Ltd.*, could apparently no longer source its mechanical components.

Since the original avometer appeared, many other companies have produced their own high-quality analogue multimeters. In North America, for example, *Simpson* and *Conway* are well-known for their precision multimeters. Americans and Canadians often call multimeters '**VOM meters**', the acronym for **v**olts, **o**hms and **m**illiamperes (most multimeters do not measure amperes).

### Reading analogue instruments

Figure 12.13

Figure 12.13 illustrates just how complicated some analogue instruments' scales are. Which of the scales on this multi-range voltmeter we need to use depends on the instrument's selector-switch setting.

A good-quality analogue instrument is a very accurate instrument – probably *more* accurate than an inexpensive 'consumer-level' digital instrument, but probably somewhat less accurate than a professional digital multimeter. However, analogue instruments require more skill to read and, particularly, to interpolate scale readings. By 'interpolate', we mean to be able to judge values between graduations on an instrument's scale.

For example, if we refer back to Figure 12.13, if we read the bottom scale when the pointer is in position *b*, there is no question that it is indicating 1.6 V. But what about when it's in position *c*? Well, it's a little more than halfway between 2.8 V and 2.9 V, so we can

reasonably interpolate it as being 2.87 V.

So analogue instruments are undoubtedly more difficult to read accurately, compared with digital instruments, and a certain amount of preparation is necessary in order to use them properly. *We should, therefore, all develop a habit of following these guidelines:*

1 Firstly, *all* analogue instruments are designed to give accurate readings *only* when placed in a particular (usually horizontal) position, so they should *never* be propped-up at an angle, placed vertically or held in the hands – where gravitational forces may act on the pointer to give an inaccurate reading.

2 Before measuring current or voltage, the pointer must be always checked to ensure it is pointing *exactly* at zero. If it isn't, then it must be mechanically adjusted, using a small screwdriver, by means of the **zero-adjust screw**, located just below the instrument's scale. *Don't confuse this with the zero-ohms adjustment.*

3 Multimeters usually have several scales, corresponding to the different functions and ranges available, as well as for whether we are measuring a.c. or d.c. values. This can be quite confusing, and it's important to ensure we use the correct function and scale settings when taking a reading.

4 Multimeters offer different **ranges** (e.g. voltage ranges offered might be 0 V – 125 V, 0 V – 250 V, 0 V – 500 V). The *highest* range should *always* be selected first, but *whichever scale then provides the greatest deflection should **always** be used when taking the actual reading.* This is because the accuracy of any instrument is highest at its *full-scale deflection*, and decreases towards the lower-end of the scale!

5 When taking a reading, the **parallax mirror** (which runs adjacent to the graduated scale, behind the pointer) should *always* be used. Its purpose is *to ensure that your eye is **exactly** above the pointer and not to one side or the other.* When the eye is exactly above the pointer, the reflection of the pointer is hidden by the pointer – as illustrated in Figure 12.14.

Figure 12.14

6 When the pointer lies between two graduations on the scale, we have to *interpolate* that reading. By that,

we mean that when the pointer is between graduations, we have to estimate whether the pointer is nearer to the lower or the higher graduation, and by how much, in order to place a value on that reading. This requires skill, which comes with practice.

7 After completing a resistance measurement, a multimeter should *never* be left on its resistance setting – this is because, should the test-leads become accidentally short circuited, the battery will become de-energised. And, of course: no battery, no ohmmeter!

8 When transporting an analogue multimeter, if there is no 'off' setting, then it should always be set to one of its **current** ranges – this will allow the instrument to 'self damp' through one of its shunt resistances (a low-value internal resistor placed in parallel with the meter movement for the purpose of extending its range). Any movement of the coil induces a voltage into the coil, and the direction of the resulting circulating current is such that it always opposes the movement of the coil during transit.

9 Good quality analogue multimeters are far more rugged than they may appear, but they should always be treated carefully when they are transported in order to avoid damaging their movement. One of the biggest problems occurs when the meter movement is displaced from its bearings.

Following the above steps and being able to interpolate between graduations on an analogue scale are essential skills we all need to acquire when using analogue instruments.

Now, let's move on to digital instruments which, in most respects, are far easier to use and read than analogue instruments.

## Digital measuring instruments

Figure 12.15

**Digital multimeters** (Figure 12.15) have now largely replaced analogue multimeters. In fact, with very few exceptions, analogue instruments are no longer being manufactured. The term 'digital' applies not only to an instrument's liquid-crystal readout, but also to the technology it employs internally.

Analogue instruments have long-reached the limits of their development – the accuracy of their movements and their ease of reading have become about as good as they can possibly get. Digital instruments have now overtaken analogue instruments in terms of their robustness, potential accuracy and the elimination of reading errors. Some digital multimeters can also measure additional quantities, such as frequency, capacitance, etc.

So, let's briefly compare **digital multimeters** with **analogue multimeters**:

- **Analogue multimeters** have several scales for measuring current, voltage and resistance and, often, separate scales for measuring a.c. and d.c. They are, therefore, relatively difficult to read, require practice and, consequently, are prone to mistakes by the user: such as reading from the wrong scale or wrongly interpolating a particular reading.
- **Digital multimeters**, on the other hand, have a single, simple, digital readout (often with automatic range adjustment), which is very easy to read, removing the need to interpolate and virtually eliminating the chance of making a reading error.
- **Analogue multimeters** movements are relatively delicate and must be treated with care to prevent them from becoming miscalibrated or their movement damaged.
- **Digital multimeters**, however, are tough and robust, and are far less likely to become damaged. Even relatively severe external physical damage is unlikely to affect their accuracy.
- **Analogue multimeters** require a careful set-up routine prior to taking a measurement, as explained in the previous section.
- **Digital multimeters** must be set to the appropriate *function* setting (current, voltage or resistance), but many types will then automatically select the most appropriate *range* within that function, will compensate for an aging battery and all will automatically self-zero when set to measure resistance.
- **Analogue multimeters** require a battery only to measure resistance, and will read current and voltage without a battery.
- **Digital multimeters**, though, require a battery for *all* their functions, and will *not* operate at all once the battery is exhausted.

## How digital instruments work

Understanding the operation of digital instruments requires an understanding of advanced electronics and digital logic which is well beyond the scope of this book. Instead, a *greatly* simplified explanation will have to suffice.

All analogue instruments are, essentially, **ammeters** which can be adapted to measure *voltage* and *resistance*. Digital multimeters on the other hand are, essentially, **voltmeters** which can be adapted to measure *current* and *resistance*.

There are *four* major component circuits to a basic digital voltmeter: (1) a **voltage-to-time converter** circuit which opens and closes (2) a '**gate**' (an **electronic switch**) which, in turn, controls the flow of pulses from (3) a **pulse generator** circuit to (4) a **counter** circuit. These components are illustrated in the block diagram shown in Figure 12.16.

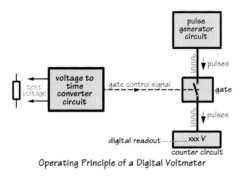

Operating Principle of a Digital Voltmeter

**Figure 12.16**

The **gate** is controlled by a digital circuit which converts the test voltage to time, with that time being directly proportional to the test voltage. The *greater the test voltage*, the *greater the time* that the gate is held open (an 'open' gate is equivalent to a 'closed' switch). The number of pulses that then pass through the gate *is proportional to the time for which it is held open* which, in turn, is proportional to the voltage being measured. So, by measuring the number of pulses passing through the open gate, the counter circuit can be calibrated to provide a digital readout directly in volts.

As already mentioned, digital instruments are all, essentially, *voltmeters*. To measure *current*, they measure the voltage drop across a very accurate internal resistance; since current is directly proportional to voltage, the readout is calibrated in amperes. To measure *resistance*, they pass a constant current through the unknown resistance and measure the resulting voltage drop, which is directly proportional

to the resistance, and the readout is calibrated in ohms.

Of course, the operating principle of an *actual* digital multimeter is significantly more complicated than this explanation – as it must be capable of measuring a.c. as well as d.c. voltages and currents over a wide range of values, as well as measuring resistance and, in some cases, other quantities too.

## Features of a digital multimeter

There is an enormous range of digital multimeters available from a great many manufacturers: from inexpensive consumer-level models to very expensive professional models. They all measure d.c. and a.c. voltages and currents, as well as resistance, but many will also have additional features, enabling them to measure frequency, capacitance, inductance, etc., and to test electronic components such as diodes and transistors.

Figure 12.17 shows the basic features that are common to most professional multimeters.

**Figure 12.17**

In this example, the black test-lead is inserted into the **common** terminal, and the red lead inserted into the appropriate function terminal: **current** (amperes), **current** (milli/microamperes) or **voltage/resistance**.

The rotary switch allows the appropriate function/range to be selected, and this selection is confirmed on the LCD display panel.

This particular instrument offers manual or automatic range selection; when 'automatic' is selected, the instrument will choose the most appropriate (most sensitive) range for any particular measurement.

This particular instrument offers audible confirmation when testing for continuity (e.g. checking whether a fuse is healthy).

Most of the preliminary set-up requirements necessary for analogue instruments (zeroing, ensuring the instrument is level, etc.) don't apply to digital instruments. Neither is there any requirement to interpolate scale readings. All of this makes digital instruments far easier to use, and less error-prone, than their analogue equivalent.

## Review your learning

Figure 12.18 shows the solution to the exercise from earlier in this chapter.

Figure 12.18

Now that we've completed this chapter, we need to examine the **objectives** listed at its start. Placing *'Can I...'* at the beginning, and a question mark at the end, of each objective turns that objective into a **test item**. If we can answer each of those test items, then we've met the objectives of this chapter.

# Chapter 13

# Cells and batteries

## Objectives

On completion of this unit, you should be able to

1 describe the function of an electrochemical cell.
2 explain the difference between a primary cell and a secondary cell.
3 describe the three components common to all electrochemical cells.
4 explain the difference between a cell and a battery.
5 explain the terms 'capacity' and 'discharge rate', as they apply to cells and batteries.
6 briefly explain the chemical process by which a simple Voltaic cell is able to separate charge in order to provide a potential difference between its plates.
7 briefly describe the construction of a Dry Leclanché cell.
8 briefly explain the advantages of miniature cells.
9 describe each of the following, as they apply to a lead-acid cell, in terms of its plates, electrolyte and relative density:
  a fully charged state
  b de-energising state
  c fully de-energised state
  d re-energising state.
10 briefly explain how a lead-acid cell's relative density is a guide to its state of charge.
11 briefly explain the disadvantages of nickel-cadmium cells, and how these are overcome using lithium-ion cells.
12 outline what is meant by the 'responsible disposal of cells and batteries'.
13 explain how a fuel cell differs from a chemical cell.
14 list and explain the functions of the four basic components of a fuel cell.
15 outline the basic principle of operation of a hydrogen/oxygen fuel cell.
16 identify the main reasons for the lack of progress in the development and acceptance of fuel cell technology.
17 determine the potential difference and internal resistance of identical
  a cells connected in series
  b cells connected in parallel
  c cells connected in series-parallel.

## Introduction

You may have heard the joke in which an American scientist is credited with inventing an 'electric rocket' which, he claims, could easily reach the moon, yet would cost the taxpayer only $5. Unfortunately, the cost of the power cord would be around $200-million!

Well, thanks to **electrochemical cells**, we *can* use many of our electrical tools, small appliances, multi-media equipment and other gadgets without having to keep them permanently plugged into our wall sockets! Although, of course, we will still require the wall socket to regularly *re-energise* those cells.

After electromagnetism, **electrochemical action** provides us with our *second most important source of electrical energy*. As well as powering our mobile phones and other electrical gadgets, **cells** and **batteries** are used to start our cars, propel submarines, provide emergency lighting, and a whole raft of other applications too numerous to mention.

In this chapter, we are going to learn about the basic behaviour of an electrochemical cell, and examine a selection of these cells and batteries together with their application in electrical engineering. Also, in this chapter, to avoid any confusion, we will consistently describe current in external circuits in terms of **electron flow**, *not* conventional flow. The use of conventional flow in this context will be unnecessarily confusing!

As we learnt in the chapter on *potential and potential difference*, an electrochemical cell uses a chemical reaction to release the energy necessary to separate charges, thus creating a potential difference across its terminals. This potential difference is then responsible for causing a drift of electric charges (free electrons) around the external circuit. These charges, of course, already exist in the conductors of the external circuit, and are not 'injected into it' by the cell! So what an electrochemical cell *does not* do, is to 'store electric charges' which it then 'pumps around' its external circuit. It is, therefore, techically incorrect to say that a cell *'stores charge'*, or that a cell *'discharges'*, or that a secondary cell can be *'recharged'*. Despite this, these terms are widely used when referring to the behaviour of cells and batteries. In this chapter, we are going to try and avoid this, by using more appropriate terms – e.g. we will say that a cell *'stores energy'*, that a cell *'de-energises'*, that a secondary cell can be *'re-energised'* and so on.

Perhaps, in school or on television, you have seen a lemon being used to demonstrate the behaviour of an electrochemical cell? By inserting two dissimilar metals, such as a galvanised nail and a copper nail, into the lemon, the resulting chemical reaction between each nail and the citric acid inside the lemon results in a small potential difference appearing across the two nails which can be measured with a voltmeter.

The same result would be obtained if, for example, we used other fruit, or even a potato, instead of the lemon. It *isn't* the lemon or the potato that is converting the energy necessary to create the resulting potential difference but, rather, *the chemical reaction that takes place within the dissimilar metals when they come into contact with an acid or alkaline.*

The 'lemon cell' comprises the same *three* components that must be present in *all* types of chemical cell: two **dissimilar electrodes** and an **electrolyte** (a dilute acid or alkaline).

The **circuit symbols** for cells and batteries are illustrated in Figure 13.1. The longer stroke always represents the positive electrode.

cell                    battery

*Circuit symbols for cells and batteries*

Figure 13.1

## Types of electrochemical cell

Electrochemical cells are either **'galvanic'** or **'electrolytic'**.

A **'galvanic cell'** creates a potential difference across its terminals, through the process of **charge separation** by the release of energy due to the chemical reaction between its electrodes and the electrolyte. Charge separation was explained in the earlier chapter on *potential difference*, and you may wish to review that chapter before continuing.

An **'electrolytic cell'**, on the other hand, uses electricity to decompose its electrolyte, a process called 'electrolysis', usually for the purpose of electroplating – such as silver or chromium plating.

Put simply, a 'galvanic cell' *produces* electrical energy, whereas an 'electrolytic cell' *uses* or *absorbs* electrical energy.

## Primary and secondary cells

Galvanic cells are broadly classified as **'primary cells'** and **'secondary cells'**, where

- **primary cells** are those in which their electrochemical reaction is irreversible, which means that they *cannot be re-energised* and, so, are disposable.
- **secondary cells** are those in which their electrochemical reaction is reversible, which allows them to be *fully re-energised* and, so, are re-useable.

While **primary cells** are *'galvanic* cells', **secondary cells** behave as *'galvanic* cells' when they are being *de-energised*, but as *'electrolytic* cells' when they are being *re-energised*.

Secondary cells (formerly called 'accumulators') are generally more economical than primary cells, despite their initially higher cost together with the cost of a recharging unit*, but that cost can be spread out over numerous discharge/recharge cycles before they finally fail. This makes secondary cells ideal for portable tools, video camcorders, etc.

*Strictly speaking, we should talk about a 're-energising unit' or a 'battery re-energiser', rather than a 'recharging unit' or 'battery recharger'. However, the terms 'recharging unit' or 'battery recharger' are so widely used that it would be impossible to change the terminology at this stage.

While *all* cells 'self de-energise' to some degree, when they are not in use, this is significantly lower with primary cells than it is with most secondary cells, meaning that primary cells usually have a very much longer shelf life than energised secondary cells. This makes disposable primary cells the better choice for applications where they see just occasional use – such as in torches (flashlights).

There are numerous different types of cell within each of these two general classifications, with some of the more common listed in Figure 13.2.

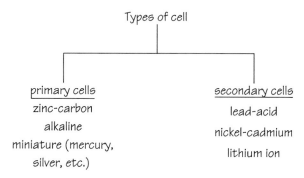

Figure 13.2

We'll examine some of these cells, in various degrees of depth, later in this chapter.

## Terminology

Before embarking on our investigation into the behaviour of primary and secondary cells, we need to understand some of the terminology and specifications that are common to both classes of cell.

## Anode and cathode

Not surprisingly, a great deal of confusion surrounds the use of the terms '**anode**' and '**cathode**', when naming a cell's electrodes. This is due to a conflict between the ways in which these terms were applied in the past, by electrical engineers and by chemists. A conflict, incidentally, which the chemists eventually won!

Accordingly, a very brief explanation of the terms 'anode' and 'cathode', *as they relate to electrochemical cells* (the terms are used differently for electronic devices!) won't go amiss! Even though we will be avoiding these terms in this chapter, you *will* come across them and you should be aware that, from the electrical point of view, *their definitions have changed over recent years*, rendering their use in older textbooks obsolete.

Contrary to popular belief amongst those who were brought up on older electrical textbooks, regardless of the type of cell (electrolytic *or* galvanic), by definition free electrons *always* travel through the external circuit **from the anode to the cathode**. It's as simple as that!

> Free electrons *always* travel through the **external** circuit **from a cell's anode to its cathode**.

So the terms 'anode' and 'cathode' are based on the *direction of electron flow*, and **not** on the polarity of the electrodes!

In other words, a cell's 'anode' and 'cathode' *change*, according to whether a cell is de-energising or is being energised – as shown in Figure 13.3. For a **galvanic cell** (a de-energising cell), *the negative plate is the anode* and *the positive plate is the cathode* whereas, for an **electrolytic cell** (an energising cell – i.e. with the current flowing in the opposite direction), *the positive plate becomes the anode* and *the negative plate becomes the cathode*.

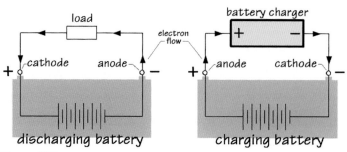

Figure 13.3

Be aware that most older textbooks invariably define the 'cathode' as being the *negative* plate and the 'anode' as the *positive* plate. *This is no longer the case!*

Because of the confusion over the use of the words 'anode' and 'cathode', we will avoid using these terms throughout the rest of this chapter. Instead, we will be referring to the electrodes as being either 'negative' or 'positive' plates.

## Capacity and discharge rate

The **capacity** of a cell is a measure of *its ability to deliver current to a load*. Capacity is related to the quantity of active material the cell contains, the surface area of its plates, and its temperature.

The **capacity** of any particular cell is determined experimentally by the manufacturer, by de-energising the cell at a constant current without exceeding a safe temperature, until the cell reaches its 'cut-off' voltage (i.e. the cell's minimum voltage, below which the cell is considered to be fully de-energised) and, then, multiplying that current by the time taken to reach that cut-off voltage. A cell's capacity, therefore, is the product of current and time and, so, is expressed in **ampere hours (A·h)**. The capacity of smaller cells is expressed in milliampere hours (mA·h).

A cell's capacity is dependent upon the temperature at which that cell is operating, and capacity is normally expressed at a temperature of 25°C.

From the above explanation of capacity, it would *seem* that a lead-acid car battery, for example, with a capacity of, say, 200 A·h, is *theoretically* capable of delivering *any* combination of current and time, whose product is 200 A·h, for example:

either: 1 A for a period of 200 h
or: 2 A for a period of 100 h
or: 10 A for a period of 20 h
or: 20 A for a period of 10 h
or even: 2000 A for a period of 6 min!
. . . and so on!

'Theoretically' maybe! But, in practice, we *have* to take care; excessively high currents cause excessively high temperatures which can actually melt a battery's electrodes or even cause a battery to explode! And 2000 A is most certainly an excessively high current! So, how do we know what the maximum 'safe' continuous de-energising current should be for a particular cell or battery?

The answer is that we must refer to another part of a battery's specification: its **discharge rate** (or, more accurately, '**de-energising rate**'), expressed in **hours**. We can use this figure, together with the corresponding capacity, to determine a cell or battery's maximum, *safe*, continuous de-energising current – that is, the maximum continuous current that can be supplied to a load without overheating the battery.

> A cell's '**discharge rate**' determines the maximum value of continuous current that cell is capable of supplying to its load without overheating.

For lead-acid cells, discharge rates are typically expressed as '**8 h**', '**10 h**' or as '**20 h**'. Often, the discharge rate is shown on specification sheets in the form: '**C/8**', '**C/10**' or '**C/20**', where '**C**' represents the capacity of the cell.

For example, the motor vehicle's lead-acid battery that we described earlier may be quoted as having a capacity/discharge rate of '**200/10**', which repre-sents a capacity of 200 A·h at a discharge rate of 10 h, which means that it is designed to supply a *maximum* continuous de-energising current of (200 ÷ 10 =) 20 A for 10 h before the cell reaches its cut-off voltage.

An alternative way of defining a cell's 'capacity' (mainly used with larger cells or batteries) is by expressing it in **watt hours (W·h)** although, strictly speaking, 'capacity' is the *wrong* word to use in this context because a watt hour is a measure of the **energy**, not of the charge, a cell can supply.

To convert the capacity of a cell, expressed in ampere hours, into the energy it can supply in watt hours, we can use the following equation:

$$\left[ \begin{array}{l} \text{energy (in watt hours)} = \text{capacity} \\ \text{(in ampere hours)} \times \text{cell voltage (in volts)} \end{array} \right]$$

For example, a 12-V lead-acid battery having a capacity of 250 A·h, will be able to supply 3000 W·h or 3 kW·h (i.e. 250 A·h ×12 V).

Now that we have learnt some of the terminology involved, let's move on and study the cells themselves.

## Primary cells

### Simple Voltaic cell

In this section, we will start by briefly examining the operation of what is known as a '**simple Voltaic cell**', whose operation is representative of the chemical process by which *all* primary cells operate, albeit with different combinations of chemicals.

The important thing to understand, not only about this particular cell, but about all cells, is that they provide the energy required to increase the potential of a charge passing though that cell. You may wish to review the chapter on *potential and potential difference*, before proceeding.

A 'simple **Voltaic** cell' is the name we give to a cell consisting of copper and zinc electrodes, inserted into an electrolyte of dilute sulfuric acid* – as shown in Figure 13.4.

*The **Royal Society of Chemistry**'s Nomenclature Committee recommended that, to support international harmonisation of chemical terminology, the spellings 'sulphur', 'sulphuric', etc., be changed to 'sulfur', 'sulfuric', etc., with effect from 1992.

A **chemical symbol** is used to identify a particular element. For example, the symbol **H** represents **hydrogen**, **Zn** represents **zinc**, and so on. Combinations of these symbols represent compounds, with numeric subscripts indicating the presence of two or more atoms of an element. For example, the chemical symbol for (the compound) water, is $H_2O$, which indicates a molecule comprising *two* atoms of hydrogen and *one* atom of oxygen.

The chemical symbol for pure sulfuric acid is $H_2SO_4$, indicating that it is a compound comprising **hydrogen** ($H_2$) and **sulfate** ($SO_4$). When dissolved in water,

sulfuric acid *dissociates* – that is, it *separates* into its constituent parts hydrogen and sulfate. During this process, the sulfate ($SO_4$) molecules each acquire two electrons from the hydrogen ($H_2$) molecules. As a result of this, the sulfate molecules become negatively charged sulfate ions ($SO_4^{2-}$), while the hydrogen molecules each become positively charged hydrogen ions ($2H^+$).

Remember, an **ion** is simply a charged atom or molecule – i.e. one that has become electrically unbalanced by gaining or losing one or more electrons. It's charge, positive or negative, is indicated by + or – superscript added to its chemical symbol (e.g. a negative sulphate ion, that has gained two excess electrons is therefore shown as: $SO_4^{2-}$), as indicated in the previous paragraph.

The ionisation of sulfuric acid (or, in fact, *any* acid or, in some cases, alkaline) in water leaves charged particles within the solution. So we can say that the solution, now termed an electrolyte, *has become a liquid conductor* and *capable of supporting current flow* which is, of course, defined as a drift of charged particles.

When the zinc electrode is inserted into the electrolyte, the zinc starts to dissolve. This process is caused by positive zinc ions ($Zn^{2+}$) which, attracted by the negative sulfate ions ($SO_4^{2-}$), detach themselves from the zinc electrode, each leaving behind two electrons, and move into the solution and combine with the negative sulfate ions to form neutral zinc sulfate ($ZnSO_4$) which plays no further part in the reaction. As more and more zinc ions detach, the zinc electrode starts to acquire a negative charge due to the surplus of electrons left behind on it.

This is quite a vigorous process and, after a short period of time, the zinc electrode has acquired a

Simple Voltaic Cell

Figure 13.4

sufficiently high negative charge to prevent any further loss of positive zinc ions into the electrolyte, and any further action stops. If any more positive zinc ions now try to detach themselves, they are simply attracted back to the negatively charged electrode.

At this point, if it were possible to measure the amount of negative potential of the zinc electrode, relative to the electrolyte, it would be found to amount to **–0.76 V**.

If we now turn our attention to the copper electrode, a similar, but much milder, process occurs. That is, positive copper ions ($Cu^{2+}$) dissolve into the electrolyte, each leaving two electrons behind, causing the copper electrode to eventually acquire sufficient negative charge to prevent any further positive copper ions from leaving. This time, if we were able to measure the potential of the copper electrode relative to the electrolyte, we would find that its potential is **+0.34 V**.

So the zinc electrode becomes the cell's **negative plate** while, *relative to the zinc electrode*, the copper electrode becomes the cell's '**positive' plate**.

At this point in our explanation you are no doubt wondering, if the copper electrode has acquired a *negative* charge, *why* its potential should be +0.34 V? The answer is that, while its potential is indeed negative, it is much *less negative* than the zinc electrode and, so, *relative to the zinc*, it is considered to be positive. If this has left you confused, by all means review 'potential', in the chapter on *potential and potential difference* before you continue on, but the key thing to remember is that *all potentials are relative to the point of reference from which they are measured.*

In fact, it is an oversimplification to say that the potentials of the two electrodes are 'measured with respect to the electrolyte'. Rather, by common agreement, their potentials are theoretically compared to that of a **hydrogen electrode**, whose reference potential has been arbitrarily assigned as being at zero volts. Even though *all* metal electrodes (as well as our hydrogen electrode) actually acquire *negative* potentials, those metals which are *less negative* than the hydrogen electrode are considered to be *positive with respect to the hydrogen electrode*. Figure 13.5 should make this clear.

The potentials of different metals, each measured with respect to this 'hydrogen electrode', are listed in what is known as the '**Electrochemical Series**' (Table 13.1), with what are described as the 'more-active' electrodes (e.g. lithium, at –3.02 V) towards the top of the series, and the 'less-active' electrodes (e.g. gold, at +1.68 V) towards the bottom of the series.

The *potential difference* between *any* pair of electrodes from this series is then obtained by simply

**Table 13.1**

| Conductor | Symbol | Electrode potential/volts |
|---|---|---|
| lithium | Li | –3.02 |
| potassium | K | –2.92 |
| barium | Ba | –2.90 |
| sodium | Na | –2.71 |
| aluminium | Al | –1.67 |
| zinc | Zn | –0.76 |
| chromium | Cr | –0.71 |
| iron | Fe | –0.44 |
| nickel | Ni | –0.25 |
| tin | Sn | –0.14 |
| lead | Pb | –0.13 |
| **hydrogen** | **H** | **0.00** |
| bismuth | Bi | +0.20 |
| copper | Cu | +0.34 |
| silver | Ag | +0.80 |
| mercury | Hg | +0.85 |
| gold | Au | +1.68 |

subtracting their individual potentials – and the *further apart* they are on the series, the *greater* the resulting potential difference between them.

So, for copper and zinc, the resulting potential difference between them is:

$$+0.34 - (-0.76) = \textbf{1.10 V}$$

The potential difference (voltage) between the copper and zinc electrodes of a simple Voltaic cell is **1.10 V**.

So, while the electrolyte plays a key part in the creation of a potential difference between a cell's electrodes, *it's actually the materials from which the*

| TRUE NEGATIVE | TRUE POSITIVE |

hydrogen
electrode
(reference)
0 V

true neutral

copper
electrode
+0.34 V (relative to hydrogen)

zinc
electrode
−0.76 V (relative to hydrogen)

**Figure 13.5**

*electrodes themselves are manufactured that determine the potential difference that can be achieved by any particular type of cell.*

You should also understand that this voltage represents the 'no load' voltage of the cell, which we call its **electromotive force (e.m.f.)**. When supplying a load, the terminal voltage will always fall below the value of the cell's e.m.f., as we shall learn later.

Let's move on, now, and find out what happens when our cell's electrodes are connected to an external load.

Remember that once the two electrodes have reached their respective potentials, *no further activity will take place within the cell*. The zinc electrode has become sufficiently negative to prevent any further positive zinc ions from detaching into the electrolyte, and the copper electrode has become sufficiently negative to prevent any further positive copper ions from detaching.

However, immediately the cell is connected to an external load, things begin to happen! *Electrons start to leave the more negative zinc electrode, and travel through the load, across to the less negative ('positive') copper electrode.*

As more and more electrons leave the zinc, its negative charge starts to fall, allowing more and more positive zinc ions to detach themselves and dissolve into the electrolyte. These positive zinc ions ($Zn^{2+}$) now start to *repel* the positive hydrogen ions ($2H^+$) adrift within the electrolye, *driving them towards the copper electrode.*

So, a drift of positive hydrogen ions becomes established within the electrolyte, from the zinc electrode towards the copper electrode. At the same time, there is a drift of negative sulfate ions towards the zinc electrode. In other words, *the drift of positive hydrogen ions and negative sulphate ions, in opposite directions, constitutes the electric*

*current within the electrolyte* while *the drift of free electrons constitutes the electric current through the external circuit.*

Although the current in the external circuit is by *free electron drift* and the current within the electrolyte is by *ion drift*, their magnitudes, of course, are identical – i.e. when the *external* current is, say, 1 A, the *internal* current is also 1 A. Although the charge carriers are different, it is, of course the *same* current!

As each positive hydrogen ion arrives at the copper electrode, it attracts an electron which has arrived at the copper electrode via the external circuit, forming a *neutral hydrogen atom*, and is liberated to form bubbles of hydrogen gas.

Some of this hydrogen gas simply bubbles out of the electrolyte to disperse in the atmosphere but as the action continues, more and more of these bubbles start to coat and isolate the copper electrode  – a condition we call *'polarisation'*. Polarisation eventually causes the copper electrode to behave as though it was a *hydrogen* electrode (which is why we see a 'hydrogen electrode' listed amongst metal electrodes in the *Electrochemical Series*), and the potential difference between the two electrodes (zinc and hydrogen) now falls towards 0.76 V.

So, if you were wondering how it is possible to have a 'hydrogen electrode', then this is the answer.

With this reduction in the cell's electromotive force, the current in the external circuit also starts to fall. This condition is equivalent to the resistance of the cell increasing and, in fact, we describe this condition as an *'increase in the cell's internal resistance'*.

Polarisation makes this particular cell unsuitable for most applications. In fact, for practical purposes, the simple Voltaic cell is little more than a laboratory demonstration. In more practical cells (e.g. the dry cells we use to power our torches, etc.), the problem of polarisation is avoided by adding a 'depolarising' agent to the electrolyte; this material contains oxygen which combines with the hydrogen to form water, reducing the amount of hydrogen bubbles and leaving the positive electrode clear.

Unfortunately, the addition of a depolarising agent does absolutely nothing to counter the continuing destruction of the zinc electrode (and, to a lesser extent, the copper electrode) and, eventually, the electrode becomes completely eaten up and the cell dies. This is accelerated by '**local action**', which describes internal electrolytic corrosion within the electrodes themselves due to the presence of impurities within the metals.

### Dry Leclanché cells

The 'dry cell' or, more accurately, the '**dry Leclanché**' cell (Figure 13.6), is also known as a '**carbon-zinc**' **cell**, after the materials from which its electrodes are manufactured. This is one of the most common, and inexpensive, disposable cells that we use to power our torches, electronic toys, etc.

Strictly speaking, 'dry cells' are not dry at all; rather, their electrolyte is a *paste* or *gel* (so it might be more accurate to describe it as a 'non-spillable wet cell'!), composed of ammonium chloride typically mixed with starch and flour.

Its positive 'plate' is a carbon rod located at the centre of the cell, immediately surrounded by a core of fine carbon granules mixed with manganese oxide, which act as a 'depolarising agent' to reduce the amount of hydrogen bubbles forming on the carbon electrode. The negative 'plate' is a zinc canister, into which the carbon rod and depolarising agent are inserted, with the space between filled with the electrolyte, before being completely sealed.

Dry Leclanché Cell

Figure 13.6

The electromotive force achieved by the dry carbon-zinc cell is approximately **1.5 V**, but this drops significantly if a large load current is drawn from the cell.

Unfortunately, as zinc ions detach into the electrolyte, the zinc container is gradually eaten away until, eventually, the cell begins to leak electrolyte. For this reason, dry cells that are approaching the end of their life should always be removed from the device they are powering to avoid the device from becoming damaged from any corrosion caused by the leaking electrolyte.

The capacity of an AA-size carbon-zinc cell is typically 400–1700 mA·h. The large battery manufacturer, *EverReady*, claims that its carbon cells have a shelf life of approximately one year while retaining 95–100 per cent of their capacity, or up to four years with a capacity of 65–80 per cent.

Other types of 'dry' disposable cells include the '**alkaline cell**'. This is similar in appearance to the carbon-zinc cell, but it uses an electrolyte of potassium hydroxide, which is an *alkaline* rather than an acid – hence its name. Alkaline cells have between three and five times the capacity of an equivalent carbon-zinc cell, making them the ideal alternative for powering photographic equipment, portable audio equipment, etc.

Because the electrodes of primary cells corrode when they de-energise, *they cannot be re-energised* – unlike the next class of cells we are going to study.

### Miniature cells

The **mercury** (or **mercuric oxide**) **cell** (Figure 13.7) is one of several types of miniature, or 'button', cells developed as a spin-off to the U.S. space programme in order to meet the demands for miniaturisation, and they have day-to-day applications which include powering electronic watches, hearing aids, etc. They are manufactured in a wide range of diameters and thicknesses, they have an excellent shelf life, and the materials used in the cell provide it with many times the capacity of other cells of comparable size.

Figure 13.7

The mercury cell is so-called because its positive electrode is manufactured from a mixture of mercuric-oxide and graphite. Its negative electrode is manufactured from zinc, and its electrolyte is potassium hydroxide. The entire cell is encapsulated in a two-part stainless-steel capsule, with the main part usually (but not always) providing the positive potential, and the 'cap' part providing the negative potential to the external circuit. Its e.m.f. is 1.35 V, which remains exceptionally steady throughout its life span.

As mercury is highly toxic, a more environmentally friendly alternative is the **silver-oxide** cell which, externally, looks identical to a mercury cell, but which has a slightly higher e.m.f. of 1.5 V. This cell uses silver oxide as its positive electrode and zinc as the negative electrode, with sodium hydroxide or potassium hydroxide as an electrolyte.

In common with other types of cell, these need to be disposed of responsibly, as discussed elsewhere in this chapter.

## Secondary cells

A **secondary cell** will deliver current to a load through a chemical reaction similar to that of primary cells, but its chemical reaction is *reversible*, enabling it to be restored to practically its original condition.

A secondary cell can be restored, or **'re-energised'**, simply by passing a d.c. current through the cell *in the opposite direction to its de-energising current*.

When a cell stops working, it has run out of **energy**, *not charge!* The chemical reactions that take place within a cell release **energy**, and this energy is used to separate charges in order to produce a potential difference across its terminals. So, when a secondary cell's chemical reaction stops, the cell has become 'de-energised' not 'discharged'. And we 're-energise' that cell; we don't 'recharge' it! Having said that, the terms 'discharge' and 'recharge' are so widely used, it's unlikely those terms will ever be replaced.

### Lead-acid cells

Invented in the mid-nineteenth century by a Frenchman, Gaston Planté, the **lead-acid cell** is one of the most widely used types of secondary cell, and the type used in practically all motor vehicles. More compact lead-acid cells are also used widely for other applications, such as emergency lighting, uninterruptible power supplies (UPS), etc.

A lead-acid cell has an open-circuit voltage of around 2.1 V when fully energised, falling to a little less than 2 V when fully de-energised.

### *Construction*

Prior to initial charging, the active material used on *both* electrodes is manufactured from a mixture of lead sulfate and lead oxide. During the initial charging, however, this material is converted to a hardened paste of brown-coloured **lead dioxide** (also known as 'lead peroxide') which forms the positive plate, and to a grey-coloured **porous metallic lead** (also known as 'spongy lead') which forms the negative plate.

These so-called 'active materials' are relatively weak, so they need mechanical support which is provided by an **electrode grid** (Figure 13.8) manufactured from a hard lead alloy. In addition to providing support for the active materials, the grid also acts to conduct the current between the active materials and the external load.

This combination of active material and supporting grid is called a '**plate**'.

electrode grid – prior to coating

Figure 13.8

The positive and negative plates are kept apart from each other by means of **separators** (Figure 13.9), which are a little larger in area than the plates themselves. These are sheets of insulation, typically manufactured from polyethylene, which is microporous in order to enable ions to pass between adjacent plates of opposite polarity.

electrode grids

separators

Figure 13.9

Combinations of plates and separators make up the individual cells (Figure 13.10).

Figure 13.10

A number of individual cells are inserted into a polypropylene container (Figure 13.11), connected together in series, and a lid fitted. Lead bars are used to interconnect plates of like polarity, and to provide posts by which the battery can be connected to its load. In the case of a nominal 12-V car battery, six cells are used.

cells connected together in series

Figure 13.11

### Behaviour of lead-acid cell during discharge/charging cycle

The action of a lead-acid cell during its de-energising/ re-energising cycle involves chemical reactions between the lead compounds on its electrodes, and changes to the relative density (specific gravity) of the electrolyte due to the formation of water.

As well as being reversible, the chemical reaction of a lead-acid is rather more complicated than that of a primary cell (with the behaviour at the positive electrode being particularly complicated!). Fortunately, its overall behaviour can be neatly summed up in the

form of what is known as a '**reversible chemical equation**', as follows:

$$PbO_2 + Pb + 2H_2SO_4 \rightleftharpoons 2PbSO_4 + 2H_2O$$

This equation might *look* complicated but, in fact, it is very simple. The double-headed equals-sign simply means that the process is *reversible*; the upper part (pointing to the right) indicating the 'completely de-energised' state, and the lower part (pointing to the left) indicating the 'fully energised' state.

So the left-hand side of the equation represents the situation *before* the cell de-energises. The chemical symbol, **PbO$_2$**, represents lead dioxide (the positive plate), **Pb** represents grey metallic lead (the negative plate), and **2H$_2$SO$_4$** represents the electrolyte, dilute sulfuric acid. These are the chemicals present within the cell *when it is fully energised*.

The right-hand side of the equation represents the situation *after* the cell has become fully de-energised. The chemical symbol, **2PbSO$_4$**, represents lead sulfate (a white-coloured compound) which has replaced the lead dioxide and metallic lead on the electrodes, while the chemical symbol, **2H$_2$O**, indicates the formation of water which dilutes the electrolyte. These are the chemicals present within the cell *after it has become de-energised.*

So let's move on and consider each of the four stages of the cell's condition: (a) *fully energised*, (b) *de-energising*, (c) *fully de-energised* and (d) *energising*.

### Fully energised cell

Table 13.2

| positive plate: | Brown lead dioxide. |
|---|---|
| negative plate: | Grey porous metallic lead. |
| electrolyte: | Maximum acid content. Maximum relative density. |

positive plate          negative plate

Figure 13.12

### Cell de-energising

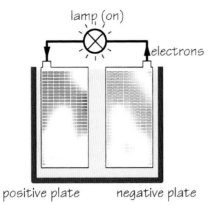

Figure 13.13

Table 13.3

| | |
|---|---|
| **positive plate:** | Brown lead dioxide being converted to (white) lead sulfate. |
| **negative plate:** | Grey porous metallic lead being converted to (white) lead sulfate. |
| **electrolyte:** | Formation of water, diluting the acidity. Relative density falling. |

### Fully de-energised cell

Figure 13.14

Table 13.4

| | |
|---|---|
| **positive plate:** | Electrode now mainly white lead sulfate. |
| **negative plate:** | Electrode now mainly white lead sulfate. |
| **electrolyte:** | Maximum water content/ minimum acid content. Minimum relative density. |

### Re-energising cell

Figure 13.15

Table 13.5

| | |
|---|---|
| **positive plate:** | White lead sulfate being converted back to brown lead dioxide. Water converted to hydrogen and oxygen gases. |
| **negative plate:** | White lead sulfate being converted back to grey porous metallic lead. Water converted to hydrogen and oxygen gases. |
| **electrolyte:** | Acid content increasing. Relative density increasing. |

### Relative density

As you can see from Tables 13.2–13.5, the **relative density** of the electrolyte is at its maximum when the cell is fully energised, and at its minimum when the cell is completely de-energised.

So, the simplest way to determine the capacity of a lead-acid cell or battery is to measure the electrolyte's **relative density**. The term 'relative density', has long replaced the older term 'specific gravity' which, in this context, is a measure of *the mass of electrolyte compared to the mass of an equal volume of pure water*.

Pure sulfuric acid has a relative density of 1.84 – which means that it has 1.84 times the mass of an equal volume of pure water. When sulfuric acid is prepared as an electrolyte, by diluting it with water (35% acid : 65% water), its relative density is typically about **1.27**. During de-energising, the electrolyte becomes even more diluted, due to the formation of water during the chemical reaction, and its relative density gradually falls to about **1.13** when the cell is completely de-energised.

The electrolyte's relative density (RD) is a guide to the cell's condition.

*Completely energised cell's RD = 1.27 at 15°C*
*Completely de-energised cell's RD ≈ 1.13 at 15°C*

hydrometer

Figure 13.16

Relative density is proportional to the level of discharge (but *not* the level of charge) and varies with temperature. The figures above are typical, rather than exact, and quoted for a temperature of 15°C. Battery rooms are normally equipped with a temperature-compensation chart which allows the measured relative density to be adjusted for the ambient temperature.

The relative density of an electrolyte is measured using a **hydrometer** (illustrated in Figure 13.16). Enough electrolyte is drawn into the hydrometer, using the rubber bulb, to enable the graduated float to rise. The relative density is then read off the scale engraved on the float, at the level of the electrolyte – as shown in Figure 13.17.

Figure 13.17

**Always beware of the corrosive properties of sulfuric acid!**

Figure 13.18

### Energising lead-acid cells and batteries

Lead-acid cells are re-energised, using purpose-built **battery chargers**. Since it is necessary to pass the energising current in the *opposite* direction to the cell's de-energising current, the battery charger must be connected with great care to ensure that its positive terminal is connected to the battery's positive electrode, and its negative terminal is connected to the battery's negative electrode.

Professional battery chargers, of the type you are likely to encounter in workshops or electrical substations, allow several ways in which to re-energise lead-acid cells or batteries, these include: **constant-current**, **constant-voltage** and **trickle**.

- **Constant-current re-energising.** A constant re-energising current of around 7 per cent of the figure representing the cell's capacity is a safe rule of thumb. For example, if the cell has a rated capacity of, say, 250 A·h, then we should use a current of 7 per cent of 250, or 17.5 A. The cell's relative density should then be monitored, hourly, and when it ceases to increase further, the re-energising can cease.

- **Constant-voltage re-energising**. This method will cause a high initial value of re-energising current. As the cell's potential difference, which opposes the battery charger's voltage, will gradually increase during re-energising, the current will gradually decrease. With badly sulfated cells (see below), this method may result in excessively high temperatures (in excess of 50°C), and is best avoided.
- **Trickle re-energising**. This method is also called 'float re-energising', and is carried out by connecting the cell across a source which provides just enough re-energising current to make up for any loss through de-energising. This method is widely used in electricity substations, for example, to ensure that the batteries remain at their full capacity in order to meet the occasional requirements for tripping circuit breakers, etc.

### Care and maintenance of lead-acid cells and batteries

Should you ever be tasked with maintaining lead-acid cells or batteries, you must be constantly aware of the **corrosive properties of electrolyte**, and of the **potentially explosive nature of the gas** vented from batteries during their charging process.

Accordingly, you must *always* wear eye protection, and the rubber aprons and gloves that will be provided by your employer and, of course, *never* smoke near batteries or, in particular, inside a battery room. Battery rooms are normally provided with an emergency eye-wash station, and you should always make yourself aware of where this is and how to use it.

It's important that lead-acid cells and batteries are regularly cleaned, and their terminals and connections lightly coated with petroleum jelly in order to prevent corrosion. Cleaning can be carried out by washing the outside of the batteries using a solution of baking soda, followed by clean water.

The electrolyte level should *never* be allowed to fall below the level of the plates and, whenever necessary, topped up using distilled water. Allowing the electrolyte level to remain below the plates will cause any lead sulfate to dry out and harden – a process called 'sulfation' – which is irreversible, and will reduce the effective active area of the plates, reducing the cell's capacity.

If you ever have to prepare electrolyte, then you must always follow the manufacturer's recommendations.

Typically, the ratio of the solution is 65% water to 35% acid. It's vitally important that you must *always add acid to water,* **never** *the other way around!* Failure to do so could result in rapid boiling and spitting as the reaction can be extremely vigourous, producing large quantities of heat.

### Performance ratings for lead-acid vehicle batteries

As we have already learnt, the capacity of a battery is normally expressed in **ampere hours** or in **watt hours**. However, the performance of a vehicle battery now includes two additional specifications, termed its '**cold cranking capacity**' and its '**minimum reserve capacity**'.

> **Cold-cranking current** is defined as *'that current which can be continuously delivered by a cell, at −18°C, for 30 s, after which each cell must deliver a terminal voltage of 1.2 V or higher'*.

For example, a 12-V car battery having a cold-cranking current of, say, '350 A', means that it must be able to deliver 350 A for 30 s, at −18°C, without the battery's voltage falling below 7.2 V (i.e. $6 \times 1.2 = 7.2$ V).

> **Minimum reserve capacity** is defined as *'the time, in minutes, that a cell will support a full accessory load (about 25 A), at 25°C, while sustaining a terminal voltage of 1.75 V or better'*.

You can think of the minimum reserve capacity of a car battery as being a sort of 'capacity insurance', indicating how long you have to find a service station after, say, your vehicle's alternator or fan belt fails. For example, a minimum reserve capacity of, say, '150' means that you have 150 minutes! Less, of course, if you don't switch off most of the car's accessories.

### Nickel-cadmium (Ni-Cd) cells

**Nickel-cadmium (Ni-Cd)** cells (widely referred to as 'ni-cads') are re-energisable cells that use nickel oxide hydroxide and metallic cadmium foil as their positive and negative electrodes, and potassium hydroxide as an electrolyte. The two plates, kept apart by a separator which is impregnated with the electrolyte, are rolled up rather in the same way as a 'Swiss roll'.

Fully energised Ni-Cd cells achieve an electromotive force of 1.25–1.35 V. One of the characteristics of a Ni-Cd cell is that its voltage remains relatively constant during de-energising until, eventually, it suddenly falls to around 0.9 V with little or no warning.

One of the major drawbacks with Ni-Cd cells is their so-called '**memory effect**', which results in them failing to achieve their rated capacity. This occurs when the cell is repeatedly re-energised after having been only partially de-energised, and causes the cell to appear as though it 'remembers' the lower capacity and appears to lose its charge very quickly.

Yet another drawback with Ni-Cd cells is an effect called '**voltage depression**', which is usually the result of repeated over-energising. Voltage depression describes a condition in which the cell's voltage falls far more rapidly than normal, giving the impression that it is not maintaining its capacity.

Nickel-cadmium cells have been widely used for powering a wide range of portable electric handtools, such as electric drills, etc., as well as for a great many consumer electronics applications, including camcorders, laptop computers, mobile telephones, etc.

### Lithium-ion (Li-ion) cells

The main drawbacks of nickel-cadmium cells, i.e. their 'memory effect' and 'voltage depression', have been overcome thanks to the introduction, during the early 1990s, of **lithium-ion (Li-ion)** cells and batteries.

As well as eliminating the problems of memory effect and voltage depression, lithium-ion cells are very much lighter and, physically, they are significantly smaller than other re-energisable cells of similar capacity. An interesting comparison is that a lithium-ion battery can supply around 150 W·h of energy per kilogram mass of battery, compared with just 25 W·h of energy per kilogram for a lead-acid battery! That's an enormous difference!

Although somewhat more expensive than a corresponding nickel-cadmium battery, lithium-ion batteries now have become the standard for professional and consumer electronics equipment such as camcorders, mp3 players, laptop computers, mobile telephones, etc., and they are now starting to appear as light-weight battery packs for power tools.

A lithium-ion cell's electrodes can be manufactured from various materials. Typically, the positive electrode consists of a thin aluminium foil, coated with lithium and cobalt metal oxides, and the negative electrode consists of a thin copper foil, coated in graphite. The electrodes are separated with porous polyethylene film. The electrolyte is lithium hexafluorophosphate ($LiPF_6$), dissolved in an organic solvent.

The electrodes, their separator and the electrolyte is wound ('Swiss roll' style) into a sealed container.

While re-energising, lithium ions are transferred through the electrolyte from the positive plate to the negative plate; while de-energising, lithium ions are transferred in the opposite direction. During this process, little change takes place to either the positive or the negative plate.

The average e.m.f. for a lithium-ion cell is 3.6 V (it varies from 4.2 V down to about 3 V); which is equivalent to that of *three* nickel-cadmium cells, thus reducing the number of cells required for similar applications.

Lithium-ion cells require their own, dedicated, two-stage charging unit. Initially, the charger limits the energising current, until the voltage across the electrodes reaches 4.2 V, at which point it then changes to a constant voltage operation as the current reduces towards zero. Under normal operating conditions, these cells are capable of between 300 and 500 energise/de-energise cycles.

At the time of writing (December 2017), the world's largest lithium 'battery' has recently been constructed near Adelaide in South Australia. The **129-MW·h** battery or, more accurately, a huge '**battery farm**', which was constructed by the electric-vehicle wind-turbine, Tesla Inc., is paired with an adjacent wind-turbine farm which supplies energy to the lithium storage units.

The battery farm occupies an area roughly equivalent to that of a soccer pitch, with each energy unit measuring 2.1 m in height, 1.3 m long, and 0.8 m wide, and having a mass of 1200 kg, and the complete farm is said to be capable of supply energy at the rate of 100 MW.

To put the size of this battery farm into context, the manufacturer claims that it has 'the capacity to supply enough energy to supply 30 000 homes for up to an hour, in the event of a severe blackout'. However, its primary purpose is to support the stability of the grid during periods of high demand. South Australia, which relies heavily on solar and wind-generated energy has been seeking ways to support its fragile energy grid since it suffered a major blackout, after a storm in 2016, followed by a series of smaller ones since then.

### Fuel cells

Just like the cells already described in this chapter, a **fuel cell** is an electrochemical energy-conversion device but, *unlike* those cells, it depends on a *constant supply of fuel (generally hydrogen and oxygen)* which it converts into water and, during that process, produces electricity. Because the only waste product is water, fuel cells are considered to be 'environmentally friendly', and a potential source of energy for many applications, including electric vehicles.

As fuel cells do not rely on a chemical electrolyte for their operation, they do not require re-energising and will provide a continuous supply of electrical energy, as long as they supplied with fuel.

We tend to think of fuel cells as 'new technology' but, surprisingly, fuel cells are *not* new at all. In fact the first fuel cell, credited to the Welsh judge and physicist, Sir William Grove (1811–1896), was developed as long ago as 1839. Grove already knew that, by passing an electric current through water, its molecules could be separated to produce hydrogen and oxygen gas and, so, he thought that it might be somehow possible to *reverse* that process, and produce electricity and water. He was correct, and the result of his experiments was a primitive device which he called a 'gas voltaic cell' – a name which, around 50 years or so later, was changed to 'fuel cell'.

There *are four* basic components to any fuel cell:

- **Anode**. This is the *negative* electrode of the cell. It has microscopic channels etched into its surface that distributes the hydrogen gas over the surface of a catalyst. It also conducts the electrons, liberated from the hydrogen molecules, to the external load.
- **Cathode**. This is the *positive* electrode of the cell. It, too, has microscopic channels etched into its surface which distributes oxygen over the surface of the catalyst. It 'receives' electrons back from the external circuit, where they recombine with the hydrogen ions, as well as the oxygen, to form water.
- **Catalyst**. A 'catalyst' is defined as any material that accelerates the rate of a chemical reaction without, itself, being consumed by that reaction. In this case, it is a porous material, usually manufactured from tiny particles of platinum deposited onto one side of a layer of carbon, which encourages the reaction of hydrogen and oxygen to take place at the electrodes.
- **Electrolyte**. This term is *not* used in the same sense as it is in a conventional electrochemical cell. Rather, it's a *membrane* or 'selective barrier' which, in this case, allows the passage of positively-charged hydrogen ions from the anode to the cathode, while blocking the passage of electrons.

## Operation

Figure 13.19 shows the basic construction of a fuel cell. Hydrogen molecules ($H_2$) enter the cell, under pressure,

at the anode side of the cell. Here, they come into contact with the platinum surface of the catalyst, where they separate into two positively-charged hydrogen ions ($H^+$) and two electrons (e-). The electrons are conducted to the external load via the anode.

**Figure 13.19**

At the same time, oxygen molecules ($O_2$) enter at the **cathode** side of the cell, where they are forced through the catalyst, and separate into two, negatively-charged, oxygen ions ($O^-$). Each of these attracts two positively-charged hydrogen ions through the electrolyte membrane, with which they combine – together with two electrons returning from the external circuit – to form water ($H_2O$).

The potential difference developed between the cathode and anode by this process is relatively low. For a PEMFC-type fuel cell (see below), it is typically less than 1 V so, in order to achieve a useful potential difference, several identical fuel cells must be connected in series to form, not a battery, but what is termed a '**fuel-cell stack**'. Unfortunately, fuel cells tend to have a relatively-high internal resistance and, so, any heavy electrical load can cause a significant internal voltage drop and reduction in terminal voltage.

## Types of fuel cell

There are various types of fuel cell, and they are *all* classified according to the **type of electrolyte** they use. These include:

- Alkaline fuel cell (AFC)
- Direct-methanol fuel cell (DMFC)
- Molten-carbonate fuel cell (MCFC)
- Phosphoric-acid fuel cell (PAFC)

- Polymer-exchange membrane fuel cell (PEMFC)
- Solid-oxide fuel cell (SOFC)

Further information on these is well-beyond the scope of this book. However, the oldest of these is the AFC type, dates back to the 1960s, and was developed for the US space programme. The characteristics of these different types of fuel cell differ in terms of their energy per unit mass, operating temperature, efficiency, etc, making some better for certain applications than for others. For example, the PEMFC type appears to be particularly suited for use by electric vehicles, while most of the others appear more suited as static power supplies – eventually, perhaps, for off-grid residential use.

## Conclusion

At the turn of the century, it was widely-believed that the fuel cell would transform the world: replacing the need to burn fossil fuels, and providing a clean and inexhaustible supply of 'green' energy – particularly for electric vehicles. But, unfortunately, this hasn't proved to be the case.

In fact, it's likely to be some time before we start to see the widespread use of fuel cells and, presently, the *biggest* problem is their **cost** – normally expressed in 'pounds (or dollars, etc.) per kilowatt hour'.

Although hydrogen is a plentiful (it's the third most-abundant element on the earth's surface) and clean fuel, it is considered to be 'energy neutral' – that is, it takes almost as much energy to *produce* it as it *delivers*!

So, at the moment, fuel cells are prohibitively expensive to manufacture and to operate. In fact it is claimed that the fuel-cell industry has yet to turn a profit! At the time of writing this book (2017), no fuel cell is cheap and efficient enough to replace the traditional ways of producing electrical energy.

There is *some* good news, though. For example, the German and Italian navies are now both operating diesel submarines which use fuel-cell stacks in place of conventional batteries, enabling those boats to operate for weeks, silently, underwater – unlike conventional diesel boats, which must re-surface regularly in order to recharge their batteries. There is also the added impetus of the need to produce efficient and inexpensive batteries to power electric vehicles.

## Responsible disposal of cells and batteries

As you are aware, cells and batteries contain hazardous and toxic materials, so it is important that we dispose of them responsibly in order to reduce any risk to the environment or to human health. This means *not* disposing of them with other general municipal waste that finds its way into landfill sites but, instead, making use of the **battery collection points** that are now available at sales points.

In order to encourage users to dispose of cells and batteries responsibly, under the European Battery Directive 2006/66/EC, manufacturers are now required to bear the cost of collecting, treating and recycling industrial, automotive and portable batteries, together with the costs of any publicity campaigns intended to inform the public of these arrangements. Countries not covered by this European directive have their own corresponding requirements which amount to the same thing.

This includes the requirement to provide facilities at sales points, where cells and batteries can be safely discarded for collection and safe disposal under the Directive.

Manufacturers are also required to provide visible, legible and indelible markings on their batteries and/or packaging which, among other requirements, displays the **WEEE (Waste Electronic and Electrical Equipment Directive 2008***)* pictogram, together with the relevant chemical symbols, as illustrated in Figure 13.20.

Pb/Hg/Cd

Figure 13.20

This pictogram also appears on any electrical product that contains embedded battery packs which are intended to last for the life of that product (e.g. electric toothbrushes), and indicates that the product itself must also be disposed of separately from municipal waste.

## Batteries

As we have already learned, a **battery** is simply *a number of cells connected together*. In most cases, cells are connected either in **series** or in **parallel**, although **series-parallel** connections, of course, are also possible.

It is *not* recommended that we connect different types of cell together, or to mix partially de-energised cells with fully energised cells, so in this section, we will assume that each cell is of *the same type* and has *the same level of energy.*

## Cells in series

Cells are connected in **series** when the positive terminal of one cell is connected to the negative terminal of the next, as illustrated in Figure 13.21.

**Figure 13.21**

We connect cells in series when we want to increase the available e.m.f. The total e.m.f. of cells connected in series is the sum of their individual e.m.f.s.

$$E_{total} = E_1 + E_2 + E_3 + etc.$$

Of course, *we can only achieve e.m.f.s that are multiples of that of an individual cell* – we *cannot*, for example, achieve an e.m.f. of, say, 10 V, using 1.5-V dry cells! We could only achieve 10 V by using a completely *different type* of cell, such as five 2-V lead-acid cells.

The total internal resistance of cells connected in series is the sum of their individual internal resistances, just like resistors in series:

$$R_{total} = R_1 + R_2 + R_3 + etc.$$

The total discharge current available from cells connected in series must not exceed the discharge current delivered by an individual cell.

## Cells in parallel

Cells are connected in parallel when all their positive terminals are connected together and all their negative terminals are connected together, as illustrated in Figure 13.22.

$$E = E_1 = E_2 = E_3$$

**Figure 13.22**

We connect cells in parallel when we want to maintain the same e.m.f. as a single cell, but wish to increase the available discharge current.

$$E_{total} = E_1 = E_2 = E_3 = etc.$$

The maximum discharge current available from cells connected in parallel is the sum of the currents deliverable by each cell.

The total internal resistance is determined in exactly the same way as we determine the total resistance of resistors in parallel:

$$\frac{1}{R_{total}} = \frac{1}{R_1} + \frac{1}{R_2} + \frac{1}{R_3} + etc.$$

The alternative, and quicker, method is to simply divide the internal resistance of one cell by the total number of cells:

$$total\ internal\ resistance = \frac{internal\ resistance\ of\ one\ cell}{number\ of\ cells}$$

## Cells in series-parallel

If we want to increase *both* the terminal e.m.f. *and* the amount of discharge current deliverably, then we could connect (identical) cells in **series-parallel**, as illustrated in Figure 13.23:

**Figure 13.23**

In the example in Figure 13.23, the total e.m.f. will be three times the e.m.f. of an individal cell, and the complete battery will be capable of delivering a discharge current that is three times that of any individual branch cell (which, in turn, is that of any individual cell).

The total internal resistance of this arrangement can be determined in exactly the same way as we learnt to determine the total resistance of any series-parallel resistive circuit. In this example, the total internal resistance will be one-third of the internal resistance of one branch.

## Review your learning

Now that we've completed this chapter, we need to examine the **objectives** listed at its start. Placing '*Can I...*' at the beginning, and a question mark at the end, of each objective turns that objective into a **test item**. If we can answer each of those test items, then we've met the objectives of this chapter.

---

**Online resources**

The companion website to this book contains further resources relating to this chapter. The website can be accessed via the following link:

www.routledge.com/cw/waygood

# Chapter 14

# Internal resistance

## Objectives

On completion of this chapter, you should be able to

1 explain what is meant by the *internal resistance* of a  voltage source.
2 sketch the '*equivalent circuit*' of a voltage source.
3 explain the difference between a voltage-source's
   a electromotive force
   b internal voltage drop and
   c terminal voltage.
4 explain the relationship between a voltage source's
   a electromotive force
   b internal voltage drop and
   c terminal voltage.
5 briefly outline the experimental method of determining the internal resistance of a cell.
6 solve simple problems on the effects of a voltage-source's internal resistance.

## Internal resistance explained

The load current supplied by *any* voltage source, whether it be a cell or battery, a generator, or a transformer, as well as passing through the load must, of course, *also pass through the voltage source itself.*

This is because the voltage source is part of the complete circuit around which the load current flows. In other words, the 'inside' of the voltage source is *in series* with its external load.

So what exactly do we mean by the 'inside' of a voltage source?

Well, in the case of a generator, for example, the load current must also pass through the windings (coils) in which the voltage is generated by the rotating machine. Because windings are wound from copper wire, this 'internal current' is, of course, a *drift of free electrons* – just as it is through the external (load) circuit (Figure 14.1).

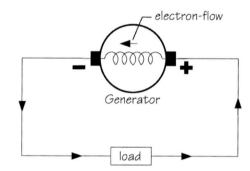

**Figure 14.1**

In the case of a cell or battery, the load current passes through the electrolyte in which its electrodes are immersed. However, this current is *a drift of ions* (charged atoms), rather than of free electrons, and may flow in the *opposite* direction (or even different ions flowing in *both* directions!) to the electrons in its external circuit (although this direction is unimportant to what follows) – as shown in Figure 14.2.

Whether the 'internal current' is a drift of free electrons *or* a drift of ions *isn't really important.* It's simpler just to think of it as being 'current'! What *is* important is that, internally, this current is *opposed* by what we call the 'internal resistance' $(R_i)$ of the voltage source – as shown in Figure 14.3.

An Introduction to Electrical Science, Waygood, ISBN 9780815391821, 2019. © Taylor & Francis

Figure 14.2

Figure 14.3

In the case of a **generator**, the internal resistance is *the resistance of the windings* within which the voltage is generated. In the case of a **cell** or **battery**, the internal resistance is made up of the combined effect of *two* components: the first is due to the resistivity of the internal conducting parts (the electrodes and any metallic connections), and the second, called 'ionic resistance', is due to the electrochemical opposition within the electrolyte. The first increases as the cell's electrodes dissolve (in the case of non-rechargeable cells) because their effective cross-sectional area reduces; the second increases as the cell de-energises.

The internal resistance of a fully energised, healthy cell is typically expressed in milliohms but, as the cell deteriorates, its internal resistance may rise markedly. A cell's internal resistance will also vary according to its state of charge.

The internal resistance of a generator's windings is also low, but will increase somewhat when the windings are hot (see the chapter on the *effect of temperature on resistance*), whenever the machine is running for prolonged periods.

The nature of a voltage source's internal resistance is unimportant; what *is* important is that it *exists* and it affects the behaviour of the voltage source as follows.

The effect of the voltage source's internal resistance is to *cause an internal voltage drop (IR) to occur within the voltage source itself.* The direction of this voltage drop is such that it **always** *acts to oppose the electromotive force of the voltage source, and to reduce the value of terminal voltage applied to the load.*

The *larger* the load current, the *larger* the corresponding internal voltage drop, and the *lower* the terminal voltage applied to the load.

This effect is sometimes referred to as the '**lost volts**' of a voltage source.

We have all experienced this effect whenever we have inadvertently switched on our car's headlights *before* starting the engine. The very large load current drawn by the starter motor causes a correspondingly large internal voltage drop to occur within the battery, resulting in a significant drop in its terminal voltage and a corresponding reduction in the brightness of the headlights (because lamps operate at their rated power *only* when subjected to their rated voltage – in this case, 12 V).

Of course, if there is no load connected to the voltage source, then no load current will flow, and there will be *no internal voltage drop*. The voltage appearing across the voltage source's terminals under this condition is called its **open-circuit voltage** or **no-load voltage** – which exactly equals the **electromotive force** (e.m.f.) of that source.

> A voltage source's **open-circuit**, or **no-load**, **voltage** corresponds to its **electromotive force**.

## Representing internal resistance in a schematic diagram

In the series of three **schematic diagrams** that follow, for simplicity, we will use a chemical **cell** to represent a voltage source, but we could also have used a generator or *any* other voltage source.

In Figure 14.4, we show where the internal resistance *(R)* really is located – between the plates, or 'electrodes', of the cell. The internal resistance is, of course, in *series* with the load, because the load current passes through both. The grey area represents the cell's container.

Figure 14.4

Of course, we don't *have* to show the internal resistance between the plates of the cell (even though that's where it *actually* is!). So, in Figure 14.5, we have moved it from between the plates, so that it sits *alongside* the cell – electrically, this is quite correct, even though it no longer represents its true *physical* location. The circuit symbol for the cell now represents an '**ideal**' cell (one having absolutely *no* internal resistance) producing its full electromotive force *(E)*, while the internal voltage drop *(IR$_i$)* now appears across the 'separate' internal resistance – acting in the *opposite* direction to the e.m.f. But remember, this is only a graphic representation (we call it an '**equivalent circuit**') – the internal voltage drop still *actually* occurs between the plates of the cell, not outside of them.

Figure 14.5

In fact, we can move the internal resistance to just about *anywhere around the circuit*, just as long as it remains *in series with the load*. So, in Figure 14.6, we have moved it even further around so that it is now adjacent to the load – again, this is electrically correct even though it is nowhere near its actual physical location.

We've moved the internal resistance to this particular position because it helps clarify the relationships between the cell's **electromotive force *(E)*,** its **internal**

Figure 14.6

**voltage drop *(IR$_i$)*,** and the **terminal voltage *(U)*** which actually appears across the load, that is:

$$U = E - (IR_i)$$

In other words, the voltage source's terminal voltage *(U), when supplying a load,* is its electromotive force *(E), less* its internal voltage drop *(IR$_i$).*

> The *shaded* part of Figure 14.6 is called the **equivalent circuit** of the voltage source, and consists of an '**ideal**' **voltage source** (i.e. one having no internal resistance), in series with an **internal resistance**.

Figure 14.7 shows a graphical representation of the drop in the terminal voltage for increasing load current. The greater the load current, the greater the internal voltage drop, and the lower the resulting terminal voltage.

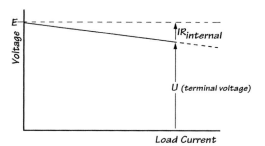

**Effect of Internal Resistance on Terminal Voltage**

Figure 14.7

Alternatively, we could calculate the terminal voltage by simply multiplying the load current by the resistance of the load:

$$U = I R_{load}$$

The **load current *(I)*** itself is determined by simply dividing the cell's **electromotive force *(E)*** by the total resistance of the circuit – i.e. by the sum of the **load resistance** $(R_L)$ and **internal resistance** $(R_i)$:

$$I = \frac{E}{\left(R_L + R_i\right)}$$

To simplify matters, we will ignore the resistance of any wires connecting the voltage source to the load but, for long conductor runs, the conductor's voltage drop *would*, of course, need to be taken into account to determine the load current and terminal voltage.

What should now be obvious is that the *larger* the load current *(I)*, the *larger* the internal voltage drop *(IR_i)*, and the *smaller* the resulting terminal voltage *(U)* appearing across the load.

The *only* time that the cell's terminal voltage will equal its electromotive force is when the load is disconnected, and no load current is flowing to cause an internal voltage drop – that is:

$$U = E - (IR_i)$$
$$U = E - (0 \times R_i)$$
$$U = E$$

. . . and this is the reason why the cell's **electromotive force** is also termed its '**open-circuit voltage**' or '**no-load voltage**'.

> By rearranging the equation, $U = E - IR_i$, we could easily determine the internal resistance of a cell, that is:
>
> $$R_i = \frac{E - U}{I}$$

Now would be a very good time to define exactly what is meant by the term 'electromotive force'. From Kirchhoff's Voltage Law, we can define **electromotive force** as follows:

> In accordance with Kirchhoff's Voltage Law, an **electromotive force** is equal to the sum of the voltage drops around any closed loop – *including the internal voltage drop within the source itself.*

We can express this definition, mathematically, as:

$$E = I_{load} \left(R_{load} + R_{internal}\right)$$

## Determining the internal resistance of a cell or battery

Voltmeters draw a tiny amount of current to operate, so if we connect a voltmeter across a cell or battery, which isn't connected to a load, then any resulting internal voltage drop is quite insignificant. So it is possible to directly measure the e.m.f. or **no-load voltage** *(E)* of a cell or battery, by simply measuring it with a voltmeter.

And we can use the same voltmeter to determine the **terminal voltage** *(U)* of a cell or battery *whenever it is supplying a load current*.

These two voltages *(E and U)* allow us to determine the **internal resistance** of a cell or battery, by applying the equation derived as follows:

$$I = \frac{U}{R_L} \quad \text{and} \quad I = \frac{(E-U)}{R_i}$$

$$\frac{U}{R_L} = \frac{(E-U)}{R_i}$$

$$UR_i = R_L\left(E-U\right)$$

$$R_i = \frac{R_L\left(E-U\right)}{U}$$

To experimentally determine its internal resistance, we need to connect the cell or battery as shown in Figure 14.8.

step 1: Measure E          step 2: Measure U

step 3:  $R_i = \dfrac{R_{load}\ (E - U)}{U}$

Figure 14.8

**Step 1:** With no load connected to the cell, a voltmeter will enable us to measure its no-load voltage or e.m.f. *(E)*.

**Step 2:** Then, we can connect a load resistor of, say, 1 Ω resistance, to the cell and measure the resultant terminal voltage (*U*), and...

**Step 3:** Determine the internal resistance using the equation:

$$R_i = \frac{R_L\left(E-U\right)}{U}$$

This process can be repeated, using resistors of increasing resistance (e.g. 2, 3, 4, 5 Ω, etc.), and the average value of the resulting internal resistance can be determined.

**Worked example 1**  A battery has an electromotive force of 9 V and an internal resistance of 600 mΩ. What will be its terminal voltage when connected to (a) a 20-Ω load, (b) a 10-Ω load and (c) no load?

**Solution**  First, sketch a fully labelled circuit diagram, indicating the values of the components (as shown in Figure 14.9).

Figure 14.9

a  With a load resistance of 20 Ω, the resulting load current will be:

$$I = \frac{E}{\left(R_L + R_i\right)} = \frac{9}{\left(20+0.6\right)} = \frac{9}{20.6} = 0.437\ \text{A}$$

Next, we can determine the terminal voltage *(U)*:

$$U = E - (IR_i) = 9 - (0.437 \times 0.6)$$
$$= 9 - 0.262 = 8.738\ \text{V (Answer a.)}$$

b  With a load resistance of 10 Ω, the load current will be:

$$I = \frac{E}{\left(R_L + R_i\right)} = \frac{9}{\left(10+0.6\right)} = \frac{9}{10.6} = 0.849\ \text{A}$$

Next, we can determine the terminal voltage *(U)*:

$$U = E - (IR_i) = 9 - (0.849 \times 0.6)$$
$$= 9 - 0.509 = 8.491\ \text{V (Answer b.)}$$

c  With *no load*, the load current will be zero, so we can determine the terminal voltage *(U)*:

$$U = E - (IR_i) = 9 - (0 \times 0.6)$$
$$= 9 - 0 = 9.0\ \text{V (Answer c.)}$$

In the above worked example, we say that the 10-Ω load is '**heavier**' than the 20-Ω load. In this context, 'heavier' *always* refers to the magnitude of the resulting *load current*. That is, *the larger the load current (or the lower the load resistance), the 'heavier' the load.*

> **Important!** A '**heavy**' load is one that draws a 'heavy' (large) current. So, a low-resistance load is a 'heavy' load, and a high-resistance load is a 'light' load.

We can also describe a load in terms of its **power**. A high-power load represents a 'heavy' load, and a low-power load represents a 'light' load. For example, a 200-W (watt) lamp represents a 'heavier' load than a 100-W lamp.

As you can see from the previous worked example, the *heavier* load (i.e. in this case, the 10-Ω load) results in the *lower* terminal voltage. In fact, a *really* heavy load may cause a *significant* fall in terminal voltage. For example, suppose we connect a 1-Ω load to the previous example.

With a load resistance of just 1 Ω, the load current will be:

$$I = \frac{E}{\left(R_L + R_i\right)} = \frac{9}{\left(1+0.6\right)} = \frac{9}{1.6} = 5.63\ \text{A}$$

And the resulting terminal voltage *(U)* will be just . . .

$$U = E - (IR_i) = 9 - (5.63 \times 0.6) = 9 - 3.38 = 5.62\ \text{V!}$$

As you can see, this is now a *substantial* drop in the battery's terminal voltage compared to its electromotive force.

The *main* problem with a battery, however, is that its internal resistance can increase markedly when the battery deteriorates and its electrodes become consumed by the chemical reaction. If a car battery fails due to a higher-than-normal internal resistance, a voltmeter will still indicate a 'normal' terminal voltage of 12 V. A voltmeter actually draws a tiny current (microamperes) and, so, the battery's internal voltage drop will also be tiny. So, in order to properly test any battery, it's necessary to simulate a typical load resistance – only then will a voltage test indicate the battery's true state. All battery voltage testers incorporate a load in their internal circuitry for this very purpose.

**Worked example 2**    Let's assume that the cell in Worked example 1 has deteriorated somewhat, and its internal resistance has risen to, say, 1.75 Ω. Let's find out what effect this will have on the terminal voltage for (a) a 20-Ω load and (b) a 10-Ω load.

*Solution*

a    With a load resistance of 20 Ω, the resulting load current will be:

$$I = \frac{E}{(R_L + R_i)} = \frac{9}{(20 + 1.75)} = \frac{9}{21.75} = 0.41\,\text{A}$$

Next, we can determine the terminal voltage *(U)*:

$$U = E - (IR_i) = 9 - (0.41 \times 1.75)$$
$$= 9 - 0.72 = 8.28\text{V (Answer a.)}$$

b    With a load resistance of 10 Ω, the load current will be:

$$I = \frac{E}{(R_L + R_i)} = \frac{9}{(10 + 1.75)} = \frac{9}{11.75} = 0.76\,\text{A}$$

Next, we can determine the terminal voltage *(U)*:

$$U = E - (IR_i) = 9 - (0.76 \times 1.75)$$
$$= 9 - 1.33 = 7.67\text{ V (Answer b.)}$$

**Worked example 3**    A battery has a no-load terminal voltage of 12 V. When it supplies a load of 6 Ω, its terminal voltage is found to be 11.25 V. Determine the internal resistance of the battery.

*Solution*

$$R_i = \frac{R_L(E - U)}{U} = \frac{6(12 - 11.25)}{11.24} = 0.40\,\Omega\,\text{(Answer)}$$

## Practical values of internal resistance

### Cells

It's useful to have some idea of the typical values of internal resistance of various common **cells**:

- A non re-energiseable **alkaline** cell, for example, has an internal resistance within the range of 150–300 mΩ (depending on its physical size, e.g. AAA, AA, etc.) when new. These disposable cells are widely used in torches, smoke detectors, etc.
- A fully charged **lead-acid** cell has an internal resistance of around 50 mΩ. This very low value of internal resistance makes it an ideal cell for the manufacture of car batteries which must supply very heavy starting currents.
- A new, fully charged **nickel-cadmium (NiCad)** cell has an internal resistance of around 75 mΩ. NiCads are widely used in portable tools, etc., but are being replaced by lithium-ion types.
- A new, fully charged **lithium-ion** cell has an internal resistance of around 320 mΩ, which increases to around 340 mΩ after around a thousand charging cycles. These cells are now widely used for powering laptop computers, camcorders, mobile telephones, etc.

Remember that these figures relate to *cells*, not batteries. Batteries are groups of cells connected together in series, parallel or series-parallel. So the internal resistance of batteries must be determined from the number of cells and the ways in which they are connected. For example, a car battery consists of six lead-acid cells, connected in series, giving a total of approximately (6×50 =) 350 mΩ.

### Generators

It's *not* possible to give equivalent figures for the internal resistance of **generators**, as there are simply too many factors involved: the type of winding configuration, the length of windings' conductor, its cross-sectional area, the machine's operating temperature and so on. Each generator needs to be assessed individually.

We will be examining the effect of a generator's internal resistance when we study generators in a later chapter.

## Summary

- All voltage sources have an ***internal resistance (R_i)***.
- A voltage source's internal resistance is *in series* with any connected **load resistance (R_L)**.

- The **equivalent circuit** of a voltage source comprises an 'ideal' voltage source, producing an electromotive force *(E)*, connected *in series* with its internal resistance.
- With a load connected, the **load current** will be the 'ideal' voltage source's electromotive force *(E)* divided by the sum of the internal resistance *(R$_i$)* and the load resistance *(R$_L$)*, that is:

$$I = \frac{E}{R_i + R_L}$$

- The load current causes an **internal voltage drop** *(IR$_i$)* to occur across the voltage source's internal resistance.
- The voltage appearing across the load, is called the **terminal voltage** *(U)*, and is the *difference* between the voltage source's **electromotive force** *(E)* and its **internal voltage drop** *(IR$_i$)*, that is:

$$U = E - (IR_i)$$

- For any given load resistance, *the greater the internal resistance, the lower the voltage source's terminal voltage.*
- For any given internal resistance, *the lower the load resistance, the lower the voltage source's terminal voltage.*
- When there is no load connected to the voltage source, the terminal voltage will equal its electromotive force, because there is no internal voltage drop.

## Review your learning

Now that we've completed this chapter, we need to examine the **objectives** listed at its start. Placing '*Can I...*' at the beginning, and a question mark at the end, of each objective turns that objective into a **test item**. If we can answer each of those test items, then we've met the objectives of this chapter.

**Online resources**

The companion website to this book contains further resources relating to this chapter. The website can be accessed via the following link:
www.routledge.com/cw/waygood

# Chapter 15

# Energy, work, heat and power

## Objectives

On completion of this chapter, you should be able to

1 define each of the following terms:
   a energy
   b work
   c heat
   d power.
2 specify the SI unit of measurement for each of the following:
   a energy
   b work
   c heat
   d power.
3 state the fundamental equation for the work done by an electric circuit.
4 state the fundamental equation for the power of an electric circuit.
5 derive alternative equations for the work done by, and the power of, an electric circuit.
6 define the term 'efficiency'.
7 solve problems on energy, work, power and efficiency.
8 explain how electricity supply companies bill their residential consumers.
9 read an analogue energy meter.
10 summarise the claimed advantages of smart energy meters compared with conventional energy meters.
11 describe the relationship between electrical energy and heat.
12 solve problems on work and heat.

**Note**: unfortunately, quantities used in thermodynamics share the same symbols as completely different quantities used in electrical engineering. So, in this unit, you should be aware that

- in **electricity**: $U$ = potential difference; $Q$ = electric charge
- in **thermodynamics**: $U$ = internal energy; $Q$ = heat

## Introduction

The study of electricity is really the study of **energy** and of **energy conversion**. Unfortunately, the word 'energy', together with the related terms, 'work', 'heat' and 'power', have 'everyday meanings' and are frequently used interchangeably by the layman. But, for those of us who are studying electrical science, it is important that we understand the *scientific* meanings of these terms.

## Energy, work and heat

### Energy

The word, '**energy**', is derived from the Greek, '*energeia*': a word that first appeared as long ago as the fourth century BC in a work by the Greek philosopher, Aristotle. However, it is thought to have been used in its modern sense only since the beginning of the nineteenth century.

'**Energy**' is probably one of the most fundamental concepts in physics, but also one of the hardest to define. In fact, it's one of those rare scientific terms where it's very much easier to explain *what it does* or *how it behaves*, rather than *what it is*.

That's why practically all technical dictionaries and textbooks define **energy** simply as '*the ability to do work*'.

> Energy is defined as '*the ability to do work*'.

But this definition doesn't address the question of *what energy actually is!* And it also assumes we know what is meant by 'work'! So it is a rather unsatisfactory definition.

So, what is energy? Well, *whatever* it is, we know that it never appears from nowhere nor does it disappear into nothingness! We call this, 'the First Law of Thermodynamics', which states that '*the total energy of an isolated system is constant; it can be transformed from one form to another, but can be neither created nor destroyed*'.

As good an explanation of this was presented during a lecture to undergraduate students, in 1961, by the American Nobel Laureate, Richard Feynman (1918–1988):

> *There is a fact, or if you wish, a 'law', governing those natural phenomena that are known to date. There is no known exception to this law – it is exact so far we know. The law is called 'the Conservation of Energy'; it states that there is a certain quantity, which we call 'energy' that does not change in those many changes which nature undergoes. That is a most abstract idea, because it is a mathematical principle; it says that there is a numerical quantity, which does not change when something happens. It is not a description of a mechanism, or anything concrete; it is just a strange fact that we can calculate some number, and when we finish watching nature go through her tricks and calculate the number again, it is the same.*

What Dr Feynman is, essentially, saying is that energy is a **quantity**, *not* an **entity**.

> Energy is a **quantity**, *not* an **entity**.

So, energy isn't an entity (i.e. 'stuff'). We cannot see it, touch it, or smell it! Rather, it's an *abstract* thing: it doesn't exist in any tangible form.

Although scientists appear unable to offer a clear understandable definition of what **energy** *is*, what they *do* tell us is that this 'quantity' can be categorised as variations of either '**potential energy**' or '**kinetic energy**'.

We can think of **potential energy** as '*energy waiting to be used*', and includes chemical, gravitational, and nuclear energies. We can think of **kinetic energy** as '*energy an object acquires through movement*', and includes electrical, light, and sound energies.

For example, an object suspended above the ground is said to possess **potential energy**. That same object, as it falls under the influence of gravity, *loses* that potential energy and *acquires* exactly the same amount of **kinetic energy**.

And *all* bodies possess a form of kinetic energy that we call '**internal energy**'. Internal energy *used* to be called 'heat energy' or 'thermal energy', but both terms are now obsolete.

**Internal energy** describes the sum-total of *all* the energies associated with the atomic structure of a body, in particular the energy associated with the vibration of the atomic particles due to their **temperature**. As we shall learn, later, the internal energy of a body is closely linked to its temperature. As a body's internal energy increases, so does its temperature, and *vice versa*.

Scientists further categorise energy as being either '**stored energy**' or '**transitional energy**':

- '**Stored energy**' describes the energy which *is stored within a body*. 'Potential energy' and 'kinetic energy', in all their forms, are examples of 'stored energy'.
- '**Transitional energy**', on the other hand, describes energy that is *being transferred in either direction between a body and its surroundings*. 'Work' and 'heat' are categorised as 'transitional energy'.

So, while a body can *possess* potential energy and kinetic energy, it *cannot* 'possess' **work** or **heat**. Work and heat *only exist until they cross the boundaries of that body*.

An analogy of **transitional energy** (i.e. 'work' and 'heat') is **rain**. Water can exist as *water vapour* in the form of a *cloud*, or as *liquid water* in the form of a *pool*. When water is in transit between being a cloud and becoming a pool, we call it '*rain*'. As soon as the rain 'crosses the boundary' of the pool, it ceases to be 'rain' – as illustrated in Figure 15.1.

Analogy between WORK, HEAT, and RAIN

Figure 15.1

Note that **work** can transit in *any* direction (e.g. from potential to kinetic energy, or *vice versa*), but **heat** can *only* transit from a higher to a lower temperature.

To summarise, '**energy**' is a difficult concept to understand or to explain. So, instead of defining energy in terms of *what it is*, we have to define it in terms of *what it does*, hence we say that '*energy is the ability to do work*'.

So what, then, is '**work**'?

## Work

**Work** (symbol: $W$) is usually defined as '*force multiplied by the distance moved in the direction of that force*'.

Although, from the electrical point of view, this can be explained in terms of the force acting on an electric charge moving through a uniform electric field, a far more understandable definition is the one used in the study of **thermodynamics** (heat science):

> **Work** is '*the process of converting one form of energy into another*'.

Scientists would express this in a slightly different way. They would say that **work** is '*energy in transit between one form of energy and another*'.

So we can think of **work** either in terms of it being the '*process*' of energy conversion, or as '*transitional energy*'. Both explanations are correct.

A generator is a machine which *converts* kinetic energy into electrical energy – so that generator is doing *work*. An electric motor *converts* electrical energy into kinetic energy – so it, too, is doing *work*.

*Any* machine that *changes one form of energy into another* is doing **work**.

## Heat

Like work, heat is classified as 'transitional energy', and describes the transfer of energy which takes place from a body at a higher temperature to one at a lower temperature. We can say:

> **Heat** is '*the process of transferring energy from a warmer body to a cooler body*'.

Again, scientists would express this in a slightly different way. They would say that **heat** is '*energy in transit from a body at a higher temperature to one at a lower temperature*' (*never* the other way around!).

So we can think of **heat** either in terms of it being a '*process*' of energy transfer, or as '*transitional energy*'. Both are correct.

## Relationship between work and heat

So there are *two* ways of 'manipulating' energy. We can **change** *it from one form into another*, through **work**. Or we can **transfer** *it from a warmer body to a cooler body* through **heat** transfer.

Let's look at a simple, everyday, example.

When we switch on an electric kettle, the **internal energy** of the water (as well, of course, as of the kettle itself) starts to increase. We know this is happening, because the temperature of the water/kettle starts to increase.

So the supplied **electrical energy** is being *converted* into **internal energy**.

As *energy conversion* is taking place, **work** is being done on the water.

As the water's internal energy continues to increase, it causes the temperature of that water to rise above the room's ambient temperature. As a result, some of that energy will now be *transferred* (lost) from the hot water into the cooler surroundings.

Because this energy is being transferred away from the water due to a *temperature difference*, we describe it as **heat** transfer.

The increase in the water's internal energy, then, is the *difference* between the work done *on* the water and the heat transfer away *from* the water. That is:

*increase in internal energy = work  heat*

Note, we cannot measure the absolute amount of internal energy in the water, only its *change* – in this case, the amount of *increase*. We will learn more about this relationship later in this chapter.

## Joule heating

'**Joule heating**', which is also known as '**resistance heating**' or '**ohmic heating**', describes the process which results in an increase in a conductor's temperature due to an electric current. Sometimes, as in the example of the electric kettle, this is desirable. But, in the case of an electric wire or conductor, this represents an energy *loss* through heat transfer into the surroundings which, of course, is undesirable but, to a large extent, unavoidable.

Joule heating is the result of collisions between a conductor's free electrons and its fixed atoms/ions. The sudden loss of the electrons' kinetic energy causes a corresponding increase in the conductor's internal energy and, because the two are linked, a corresponding *increase in its temperature*.

The kinetic energy of any object is proportional to the square of its velocity which, in the case of free electrons, corresponds to *the square of their drift velocity*. As we learnt in a previous chapter, the **drift velocity** increases significantly in wires having a smaller cross-section — which explains why, for example, the temperature of a fuse wire, whose cross-sectional area is far lower than the circuit conductor which it protects, increases so much faster than that conductor and reaches its melting point sooner.

## Power

Suppose two vehicles, say, an SUV and a compact car, are each supplied with, say, exactly one litre of fuel. Both vehicles would do **work** as they convert the potential (chemical) energy supplied by that litre of fuel into kinetic energy, as they move along a road. However, the SUV would use up its supply of energy (i.e. its fuel) much faster than the compact car and, therefore, not travel as far before running out of fuel. This is because the SUV has a much more **powerful** engine than the car, and expends its source of energy at a much higher *rate*.

> Power is *the rate of doing work or of heat transfer*.

**Power** (symbol: *P*) is simply a *rate* (i.e. energy divided by time), in just the same way as **speed** (i.e. distance divided by time) is a *rate*. There's no such thing as different *types* of speed and, in the same way,

there's no such thing as different *types* of power! So, it's technically incorrect to talk about 'electrical power', 'mechanical power', etc. – these terms are just as meaningless as trying to describe different types of velocity.

And, for the same reason as we cannot 'consume' velocity, we cannot 'consume' power either. Power is simply a 'rate'; it isn't some sort of 'stuff', so, it's technically incorrect to talk about 'power consumption' for example! What is being consumed is energy; power merely tells us the *rate* at which that energy is being consumed… or transferred… or gained… or lost.

So, for example, when engineers *say* 'power loss', what they actually *mean* is the 'rate of energy loss'.

## Units of measurement

### Energy, work and heat: the 'joule'

As we've already learnt, work and heat are simply considered to be *transitional forms of energy* — therefore, **work**, **heat**, together with all other forms of **energy** *all* share the same SI unit of measurement, the **joule** (symbol: **J**), named after the English scientist, James Prescott Joule.

Another unit of measurement of energy or work that you will come across is the **kilowatt hour** (symbol: **kW·h**), also known as the '**unit**' (after '**Board of Trade Unit**' – the Board of Trade, at one time, governed how much energy companies could charge their consumers for energy). Because the joule is actually a *very* small unit of measurement, power utility companies charge their customers for the energy they use in kilowatt hours. A kilowatt hour is simply a 'big' version of the joule. How much bigger? Well, as we shall learn later, it's equivalent to 3.6 MJ.

### Power: the watt

As **power** is the *rate of doing work* (or of *heat transfer*), it follows that its unit of measurement is the **joule per second**. However, in common with many other SI derived units, this unit is given a special name: the **watt** (symbol: **W**), named in honour of the Scottish engineer, James Watt. So a 'watt' is just another way of saying, 'joule per second'.

You should be aware that the imperial unit of measurement of power is the **horsepower** (symbol: **HP**). Although the horsepower has been phased out in the United Kingdom since the adoption of SI, you may come across older electric motors whose nameplates specify their rated power output in horsepower, rather than in watts. To convert horsepower to watts:

$$1\ HP = 746\ W$$

You have probably noticed that car engines are now rated in kilowatts, rather than in horsepower. Some books refer to 'horsepower' being used to measure 'mechanical power' but, as we have learnt, there is really no such thing as 'mechanical power', 'electrical power', etc!

There is absolutely no reason why you shouldn't measure the power of an electric fire in horsepower – in fact, a 'three-bar 3-kW electric fire' could just as correctly be called a 'three-bar 4-horsepower electric fire'!

## Energy, work, heat and power in electric circuits

In Chapter 5, we learnt that the volt is defined as *'the potential difference between two points such that the energy used in conveying a charge of one coulomb from one point to the other is one joule'*. This can be expressed as:

$$E = \frac{W}{Q}$$

where:

$E$ = voltage, in volts

$W$ = work, in joules

$Q$ = electric charge, in coulombs

Rearranging this equation, to make $W$ the subject:

$$W = EQ$$

but we have also learnt that $Q = It$, so, substituting for $Q$ in equation (1), we have:

$$W = EIt \qquad \text{—equation (1)}$$

where:

$W$ = work, in joules

$E$ = voltage, in volts

$I$ = current, in amperes

t = time, in seconds

The above equation is the fundamental equation for the work done, or energy expended, by an electric circuit.

As **power** is defined as *the rate of doing work*, we can express this as:

$$P = \frac{W}{t}$$

... substituting the fundamental equation for work into equation (2), we have:

$$P = \frac{EIt}{t}$$

which is simplified to:

$$P = EI \qquad \text{—equation (2)}$$

where:

$P$ = power, in watts

$E$ = voltage, in volts

$I$ = current, in amperes

As resistance is the *ratio of voltage to current*, we can derive alternative equations for both work and power:

$$W = I^2Rt \qquad\qquad P = I^2R$$

Again, from Ohm's Law, $I = \frac{E}{R}$. Substituting for $I$, in eq.1:

$$W = E\left(\frac{E}{R}\right)t$$

$$W = \frac{E^2}{R}t$$

Again, from Ohm's Law, $I = \frac{E}{R}$. Substituting for $I$, in eq.2:

$$P = E\left(\frac{E}{R}\right)$$

$$P = \frac{E^2}{R}$$

## Summary of equations for work and power

For **work**:

$$W = EIt \qquad W = I^2Rt \qquad W = \frac{E^2}{R}t \qquad W = Pt$$

... and for power:

$$\boxed{P = EI} \quad \boxed{P = I^2 R} \quad \boxed{P = \frac{E^2}{R}} \quad \boxed{P = \frac{W}{t}}$$

**Worked example 1** Calculate the work done by a resistor when connected across a 230-V supply, if it draws a current of 10 A for 1 min.

**Solution** **Important.** Don't forget to convert minutes to seconds.

$$W = E\,I\,t$$
$$= 230 \times 10 \times (1 \times 60)$$
$$= 138\,000 \text{ J (Answer)}$$

**Worked example 2** Calculate the current drawn by a 100-W lamp from a 230-V supply.

**Solution**

Since $P = EI$

then $I = \dfrac{P}{E} = \dfrac{100}{230} = 0.435$ A (Answer)

**Worked example 3** Calculate the rate at which energy is lost when a conductor of resistance 0.5 $\Omega$ carries a current of 13 A.

**Solution** **Note:** in this question, we are asked to calculate the *rate* at which energy is lost – this is the same as 'calculate the *power* loss'.

$$P = I^2 R = 13^2 \times 0.5 = 84.5 \text{ W (Answer)}$$

**Worked example 4** The voltage drop across a resistance of 200 $\Omega$ is found to be 12 V. Calculate (a) the power developed by the resistor, and (b) the work done by the resistor in 30 s.

**Solution**

a $P = \dfrac{E^2}{R} = \dfrac{12^2}{200} = 0.72$ W (Answer a.)

b $W = Pt = 0.72 \times 30 = 21.6$ J (Answer b.)

Be aware that *some* electrical textbooks may use the symbol '*W*' for '**power**'. This has been done (ill advisedly!) because '**W**' is the symbol for the 'watt'. This can lead to confusion, as

'*W*' (in italics) actually represents '**work**', *not* 'power'!!!

So this is a confusing and extremely slipshod method of writing any equation, because you should *never ever* write an equation in terms of the *units* used. After all, you would *never* (or shouldn't!) write, **V = Ω×A**, so *why* would you do it for power?

In equations, *always* use the symbol for a 'quantity' (e.g. *P* for '**power**') never the symbol for that quantity's' unit of measurement' (e.g. **W** for '**watt**'). That is:

**P = 300 W *never* W = 300 W**

## Misconceptions about energy transfer

Many students believe that *electrons* deliver energy from the supply to the load, after which they return to the supply, depleted of their energy, in order to 'collect' more energy which they deliver the next time they move around the circuit! *This 'conveyor belt' model of energy transfer is quite wrong,* as a brief review of the behaviour of free electrons will reveal!

You will recall from the earlier chapter on the *electron theory of electricity* that the free electrons in a metal conductor are in a constant state of vigorous, chaotic, motion and, whenever a potential difference is applied across that conductor, the electrons drift, *very slowly,* along the conductor (in the order of millimetres per hour!). In the case of alternating current, the electrons don't even move through the circuit at all; they merely vibrate backwards and forwards as the alternating supply voltage continuously changes its magnitude and direction. From this, *it should be clear that the idea of free electrons acting like an 'energy conveyor belt', is a complete misconception.*

So, how *does* energy transfer between a circuit's supply and its load? The simple answer is that, although we know how it *doesn't* supply energy, no one truly knows how it *does*, although there are various theories.

For most of these theories, the 'flow' of energy from the supply to the load in an electrical circuit can only be satisfactorily explained using post-graduate mathematics.

Perhaps the most well-known of these theories is one that describes the combination of a circuit's magnetic and electric fields to create what is known as the 'Poynting field', named in honour of the British physicist, John Poynting (1852–1914), which causes energy to 'flow' in a direction perpendicular to both these fields, and causes that energy to be *transferred through empty space around – not through – the conductors.*

Unfortunately, there is very little point in discussing this theory very much further, as the mathematics involved is extremely abstract, and the process simply cannot be described in terms that a non-physicist can easily understand. However, Figure 15.2, which has been extremely simplified, attempts to convey the general idea.

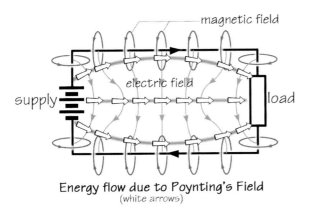

**Energy flow due to Poynting's Field**
(white arrows)

Figure 15.2

In Figure 15.2, a pair of conductors connects a battery to a load, and (conventional) current flows in a clockwise direction around the circuit, as shown. The conductors are surrounded along their entire length by a magnetic field, and an electric field stretches between the upper (positive) conductor and the lower (negative) conductor. The white arrows are perpendicular to both the magnetic *and* the electric fields, and represent the Poynting field which acts to transfer energy *entirely through the empty space that surrounds the conductors between the supply and the load.*

> Energy transfer from supply to load is via a combination of the electric and magnetic fields surrounding a circuit's conductors – *not* through the conductors themselves.

## Power ratings

Electrical appliances, such as incandescent lamps, are required to have the **rated power** at which they are designed to operate displayed. The rated power is *always* shown together with the rated voltage, because an appliance *will only develop its rated power at its rated voltage.*

This information is termed **nameplate data**. In the case of lamps, for example, the nameplate data is printed either on the glass envelope or on the bayonet/screw mount, in the form: '**100 W/230 V**', etc.

If an electrical appliance, such as a heater, is operated below its rated voltage, then *the power developed can be substantially reduced from its nameplate power-rating.*

**Worked example 5** An electric heating element has a resistance of 10 Ω and is rated at 230 V. Calculate, (a) the power of the heating element, (b) the resulting power if the rated value of the supply voltage drops by 10%, and (c) the percentage power loss. Assume that there is negligible change in the heating element's cold and hot resistance.

**Solution**

(a) $P_{(a)} = \dfrac{E^2}{R} = \dfrac{230^2}{25} = 2116\,\text{W}$ —(Answer a.)

(b) A 10% drop in voltage = 207 V

$P_{(b)} = \dfrac{E^2}{R} = \dfrac{207^2}{25} = 1714\,\text{W}$ —(Answer b.)

(c) percentage drop in power = $\dfrac{P_{(a)} - P_{(b)}}{P_{(a)}} \times 100\%$

$= \dfrac{2116 - 1714}{2116} \times 100\%$

$= 19\%$—(Answer c.)

The above worked example is particularly interesting, because it shows that a 10% drop in the supply voltage can result in a 19% drop in power!

This demonstrates the importance of an electricity network company's responsibility for *maintaining the supply voltage within its legal limits.* In the UK this is 230 V (+10%/–6%) – in other words, the supply voltage must not rise beyond 253 V or fall below 216.2 V.

It also demonstrates why lamps and other residential electrical loads *must* be connected in **parallel**. Only by connecting them in parallel will they be each subject to their rated supply voltage, enabling them to operate at their rated power.

However, with some types of electrical device, the power rating indicates the '*maximum power*' at which that device is designed to operate without overheating, and not its normal '*operating power*'.

For example, unlike lamps, **resistors** can operate quite normally well below their power rating, but if they are allowed to *exceed* their power rating, *then they will likely overheat and burn out.* Knowing the power rating of a resistor allows us to determine the maximum continuous current that the resistor can carry without overheating.

**Worked example 6** Calculate the maximum continuous current that a 4.7-kΩ/1-W resistor can support without overheating.

**Solution**

$$P = I^2 R$$

$$I = \sqrt{\frac{P}{R}} = \sqrt{\frac{1}{4.7 \times 10^3}} = \sqrt{0.21 \times 10^{-3}}$$

$$= 14.8 \times 10^{-3} = 14.8 \,\text{mA}$$

## Power in series, parallel and series-parallel circuits

In the circuit in Figure 15.3, two resistors of 10 Ω and 5 Ω are connected across a supply of 30 V. Let's calculate the power of the complete circuit, and the power of each resistor.

**Figure 15.3**

total resistance: $R_T = R_1 + R_2 = 10 + 5 = 15 \,\Omega$

circuit current: $I = \dfrac{E}{R_T} = \dfrac{30}{15} = 2 \,\text{A}$

circuit power: $P = EI = 30 \times 2 = 60 \,\text{W}$

power of $R_1$: $P_1 = I^2 R_1 = 2^2 \times 10 = 40 \,\text{W}$

power of $R_2$: $P_2 = I^2 R_2 = 2^2 \times 5 = 20 \,\text{W}$

Let's look at the relationship between the power developed by the individual resistors compared to the power of the complete circuit. It's plain to see that, for a series circuit, *the total power is the sum of the power of each resistor.*

Now, let's look at a parallel circuit, in Figure 15.4, comprising two resistors, each of 10 Ω, connected across a supply of 30 V. Let's calculate the power of the complete circuit, and the power of each resistor.

**Figure 15.4**

$$\frac{1}{R_T} = \frac{1}{R_1} + \frac{1}{R_2}$$

$$\frac{1}{R_T} = \frac{1}{10} + \frac{1}{10}$$

total resistance: $\dfrac{1}{R_T} = \dfrac{2}{10}$

$$R_T = \frac{10}{2} = 5 \,\Omega$$

power of the circuit: $P = \dfrac{E^2}{R_T} = \dfrac{30^2}{5} = 180 \,\text{W}$

power of $R_1$: $P_1 = \dfrac{E^2}{R_1} = \dfrac{30^2}{10} = 90 \,\text{W}$

power of $R_2$: $P_2 = \dfrac{E^2}{R_2} = \dfrac{30^2}{10} = 90 \,\text{W}$

Once again, *the total power of the circuit is equal to the sum of the power of each resistor.*

If we were to repeat this for a *series-parallel* circuit or, in fact, for *any type of circuit*, we would obtain the *same* result!

So, we can state the general rule that:

The total power of *any* type of circuit is the sum of the powers developed by that circuit's individual components.

## Connecting lamps in series: oh, what a surprise!

If two incandescent lamps, with *identical voltage ratings*, but *different power ratings*, were to be connected in *parallel* across their rated voltage supply, then it is obvious that the more-powerful lamp would be the brighter of the two.

But what would happen if, instead, they were connected in *series* with each other across the same supply?

Well, we'd expect the lamps to operate much less brightly, of course. But would the lamp with the higher power rating still be the brighter of the two? Intuitively, it would be reasonable to expect that this would be so.

But, on this occasion, our intuition would let us down!

This topic is only included in this chapter because it usually comes as a complete surprise to most students that, when lamps of the same voltage rating but *different power ratings*, are connected in series, it's the lamp with the lower power rating that will end up being the brighter of the two!

So *why* should this be so?

Well, let's take the example of a **230-V/100-W** lamp and a **230-V/200-W** lamp.

First of all, let's calculate the 'hot' or operating resistance of each lamp when each is subject to their rated voltage:

| **For the 100-W lamp:** | **For the 200-W lamp:** |
|---|---|
| $$P = \frac{E^2}{R}$$ | $$P = \frac{E^2}{R}$$ |
| $$R = \frac{E^2}{P} = \frac{230^2}{100} = 529\,\Omega$$ | $$R = \frac{E^2}{P} = \frac{230^2}{200} = 264.5\,\Omega$$ |

So, the first thing that often comes as a surprise to many students is that, when operated individually, the hot resistance of the lamp with the lower power rating, is higher than that of the lamp with the higher power rating.

Many students expect it to be the other way around!

The cold resistance of each lamp will be between 15 – 20 times lower than their hot resistance. But the cold resistance of the lamp with the lower power rating will, of course, still be higher than that of the lamp with the higher power rating.

Figure 15.5

Now, if we connect the two lamps in series across a 230-V supply, as shown in Figure 15.5, the voltage drop *(U = IR)* across the lamp with the lower power rating will (because of its higher resistance) be *higher* than that across the lamp with the higher power rating and it will, therefore, *be the brighter of the two*. Being brighter, of course, means that it is hotter, so its resistance increases yet further: further increasing the voltage drop across it, making it brighter still, compared with the other lamp!

The lamp will not operate at its *rated* brightness, of course, because it is not being subject to its rated voltage; it will merely be *brighter* than the second lamp.

## Efficiency

Not *all* of the energy supplied to a load goes into doing useful work. For example, some of the energy supplied to an electric motor is lost overcoming *friction*, *windage* (air resistance), *vibration* and *heat transfer* into the surrounding atmosphere. So the useful work available from a motor is *always* somewhat less than the electrical energy supplied to the motor! Rotating machines, such as motors (**M**), are always less efficient that stationary machines, such as transformers, due to the friction of their bearings, windage losses, etc.

In practice, we talk about '**input power**', '**output power**' and '**power losses**'. But, as we have learnt, 'power' is simply a 'rate' so it is, strictly speaking, incorrect to talk about power in these ways, so a word of explanation of what we *really* mean, here, is necessary.

Figure 15.6

- By '**input power**', what we *really* mean is *the rate at which (electrical) energy is being supplied to the load.*
- By '**output power**', what we *really* mean is *the rate at which useful energy* (in the case of a motor, *kinetic energy) is utilised by the load.*
- By '**power loss**', what we *really* mean is *the rate at which energy is lost* overcoming friction, windage, heat transfer, etc.

Motors are *always* rated according to their **output power**, as this is what is available from the motor to drive their load and, so, it is their output power that matters, *not* their input power. The output power of a machine is often called its 'brake power' – in fact, in North America, the term 'brake horsepower' is often used to describe the output power of a machine. At present, the *output* power of an American motor is *always* expressed in horsepower, whereas its input power is always expressed in watts.

The ratio: *output power divided by input power* is termed **efficiency** (symbol: $\eta$ —the Greek letter '*eta*').

$$\eta = \frac{P_{output}}{P_{input}}$$

Efficiency is simply a *ratio* and, so, it does *not* have any units of measurement. Instead, it is expressed either as a **per unit** value (e.g. '**0.85**') or, more commonly, as a **percentage** value (e.g. '**85%**'). In the above equation, efficiency is a 'per-unit' value. To change a per-unit value to a percentage value, we must multiply by 100:

$$percentage = (per \ unit \times 100)\%$$

There is simply no such thing as 'lossless' energy transfer, so efficiency can *never* reach unity (or 100%). Even the most efficient electrical machine, the transformer, can only achieve a full-load efficiency of around 0.95–0.98 p.u. (or 95%–98%).

**Worked example 7** What is the efficiency, expressed both as a per-unit value and as a percentage value, of an electric motor whose output power is 3900 W when its input power is measured as 5000 W?

**Solution**

$$\eta = \frac{P_{output}}{P_{input}} = \frac{3900}{5000} = 0.78 \ \text{p.u.(Answer)}$$

or $0.78 \times 100 = 78\%$ (Answer)

**Worked example 8** A 230-V electric motor, drawing a current of 12 A, is 75% efficient. Calculate its input and output powers.

**Solution**

Input Power, $P_{input} = EI = 230 \times 12 = 2760$ W

$$\text{Since } \eta = \frac{P_{output}}{P_{input}}$$

$$P_{output} = \eta P_{input} = \frac{75}{100} \times 2760 = 2070 \text{ W (Answer)}$$

**Worked example 9** A 400-V electric machine, operating at an efficiency of 80%, has an output power of 3 kW. Calculate the current supplied to the machine.

**Solution** In this problem, we have to first find the machine's input power, from which we can then find the current supplied.

$$\text{Since } \eta = \frac{P_{output}}{P_{input}}, \text{t}$$

$$\text{then } P_{input} = \frac{P_{output}}{\eta} = \frac{3000}{0.80} = 3750 \text{ W}$$

Since $P_{input} = EI$,

$$\text{then } I = \frac{P_{input}}{E} = \frac{3750}{400} = 9.38 \text{ A (Answer)}$$

## Maximum Power-Transfer Theorem

In the earlier chapter on *internal resistance*, we learned that the effect of a voltage source's **internal resistance** is *to reduce the terminal voltage of that source* whenever it supplies current to a load. We might also ask ourselves whether internal resistance affects any other circuit behaviour.

This was a question that interested a German engineer by the name of Moritz von Jacobi (1801 – 1874) and, after applying some thought together with 'a great deal of common sense', he concluded that:

*Maximum power is transferred when the internal resistance of the source equals the resistance of the load, when the external resistance can be varied, and the internal resistance is constant.*

As 'power' is simply a 'rate', and not something tangible that can be moved around, a more technically-accurate definition for 'Jacobi's Law' or, as it is better known, the 'Maximum Power-Transfer Theorem', would be:

> *'The maximum rate of energy transfer from a source to its load occurs when the internal resistance of the source equals the resistance of the load, when the external resistance can be varied, and the internal resistance is constant.'*

The 'Maximum Power-Transfer Theorem' can be easily proved by simply performing calculations on a circuit, such as that illustrated in Figure 15.7, for several different values of load resistance which are below, equal to, and above the value of the cell's internal resistance.

**Figure 15.7**

Table 15.1 shows the results of calculating the rate at which energy transferred to the load, between load resistances of 0 Ω to 10 Ω. The first column shows those values of load resistance. The second column shows the resulting load current, calculated as follows:

$$I_{\text{load}} = \frac{E}{R_{\text{internal}} + R_{\text{load}}}$$

The third column shows the rate at which energy is delivered to the load, calculated as follows:

$$P_{\text{load}} = I_{\text{load}}^2 R_{\text{load}}$$

We'll worry about the remaining (% Efficiency) column in a moment.

**Table 15.1**

| R load | I load | P load | % Efficiency |
|---|---|---|---|
| **0** | 20.0 | 0.0 | **0.0** |
| 1 | 16.7 | **277.8** | 16.7 |
| 2 | 14.3 | 408.2 | 28.6 |
| 3 | 12.5 | 468.8 | 37.5 |
| 4 | 11.1 | 493.8 | 44.4 |
| 5 | 10.0 | 500.0 | 50.0 |
| 6 | 9.1 | 495.9 | 54.5 |
| 7 | 8.3 | 486.1 | 58.3 |
| 8 | 7.7 | 473. | 61.5 |
| 9 | 7.1 | 459.1 | 64.3 |
| 10 | 6.7 | 444.4 | 66.7 |

Immediately, we can see that when the resistance of the load equals the internal resistance of the source (i.e. at 5 Ω), the **maximum power**, or the *maximum rate at which energy* (500 W) *is* transferred to the load — confirming the 'Maximum Power-Transfer Theory'.

If we draw a graph of $P_{\text{load}}$ against $R_{\text{load}}$ (Figure 15.8), we can see how the power transferred to the load peaks at 5 Ω – i.e. when the load resistance is the same as the internal resistance of the supply.

**Figure 15.8**

It is important to understand that Jacobi's theorem describes the maximum *rate of energy transfer*, and not the maximum *efficiency* of transfer. This is because, at the maximum rate of energy transfer, just as much energy is lost by the internal resistance as is delivered to the load – making the transfer just 50% efficient!

If we refer back to Table 15.1, column 4 shows the percentage efficiency for each value of load resistance, calculated as follows:

$$\% \text{ Efficiency} = \left( \frac{P_{\text{load}}}{P_{\text{internal}} + P_{\text{load}}} \right) \times 100$$

So, as we can see from column 4, greater *efficiency* is only achievable by making the resistance of the load *greater* than the resistance of the source: but the magnitude of that power is lower.

## The Significance of the Maximum Power-Transfer Theorem

If the maximum rate of energy transfer can only be achieved with an efficiency of just 50%, then we might be forgiven for asking, *'What is the significance of the 'Maximum Power-Transfer Theorem'?'*

Well, in some cases, *achieving maximum power transfer is simply more important than how efficiently it is done.*

For example, when we start a vehicle's engine, it's important to ensure that the battery is delivering maximum power to the starter motor. This can be achieved by designing the starter motor such that the resistance of its windings is roughly the same as the internal resistance of the car's battery. And, in order to achieve this, *it's worth putting up with the fact that it's not a particularly efficient way of doing so.* As an aside, it's also worth mentioning that, because the internal resistance of a battery is so low, transferring maximum power to a vehicle's starter motor cannot be sustained for very long without significantly overheating both the battery and *the* starter motor!

This theorem is also widely used in electronics, where it's referred to as '**impedance matching**'. 'Impedance', as we shall learn later, is a circuit's opposition to the flow of *alternating current*, and is a combination of resistance and another form of opposition called 'reactance' which only appears in a.c. circuits. For example, if we want a sound system's amplifier to deliver maximum power to a loudspeaker, then it's important that the impedance of the loudspeaker's voice coil (i.e. the coil that drives the diaphragm) should match the output impedance of the amplifier. Again, it's more important that maximum power is transferred to the loudspeaker than it should be done efficiently.

## Residential electricity utility bills

During the 1860s, the then Chancellor of the Exchequer, William Gladstone, asked Michael Faraday, 'But, what is the use of electricity?' Faraday replied that he didn't really know, but continued 'however there is every possibility that you will soon be able to tax it!' Little did Faraday realise that his humorous response would eventually become a reality!

When an electricity supply company bills its residential consumers, it is charging them for the *energy consumed*, or the *work done by that energy.*

The supply company is not interested in the *rate* – i.e. the power – at which a *residential* consumer uses that energy, although it *does* care when it comes to industrial consumers, as we shall learn later.

As we have learnt, in SI the unit of measurement for work or energy is the joule. The joule represents a very small quantity of energy, so more practical units are the megajoule or gigajoule. However, for some reason, electricity supply companies have never adopted these units but use, instead, the **kilowatt hour (kW·h)**, which is equivalent to 3.6 MJ.

> Some students describe the kilowatt hour as being *'the amount of power consumed over a period of one hour'.* This is misleading, as you do not 'consume' power! It is far more accurate to describe it as *'amount of energy consumed, over a period of one hour, at a rate of one kilowatt'.*

In the UK, the kilowatt hour is also known as a '**unit**', which is short for '**Board of Trade unit**'. The Board of Trade no longer exists, but was at one time the government organisation charged with determining the prices that electricity authorities could charge their customers.

The kilowatt hour is derived as follows:

$$\text{power} = \frac{\text{work}}{\text{time}}$$
$$\text{so, work} = \text{power} \times \text{time}$$

Now, if we measure power in kilowatts (kW), and time in hours (h), then:

$$\text{work} = \text{kilowatts} \times \text{hours} = \text{kilowatt hour}$$

So, to calculate the work done, or energy consumed, by a load in kilowatt hours, the following special equation can be used:

> work (in kilowatt hours) = power (in kilowatts) × time (in hours)

From this, we can say that *a kilowatt hour is the work done by a load, at a rate of one kilowatt, over a period of one hour.*

**Worked example 10**   A 60-W lamp is left switched on, continuously, for a period of six hours every day. How much energy is expended over a period of one week, and how much does it cost over that period, if the electricity supply company charges 17.5 p per kilowatt hour?

**Solution**

$$\text{work (in kilowatt hours)}$$

$$= \text{power (in kilowatts)} \times \text{time (in hours)}$$

$$= \left(\frac{60}{1000}\right) \times (6 \times 7) = 2.52 \text{ kW} \cdot \text{h (Answer a)}$$

$$\text{cost} = 2.52 \times 17.5 = 44.1 \text{pence (Answer b)}$$

## Measuring energy consumption

The energy used by residential consumers is measured using an instrument called an **energy meter**.

The energy consumed during a billing period is the difference between the current meter reading and the previous meter reading.

Energy meters are the property of the electricity supply company, and it is illegal to tamper with them. They are installed at the service entry point of a property. In modern residences, they are usually installed in an external cabinet, where they can be accessed and read without the meter reader having to enter the property. Most older properties in the UK, however, have their energy meters indoors, often in relatively inaccessible places – which means that meter readers have to enter the property in order to take a reading. Not only is this inconvenient, but it requires the householders to be at home whenever the meter reader calls.

supply ═══ load          supply ═══ load

*digital energy meter*     *analogue energy meter*

Figure 15.9

Modern energy meters provide a digital display which practically anyone can read. However, a great many older properties still have energy meters with *analogue* displays, which are relatively difficult to read. As it is not unreasonable to expect anyone working in the electricity industry to be able to read an analogue energy meter, we will now learn how this is done.

Typically, analogue energy meters have six dials, one (usually coloured red) of which is only used for calibration purposes and, so, can be ignored when reading the meter. Each dial is directly gear-driven by the one to its right and, so, *alternate dials rotate in opposite directions*. The remaining five dials are read, in sequence, *from right to left*, and each dial represents kilowatt hours expressed in: **units, tens, hundreds, thousands** and **tens of thousands** – as illustrated in Figure 15.10.

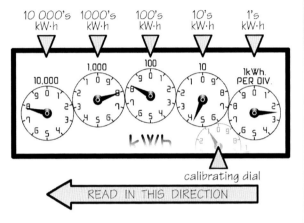

Figure 15.10

As explained, an energy meter's dials are read *from right to left*, and the figures are written down in the same sequence. When a pointer is between two numbers, the *lower* number is read, as shown in the example in Figure 15.11.

Figure 15.11

Not all readings are as obvious as the example shown in Figure 15.11.

Is this pointing at 7?
or is it just ahead of 7?
or is it just before 7?

Is this pointing at 5?
or is it just ahead of 5?
or is it just before 5?

Figure 15.12

Consider the example in Figure 15.12 .
Front right to left:

- **dial 1**: reads **0**.
- **dial 2**: reads **5**, because dial 1 has just started a new revolution.
- **dial 3**: reads **9**.
- **dial 4**: reads **6**, because dial 3 has not yet completed a full revolution
- **dial 5**: reads **2**.

So the energy meter reads: **26 950 kWh**.

### Self-test exercise

For each of the following, enter the meter readings in the spaces provided in Figure 15.13.

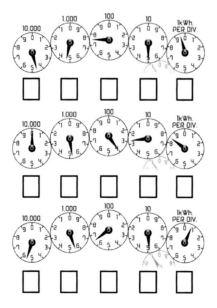

Figure 15.13

Answers are given at the end of this chapter.

It is claimed that the installation of smart meters into millions of residences across Europe, North America, New Zealand, and Australia represents one of the biggest energy-industry programmes in history. However, Smart Energy GB, the organisation tasked with promoting the installation of smart meters in the UK, has stated that (for the present, at least!), 'Smart meters aren't mandatory –if they wish, consumers can choose not to opt in to the programme'.

So, what exactly is a 'smart' meter?

A so-called **'smart' meter** is a digital energy meter, which records electrical energy consumption in intervals of one-hour or less, and wirelessly communicates that data directly to the electricity supply company, without the need for a meter reader to visit the consumer's premises. Smart 'gas meters' are also available for measuring gas consumption and, these too, report data back to the gas supplier.

Smart meters also inform the *consumer* how much energy is being consumed, together with its cost, in near real-time, via a separate **in-home display monitor**. It is argued that this will enable consumers to manage their energy consumption 'better' and save money if they can identify which of their appliances should be switched off when they are not needed.

A newspaper report described how, after moving into a new property, a consumer kept receiving higher energy bills than he could account for. It turned out that the previous occupant had placed a heating element in the cold water supply tank to prevent it from freezing during the winter, and had forgotten to disconnect it when he moved out! A display monitor could have helped the new property owner determine, through a process of elimination, that something, other than his normal loads, was causing his bills to be so high.

Traditional mechanical or electronic energy meters only measure the *quantity* of energy consumed, and provide absolutely no information whatsoever on *when* that energy is being used. Smart meters, on the other hand, are able to convey that information back to the energy supplier, enabling them to introduce pricing tariffs based, not simply upon the *quantity* of energy consumed, but also upon the *time of day* and the *season of the year* when that energy is consumed.

The arguments put forward by the supply companies for how the introduction of smart meters will benefit consumers are:

- smart meters will automatically send the meter reading to the supply company – no more visits by meter readers or estimated bills.
- smart meters are supplied together with a **smart energy monitor** which allows consumers to
  - track their energy use by day, week, month, and year.
  - compare their energy consumption with those around them.
  - see how much it costs to use energy for daily tasks, such as cooking the evening meal or running the dishwasher.
- interactive online tools will be made available, so that consumers can see *how* they are using the energy they consume.
- pay-as-you-go consumers can view their balance using the supplied energy monitor.
- faster restoration of service following a power cut, because smart meters also act as sensors which will detect a power cut instantaneously and inform the distribution network company of the area affected.

One drawback with the current generation (SMETS1) of energy meters is that they cannot necessarily be 'switched' between energy suppliers. So, if a consumer wishes to switch from one supplier to another to take advantage of a cheaper or more-suitable tariff, then the meter reverts back to being a 'dumb' meter – i.e. it will continue to record energy consumption, but it *cannot* transmit any data to the new energy supplier. Instead, a new smart meter has to be installed. This drawback has been recognised by the industry, and will be overcome with the introduction of a new generation of smart meters (SMETS2).

Of course, it would be naïve to believe that smart meters exist simply for the benefit of the consumers. There are huge benefits for the supply companies, too.

For example, the data provided by smart meters will enable supply companies to change their consumers' behaviour by, for example, charging more for the energy provided during peak hours, if they believe such tariffs can increase the efficiency of the entire energy supply system. By introducing such financial incentives for consumers to shift their demand from peak hours to off-peak hours, the effect will be to reduce the load during peak hours and 'flatten' the electricity-grid's demand curve – thus reducing the need to invest in peak-load generation plant.

There is a certain degree of mistrust on the part of consumers on the subject of smart meters, particularly if the information they transmit back to the supply companies could result, for example, in making it more expensive to use energy at peak times. There have also been some reports in the press of metering inaccuracies, faulty or hazardous installations, etc.

## Electricity and heat

### Internal energy

Everything above absolute zero temperature (zero kelvin) possesses **internal energy** (symbol: $U$). The term 'internal energy' describes the sum-total of *all* the various energies associated with molecules (in particular, the *motion* or *vibration* of those molecules) in a body, and is closely linked to temperature. Molecules are in a constant state of vibration; the higher the temperature, the greater the vibration, and the greater the internal energy.

> The term, '**internal energy**', has long-replaced the terms 'heat energy' or 'thermal energy'. It's very important to understand that these terms *are no longer used in this context*. These days, we *only* use the word '**heat**' to describe the energy transfer between a body at a higher temperature to one at a lower temperature. **It is incorrect to speak of a body 'containing heat'.**

An increase in the temperature of a body, then, will cause its internal energy to rise (and *vice versa*). However, the 'state' (e.g. solid, liquid or gas) of the body is also important. For example, ice and liquid water can co-exist at 0°C, but liquid water will *always* have a higher internal energy than ice at that particular temperature. Similarly, liquid water and steam can co-exist at 100 °C, but steam will *always* have a higher internal energy than liquid water at that temperature.

### Change in internal energy

We cannot measure the absolute ('total') internal energy of a body. We can only measure the *change* in its internal energy. The change in a body's internal energy depends upon its **mass** (symbol: *m*), the change in its **temperature** (symbol: *T*) and upon a constant called the **specific heat capacity** (symbol: *c*) of the body.

**Specific heat capacity** is a constant that depends on the material involved, and is defined as *the energy required to raise the temperature of that material by one degree*. The SI unit of measurement for specific heat capacity is the **joule per kilogram kelvin** (symbol: **J/kg·K**) or, in 'everyday' units, the **joule per kilogram degree Celsius** (symbol: **J/kg·°C**).

The relationship described above is given by the equation:

$$\Delta U = m\,c\,(T_{final} - T_{initial})$$

where:

$\Delta U$ = change in internal energy

$m$ = mass

$c$ = specific heat capacity

$T$ = temperature

**Important!** In the above equation, the Greek letter, $\Delta$ ('delta'), placed in front of the '*U*', is simply the mathematical 'shorthand' meaning '*change in*'. So, '$\Delta U$' (spoken as '*delta-U*') simply means '*change in internal energy*'.

Let's now look at an example of the relationship between work, heat and internal energy.

Whenever we switch on an electric kettle, what happens?

Well, an electric current flows through the kettle's heating element, causing electrical energy to be *converted* into internal energy. So **work** *(W)* is being done on the water (as well as on the kettle itself). This additional internal energy causes the existing internal energy of the water to rise, and this is accompanied by a rise in its temperature. As the temperature rises above the ambient temperature, energy is lost to the surrounding atmosphere through **heat** transfer *(Q)*.

In other words, the increase in the internal energy ($\Delta U$) of the water and the kettle must be *the difference between the **work done (W)** by the electricity on the water and the **heat transfer (Q)** away from the water.* We can express this as follows:

$$W - Q = \Delta U$$

If we now substitute for *U*, we have:

$$W - Q = m\,c\,(T_{final} - T_{initial})$$

where:

$W$ = work done

$Q$ = heat transfer

$m$ = mass

$c$ = specific heat capacity

$T$ = temperature

This equation can be illustrated as shown in Figure 15.14.

$$W - Q = m\,c\,(T_{final} - T_{initial})$$

**Figure 15.14**

The following worked examples demonstrate how the above equations are used.

**Worked example 11**   An electric kettle contains 1.5 kg of water at 20°C. If the kettle draws a current of 10 A from a 230-V supply, calculate the time it would take to raise the temperature of the water to boiling point (100°C), if the average heat loss during this period is 50 kJ. The specific heat capacity of water is 4190 J/kg°C (ignore the work done on the kettle itself).

**Solution**   It's *always* a good idea to start by making a sketch, showing all the information given in the problem (Figure 15.15).

**Figure 15.15**

$$W - Q = mc(T_{final} - T_{initial})$$

But we know that the work done $W = E I t$, so:

$$(EIt) - 50\,000 = 1.5 \times 4190 \times (100 - 20)$$
$$(230 \times 10 \times t) - 50\,000 = 502\,800$$
$$2300t = 502\,800 + 50\,000$$
$$t = \frac{552800}{2300} \simeq 240\,\text{s}$$

As there are 60 s in one minute,

$$t = \frac{240}{60} = 4\,\text{min (Answer)}$$

**Worked example 12**    Repeat the previous worked example but this time take into account the metal from which the kettle itself is manufactured. Assuming the kettle itself has a mass of 0.5 kg and has a specific heat capacity of 500 J/kg·°C.

**Solution**    This time, we have to add together the change in internal energy of *both the water* **and** *the kettle*.

$$W - Q = \left[ m\,c\,(T_{final} - T_{initial}) \right]_{WATER}$$
$$+ \left[ m\,c\,(T_{final} - T_{initial}) \right]_{KETTLE}$$
$$(EIt) - 50\,000 = [1.5 \times 4190 \times (100 - 20)]$$
$$+ [0.5 \times 500 \times (100 - 20)]$$
$$(230 \times 10 \times t) - 50\,000 = [6285 \times 80] + [250 \times 80]$$
$$2300 \times t = 502\,800 + 20\,000 + 50\,000$$
$$t = \frac{572\,800}{2300} = 249\,\text{s}$$

As there are 60 s in one minute,

$$t = \frac{249}{60} = 4\,\text{min}\,9\,\text{s (Answer)}$$

## Summary of important electrical equations

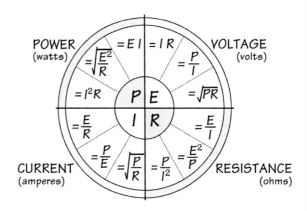

**Figure 15.16**

## Self-test exercise

By using the appropriate equations for work and power, complete Table 15.2 as it applies to the circuit. The first two have been done for you.

## Self-test exercise – answers

Answers are shown in bold at the end of this chapter.

## Misconceptions

**Power is a tangible thing. You can transfer it and manipulate it**
No. Power is simply a 'rate' – equivalent, if you like, to kilometres per hour. It simply tells us the *rate* at which energy is being transferred or consumed. When engineers talk about 'input power', 'output power', etc., what they really *mean* is the rate at which energy is being moved around.

**Electrical power is different from, for example, mechanical power**
Strictly-speaking, no. There are no 'types' of power! Power is simply a *rate* – the rate of doing work. So there is no such thing as 'electrical power', 'mechanical power', etc. You could, however, argue that '*electrical* power' means the 'rate of transfer of *electrical* energy'.

Table 15.2

|  | Work (joules) | Power (watts) | Voltage (volts) | Current (amperes) | Resistance (ohms) | Time (seconds) |
|---|---|---|---|---|---|---|
| a. | **30 000** | 1000 | 200 | **5** | **40** | 30 |
| b. | 20 000 | 500 | **50** | 10 | **5** | **40** |
| c. |  |  | 100 |  | 20 | 60 |
| d. |  |  |  | 5 | 100 | 50 |
| e. | 5000 | 100 | 20 |  |  |  |
| f. |  | 1000 |  | 4 |  | 20 |
| g. |  |  | 25 |  | 25 | 60 |
| h. | 3000 |  |  | 2 | 1000 |  |
| i. |  | 50 |  | 2 |  | 120 |
| j. | 2500 | 5000 | 200 |  |  |  |

Table 15.3

|  | Work (joules) | Power (watts) | Voltage (volts) | Current (amperes) | Resistance (ohms) | Time (seconds) |
|---|---|---|---|---|---|---|
| a. | **30 000** | 1000 | 200 | **5** | **40** | 30 |
| b. | 20 000 | 500 | 50 | 10 | 5 | 40 |
| c. | **30 000** | **500** | 100 | **5** | 20 | 60 |
| d. | **125 000** | **2500** | **500** | 5 | 100 | 50 |
| e. | 5000 | 100 | 20 | **5** | **4** | **50** |
| f. | **20 000** | 1000 | **250** | 4 | **62.5** | 20 |
| g. | **1500** | **25** | 25 | **1** | 25 | 60 |
| h. | 3000 | **4000** | **2000** | 2 | 1000 | **0.75** |
| i. | **6000** | 50 | **25** | 2 | **12.5** | 120 |
| j. | 2500 | 5000 | 200 | **25** | **8** | **5** |

**Horsepower measures mechanical power; watts measures electrical power.**

The horsepower is simply an Imperial unit of measurement for power, whereas the watt is the SI unit of measurement for exactly the same thing. We could, if we wished, describe the power of an incandescent lamp in horsepower, and a motor car's engine output in watts – in just the same way as we can use inches or millimetres to measure distance.

**Isn't the output of an electric motor measured in horsepower?**

Before the adoption of SI, the 'horsepower' was used to describe the output power of electric motors – these days, we use the watt. But, in North America, the horsepower is *still* generally used to measure the output power of motors.

**The terms power, work, and energy can be used interchangeably.**

In everyday language, this is usually the case. But in science and engineering, the difference between these terms is very important. Power is the rate at which work is done, or energy is expended. If you like, power is equivalent to 'kilometres per hour', whereas 'work or energy' is equivalent to 'kilometres travelled'.

**Energy is delivered from the supply to the load by the electrons moving through the circuit.**

No, this is impossible as, in d.c. circuits, no single electron moves fast enough to travel between the supply and the load and, in a.c. circuits, the electrons simply vibrate backwards and forwards..

**Heat is 'thermal energy'.**

Heat describes the *transfer* of energy from a warmer body to a cooler body. So 'heat' describes the *transfer* of energy – not energy itself.

**Isn't heat measured in calories? Or BTUs?**

The 'calorie' is simply an unit of measurement for energy, used in the – now archaic – 'centimetre-gram-second-ampere' (cgsA) system of measurement, that predates the SI system. And the BTU (British Thermal Unit) is an Imperial system unit of measurement. Energy, work, and heat are *all* measured in joules in SI.

**Heat and work are unrelated.**

Heat and work are very much related. The difference between the work done *on* a mass, and the heat transfer *from* that mass represents the change in internal energy of that mass.

## Review your learning

Now that we've completed this chapter, we need to examine the **objectives** listed at its start. Placing '*Can I...*' at the beginning, and a question mark at the end, of each objective turns that objective into a **test item**. If we can answer each of those test items, then we've met the objectives of this chapter.

**Online resources**

The companion website to this book contains further resources relating to this chapter. The website can be accessed via the following link:
www.routledge.com/cw/waygood

# Chapter 16

# Magnetism

## Objectives

On completion of this chapter, you should be able to

1. state the fundamental law of magnetism.
2. explain the term 'magnetic field'.
3. explain the term 'magnetic flux'.
4. explain the term 'flux density'.
5. state the direction allocated to magnetic flux.
6. sketch the pattern of the magnetic field, surrounding:
   a a bar magnet
   b a horse shoe magnet.
7. use the 'Domain Theory' to explain
   a magnetising an unmagnetised ferromagnetic material
   b why a north or south pole cannot exist in isolation
   c saturation.
8. list four methods of making a magnet.
9. explain the difference between permanent and temporary magnets.

**Important!** Throughout this chapter, for the purpose of clarity, the capitalised words 'North' and 'South' (including 'Magnetic North') refer to those **locations** on the Earth. The non-capitalised words 'north' and 'south' refer to **magnetic polarities**.

It's important to understand that the word '**poles**' has *two* meanings. In the 'geographic' sense, the 'poles' are *locations* at, or near to, the Earth's axis of rotation. In the 'magnetic' sense, 'poles' refer to *polarities of magnets*.

## Introduction

According to legend, around 900 BC, a Greek goatherd named Magnus discovered that the iron nails in his shoes were attracted to the stones in a field where he grazed his goats. The stones in Magnus's field were naturally magnetised ferrites that we know, today, as '**magnetite**' – named after 'Magnus's stones'.

As fascinating as this legend is, it is far more likely that the origin of the word 'magnetite' is derived from the region of Greece known as 'Magnesia'.

**Magnesia** is located in south-eastern Greece, and features a peninsula which partly encompasses a huge and spectacular bay on the Aegean coast, and whose modern capital is the city of Volos. Magnesia is famous as the home of a number of Greek mythical heroes – including *Jason* and *Achilles*. Jason was the heroic leader of the *Argonauts* and their quest for the 'golden fleece', while Achilles was the hero of the Trojan Wars and a central character in Homer's *'Iliad'* whose only weakness gave us the modern expression *'Achilles' heel'*.

But more relevant to this chapter, in ancient times Magnesia was a major source of large deposits of a black mineral called ferrous-ferric oxide, but more commonly known as **magnetite** – the most naturally magnetic of all the earth's minerals.

Magnetite is commonly found deposited on the Earth's surface, often in the form of octahedron (eight-faced) nodules, as illustrated in Figure 16.1. It seems possible that the composition of magnetite *may* have allowed it to become magnetised as the result of the magnetic fields that would have accompanied countless lightning strikes occurring over billions of years.

**Figure 16.1**

An Introduction to Electrical Science, Waygood, ISBN 9780815391821, 2019. © Taylor & Francis

Around the fifth century BC, the Greeks discovered that magnetite had the strange property of being able to attract and pick up small pieces of iron. Today, we call this mysterious property '**magnetism**', a word derived from a Greek word, meaning *'Magnesia stone'*.

Different sources credit both the ancient Greeks and the ancient Chinese with discovering that if a splinter of magnetite is freely suspended horizontally, so that it can rotate, it will *always* come to rest pointing in an approximately North–South direction. The first person to write about this property was an eleventh-century Chinese scientist, Shen Kua (1031–1095), who explained that it helped his countrymen improve the accuracy of their navigation, with ocean-going junks travelling from Cylon to Africa. And there is certainly a great deal of evidence to suggest that, by the twelfth century, the Chinese were using 'lodestones' (pieces of magnetite) for the purpose of navigation . . . as well as for fortune telling!

It also seems certain that the Vikings used lodestones to improve their, already excellent, navigational skills.

Interestingly, Chinese lodestones were designed to be 'South pointing', in accordance with oriental mythology.

Preserved examples of ancient Chinese lodestones include thin, slightly concave slivers of magnetite, shaped like fishes, and intended to float in small bowls of water enabling them to rotate. The peculiar, 'spoon-shaped' lodestone, illustrated in Figure 16.2, was designed to balance on its bowl, so that it could easily rotate under the influence of the earth's magnetic field, with its handle indicating South – South being more important than North in Chinese culture.

Figure 16.2

The word '**lodestone**' (as illustrated in Figure16.3) is actually a Middle-English (the English spoken during the four centuries following the Battle of Hastings) word, meaning *'leading stone'*, and was used to describe the first primitive magnetic compasses.

Figure 16.3

If an ordinary piece of steel is stroked with a piece of magnetite, it will become **magnetised** and, just like the lodestone, when freely suspended, will always come to rest with one of its ends pointing roughly towards the Earth's geographic North Pole (located in the Arctic), and with its opposite end pointing approximately towards the Earth's geographic South Pole (located in the Antarctic).

Until strong, artifical magnets could be created using electromagnetism, lodestones were the *only* source of magnetism and, as late as the seventeenth century, these weak natural magnets were still being used to 'regenerate' the failing strength of compass needles used by Royal Navy warships, by stroking them in the appropriate direction with a lodestone kept on board for that very reason. A metal needle, magnetised in this way, formed the basis of what, eventually, became the **magnetic compass** (Figure 16.4) we are familiar with today.

magnetic compass

Figure 16.4

Because of its useful navigational properties, one end of a magnet or compass needle came to be called its '**North-seeking pole**', while the other end was called its '**South-seeking pole**'. Over time, the use of the word '**seeking**' has fallen into disuse, and we now simply say: '**north pole**' and '**south pole**' which, by convention, are also used to name their *magnetic polarities*.

> The poles of a magnet, or of a compass needle, are named after the directions in which they point.

But, in fact, a compass needle doesn't actually point to the Earth's geographic or True North pole at all, but to a nearby *location* which we call 'Magnetic North'. It's very important to understand that 'Magnetic North' is the name we give to that *location* to distinguish it from True North, and is *not the magnetic polarity at that location!*

The reason that the compass acts in this way is because the Earth itself behaves as though it contains a gigantic bar magnet buried deep within its surface – with its axis slightly displaced from the Earth's axis of rotation.

Actually, it would be far more accurate to think of the source of the earth's magnetic field as a giant **natural generator**, rather than a 'bar magnet'. This 'generator' is created, around 3500 km beneath the earth's surface, within a liquid metallic layer, trapped between the earth's crust and its solid central core, where enormous temperatures and pressures result in circulating electric currents which, when combined with the earth's rotation, create a 'dynamo effect'.

In the **dynamo effect**, these natural electric currents swirl around the earth's axis, within its liquid core, following a path rather like a giant ring doughnut in shape, creating the magnetic field emanating outwards from the centre of the 'doughnut', slightly misaligned with the axis of rotation, and extending outward into space.

This 'dynamo effect' principle is essentially the same as the way in which a magnetic field emanates from the centre of a loop of wire around which an electric current is passed —as we shall learn in the later chapter on *Electromagnetism*.

To distinguish between the *locations* of the **geographic poles** (located at the earth's axis of rotation), and the *locations* of the **poles** where the magnetic field is strongest at the earth's surface, we call them '**True North**' and '**Magnetic North**', and '**True South**' and '**Magnetic South**', respectively.

It's important to understand that 'Magnetic North' and 'Magnetic South' are *locations* and **not**, as we are about to learn, the magnetic *polarities* of those locations.

Again, it's important to emphasise that 'Magnetic North' and 'Magnetic South' are *locations*, and *not* the magnetic polarities of those locations.

## Fundamental law of magnetism

The fundamental law of magnetism states that '**like poles repel, and unlike poles attract**'.

In accordance with this law then, if a compass needle's north (-seeking) pole is attracted towards the *location* we call 'Magnetic North', then the **magnetic polarity** at that location must be **south**. By the same logic, the **magnetic polarity** of the *location* we call 'Magnetic South' must be **north**.

In accordance with this law then, if a compass needle's north (-seeking) pole is attracted towards the Earth's Magnetic North, then the **magnetic polarity** of that location must be **south**.

So the Earth behaves as though it has a 'gigantic bar magnet' buried deep within its surface, with its south magnetic pole located in the Northern hemisphere, and its north magnetic pole located in the Southern hemisphere – as illustrated in Figure 16.5.

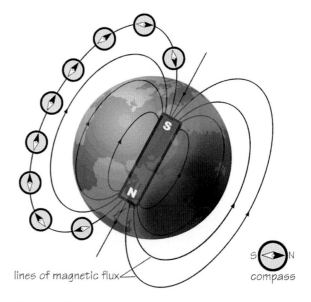

lines of magnetic flux

compass

**Figure 16.5**

This 'giant bar magnet' is surrounded by a **magnetic field**, and a needle of a compass placed *anywhere* within that field will always align itself to lie along that field, and point towards the south magnetic pole buried deep below Magnetic North.

As we have already learnt, the axis of this 'giant bar magnet' does not coincide with the Earth's axis of rotation but, instead, 'wobbles' around it – rather in the same way that a spinning top wobbles when it starts to lose momentum. As a result of this, Magnetic North is always changing its position relative to True North as, of course, is Magnetic South with True South.

The angle between True and Magnetic North (Figure 16.6) we call the **'angle of declination'** – an angle that navigators, when using compasses, must *always* compensate for when plotting their routes. But, as if to make life difficult for navigators, this angle is continuously changing! So if you were to follow your compass North today, you would arrive at a different place than you would have arrived at, say, 10–15 years ago! Accordingly, navigation charts are regularly updated to incorporate changes in the angle of declination.

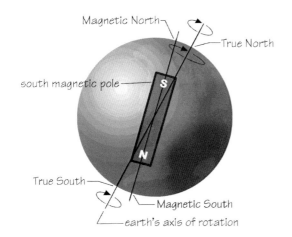

Figure 16.6

Furthermore, depending on where you are on the Earth, the angle of declination will be different; at some locations the declination will be minimal whereas, at other locations, the declination will be large.

Currently, Magnetic North is located in Canada's Arctic Ocean and is moving north-west towards Siberia at the rate of around 40–50 km per year. As well as moving in this general direction, Magnetic North also 'wanders around' its assumed route, so that on any given day, it can be some distance from where it has been predicted *to* be!

## Magnetic fields

Like gravity, magnetism is an invisible force, for which there has been *no satisfactory explanation*. However,

while we can neither see, nor explain, the nature of this force, we *can* observe and quantify its **effects**.

Unlike gravity, however, which is a force of *attraction* between masses, magnetism represents a force of *repulsion* as well as attraction, and obeys the fundamental law, which states:

**Like poles repel; unlike poles attract.**

Knowing this, if we ever need to identify the poles of an unmarked magnet we can do so using a **compass**. A compass needle's north pole (usually coloured black or red) will be *attracted* by and, therefore, identify the magnet's south pole. And, of course, the compass needle will be *repelled* by a magnet's north pole (Figure 16.7).

The *region surrounding a magnet*, in which its effects can be observed, is termed a **magnetic field**.

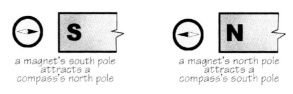

a magnet's south pole attracts a compass's north pole

a magnet's north pole attracts a compass's south pole

Figure 16.7

> A **magnetic field** is the area surrounding a magnet in which its effects may be observed.

There is actually a way in which we can apparently 'see' the shape of a magnetic field – by conducting a very simple, but classic, experiment using a magnet, a sheet of card, and some iron filings.

As, no doubt, every schoolboy knows, if iron filings are sprinkled onto a sheet of card which has been placed over a magnet, and the card is then gently tapped so that it vibrates, the filings will be seen to arrange themselves in distinctive patterns which represent the magnetic field. The filings trace out *what appear to be* 'lines of force', called **magnetic flux** (Figures 16.8 and 16.9).

What you are actually seeing, as a result of this experiment, are *not* really lines of force – it just *looks* that way, because the 'tip' of one iron filing is attracted to the 'tail' of another, causing the filings to form closed 'chains'!

However, representing a magnetic field by lines is a very useful model, first presented by the scientist Sir Michael Faraday (1791–1867).

In Faraday's 'model' of a magnetic field, the lines of magnetic flux are allocated **direction**, determined

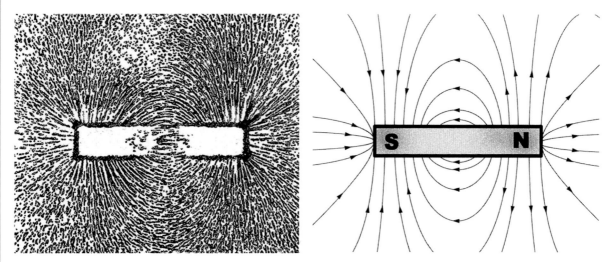

**Figure 16.8** magnetic flux surrounding a bar magnet, traced with iron filings (left), and how we represent them in a drawing (right)

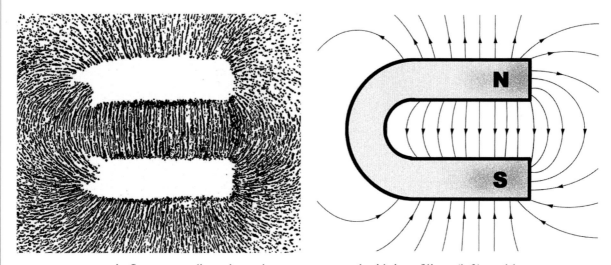

**Figure 16.9** magnetic flux surrounding a horseshoe magnet, traced with iron filings (left), and how we represent them in a drawing (right)

**Figure 16.10**

by the *direction in which a compass needle would point* when placed within the field. As a compass needle always points *along the lines of magnetic flux*, towards a magnet's *south* pole, this direction

is always **from a magnet's north pole towards its south pole**.

> We say lines of magnetic flux 'leave' the north pole of a magnet, and 'enter' the magnet at its south pole.

We can show this by moving a plotting compass (a small, transparent, compass, used for experiments) around a magnet, as illustrated in Figure 16.10.

From the above images, as well as from the directions in which the plotting compasses point, it can clearly be seen that the lines of magnetic flux 'leave' and 'enter' a magnet in the areas close to the *ends* of that magnet – indicating that the *poles are concentrated at the ends of a magnet*. This can be further demonstrated by plunging a magnet into a box of small steel tacks or iron filings; when the magnet is withdrawn, you find that the majority of the tacks or filings are clustered around the ends of the magnet with few, if any, towards its centre (Figure 16.11).

Figure 16.11

As already mentioned, Faraday is credited with first representing a magnetic field using lines of magnetic flux, and for establishing the following 'properties' of these lines. But, once again, we must continually remind ourselves, that these lines of magnetic flux *don't actually exist*; they are merely a very useful model to help us to form a 'mental picture' of the behaviour of an otherwise invisible magnetic field. However, like all models, it does have its limitations.

So, according to Faraday's 'lines of magnetic flux' model:

- lines are *elastic*, enabling them to stretch or contract
- parallel lines, acting in the *same* direction, *repel* each other
- parallel lines, acting in the *opposite* direction, *cancel* each other
- lines *never cross each other*
- lines *always take the path of least opposition*
- there is no known *'insulator'* to the lines
- the *density* of the lines is an indication of the *strength* of the magnet.

The first four of these properties are illustrated in Figures 16.12–16.15

as magnets are moved further apart the lines of magnetic flux will **stretch**

Figure 16.12

parallel lines, running in the same direction **repel** each other

Figure 16.13

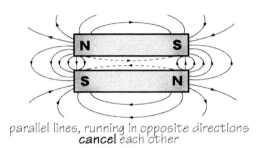

parallel lines, running in opposite directions *cancel* each other

Figure 16.14

lines of magnetic flux **cannot cross**

Figure 16.15

When we say that lines of magnetic flux *'take the path of least opposition'*, we need to understand that ferrous metals provide a *very* much easier medium in which to support the formation of flux than air – in fact, these metals are *thousands* of times better at supporting the formation of magnetic flux than air is. As a result, a piece of ferrous metal placed within a magnetic field will always *significantly* distort that field – as in the example in Figure 16.16.

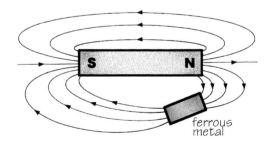

Figure 16.16

This emphasises how important it is *not* to have any ferrous metals anywhere near a compass when a directional reading is being taken, as you are very likely to obtain a false reading due to resulting distortion to the Earth's field.

> Magnetic mines explode when the metal hull of a ship distorts the Earth's magnetic field as it passes in the vicinity of the mine; they are *not* set off because their magnetic sensor is attracted towards the hull.

Although there appears to be no method of insulating magnetic flux, we *can* protect sensitive instruments from flux by **shielding** them. 'Shielding' takes advantage of the sort of distortion described above, by using a ferrous-metal container to *divert* the magnetic flux *around* the sensitive instrument – as illustrated in Figure 16.17.

**Figure 16.17**

The *intensity* of the lines of flux is an indication of the strength of a magnet. We call this **flux density** (symbol: **B**), which is defined as *'the flux per unit area'*, and we will discuss this in detail in the chapter on *magnetic circuits*. Figure 16.18 shows how the flux density varies, according to where it is measured (note that although the flux is shown in two dimensions, it is actually in three dimensions).

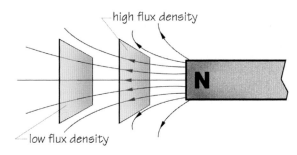

high flux density

low flux density

**Figure 16.18**

We have already referred to the fundamental law of magnetism, 'like poles repel while unlike poles attract', so let's look at how the 'properties' of the lines of magnetic flux, in our model of the magnetic field, contribute to this law.

In the case of 'unlike poles', it is the contraction of the 'elastic' lines of flux that act to pull the opposite poles together. In the case of the 'like poles', it is the repulsion between parallel lines of flux that act to push similar poles apart (Figure 16.19).

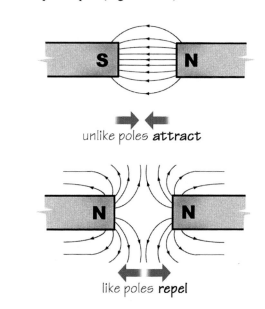

unlike poles **attract**

like poles **repel**

**Figure 16.19**

## The Domain Theory of Magnetism

Whenever a bar magnet is broken in half, two new magnets are created, no matter how many times this is repeated. This is because *no pole can exist in isolation*. All magnets exist as *dipoles* – that is, they always have *pairs* of poles, as illustrated in Figure 16.20.

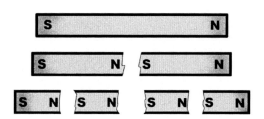

Effect of breaking a magnet in half

**Figure 16.20**

When contemplating this behaviour, the nineteenth-century German scientist Wilhelm Eduard Weber (pronounced *'Vay-ber'*) theorised that if this action could be repeated often enough, then we would ultimately arrive at a molecule, or even an atom, which would behave just like a tiny bar magnet. This 'molecular magnet' was eventually termed a '**domain**' and this, in turn, led to what has became known as the **Domain Theory of Magnetism**.

According to the **Domain Theory of Magnetism**, every electron not only orbits the nucleus of its atom, but also *spins about its own axis*. This causes each electron to generate its own magnetic field (rather like the Earth does) – causing it to behave like an incredibly small magnet. The polarity of the electron's magnetic field depends upon the direction of its spin – as illustrated in Figures 16.21–16.22.

Figure 16.21

Figure 16.22

Once again, we must continue to remind ourselves that this is only a 'model', and it is not necessarily what is *really* happening inside an atom!

In most atoms, pairs of electrons spin in *opposite* directions – effectively cancelling out each other's magnetic fields. Such atoms, therefore, exhibit no overall magnetic field. (Figure 16.23, of course, is not to scale!)

Figure 16.23

In magnetised materials, however, *more electrons spin in one direction than in the other.* For example, in Figure 16.24 (showing only one shell) four electrons spin in one direction, while only two spin in the opposite direction.

Figure 16.24

Such atoms or, more likely, molecules therefore exhibit a natural magnetic field, and are called **domains**. Magnetite is largely made up of molecules which are domains.

> A **domain** is a molecule or an atom which exhibits a magnetic field.

In the following illustrations, for the sake of simplicity, we will represent a domain as a miniature magnet, as shown in Figure 16.25.

Figure 16.25

In **unmagnetised** ferromagnetic materials, the domains arrange themselves roughly in closed 'chains', with the north pole of each domain attracted to the south pole of another domain – as illustrated in the (very much!) simplified Figure 16.26. Because these chains are closed, their poles tend to cancel out the effect of each other, and no overall magnetic effect is noticeable externally. (Note that although we have represented these domains in two dimensions, they of course actually exist in three dimensions.). Of course, the actual distribution of domains would not be as neat and as well organised as they have been illustrated here!

Figure 16.26 domains form 'chains' in unmagnetised ferromagnetic materials

In **magnetised** ferromagnetic materials, however, the domains are arranged in 'rows' (Figure 16.27) – leaving columns of unattached north poles at one end of the material, and columns of unattached south poles at the opposite end. This explains why the poles of a magnet are concentrated at each end.

Figure 16.27 domains forming 'rows' in magnetised ferromagnetic materials

The Domain Theory also explains why, if a magnet is broken in several places (Figure 16.28), *new* magnets are created, and why *poles cannot exist in isolation.*

Figure 16.28 broken magnets result in two new magnets, each with its own north and south poles

## Making magnets

The Domain Theory helps us understand what probably happens when we make magnets, and why some materials become **permanent magnets** while others become only **temporary magnets**.

In order to make an unmagnetised ferromagnetic material into a magnet, its domains must be realigned to form rows. To do this, the domains must be forced out of their natural 'chain' formations and, then, allowed to come to rest forming 'rows'. This can be brought about by using an **external magnetic field** (in some cases, simply the Earth's natural magnetic field), to influence the final direction of the domains.

Magnets can be made in several ways, *all* involving the transfer of energy while under the influence of an external magnetic field, including by

- **stroking** the material with another magnet.
- **hammering** the material, *while aligned in a North–South direction.*
- **heating** the material, and allowing it to cool *while aligned in a North–South direction.*
- **electromagnetic induction**, in which a direct current is passed through an insulated coil that has been wound around the material.

By **stroking** a ferromagnetic material with another magnet, the domains' poles are attracted by the (opposite) pole of the magnet, and 'dragged' from their natural chain formations and realigned into rows.

By **hammering** or **heating**, energy is imparted into the ferromagnetic material which results in the domain chains becoming excited and breaking up. If then left aligned in a North–South direction, the Earth's natural magnetic field will cause the domains to form rows as their energy levels return to normal. This, incidentally, explains why tools, such as screwdrivers, often become partially magnetised after rattling around in a toolbox.

Each of the methods described above are fairly *inefficient* ways of creating magnets, and the resulting magnet will be relatively weak.

Until 1820, when the Danish physicist Hans Christian Ørstedt (1777–1851) discovered the relationship between electricity and magnetism, all magnets were difficult to manufacture and they produced weak magnetic fields. But Ørstedt's discovery led to the most efficient, and practical, method of making a powerful magnet: by **electromagnetic induction**. As we will learn in the next chapter, a direct current flowing through a coil creates a very strong magnetic field. So, a current-carrying coil wound around a length of ferromagnetic material will very effectively force domain chains into rows.

## Magnetic saturation

Once all the domains that *can* be aligned into rows *have* been aligned, the magnet is said to be **saturated**, and its magnetic field has reached its maximum intensity, and *it is not possible to strengthen the magnetic field beyond this point.* We will examine this more closely in the next chapter.

## Induced magnetism

When a magnet is brought into contact with a non-magnetised piece of ferrous metal, it causes that metal itself to become magnetised. We call this effect 'induced magnetism'.

If the ferrous metal is **iron**, then it will lose its induced magnetism immediately the magnet is removed. If it is **steel**, then it will *retain* its induced magnetism after the magnet has been removed – as explained in the next section.

The part of the ferrous metal closest to the magnet's pole acquires the *opposite* polarity to that pole, while the far end of the metal acquires the same polarity – essentially elongating the original magnet. This phenomenon can be explained in terms of the domain theory, as the domains in the ferrous metal become aligned under the external influence of the permanent magnet (Figure 16.29).

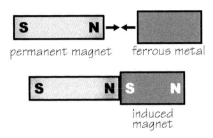

permanent magnet    ferrous metal

induced magnet

Figure 16.29

This can be repeated numerous times, as is the case in the example shown in Figure 16.30.

example of
**induced magnetism**

Figure 16.30

## Ferromagnetic materials

We will explain the term 'ferromagnetism' in the next chapter. But, in the meantime, it is acceptable to think of ferromagnetic materials as being **iron**, or alloys that contain iron, such as **steel**.

For materials such as **iron**, the chains of domains are held together very weakly, and their domains realign into rows very easily under the influence of an external magnetic field. But once that external magnetic field is removed, the domains then very easily return to their chain formations. Materials such as iron, therefore, lose most of their magnetism immediately any external magnetic field is removed, and are termed **temporary magnets**.

> **Temporary magnets are:**
>
> • easy to magnetise
> • lose their magnetism immediately an external magnetic field is removed.

For materials such as **steel**, the chains of domains hold very strongly together, and a great deal of external energy must be applied to realign them into rows. Once realigned, however, it then takes a great deal of energy to break up this new alignment. Such materials, therefore, retain most of their magnetism once any external magnetic field is removed, and are termed 'permanent magnets'.

Interestingly, some of the strongest permanent magnets are manufactured from alloys of iron together with materials which, themselves, are magnetically very weak, including aluminium and copper. For example, 'alnico' (a trade name for a metal alloy used to manufacture magnets) is an alloy of iron, aluminium, cobolt, copper and nickel. One of the very strongest magnets (over twenty times more powerful than alnico) is manufactured from an alloy of cobolt and platinum, and contains no iron at all.

> **Permanent magnets**
>
> • are difficult to magnetise
> • retain their magnetism after the external magnetic field has been removed
> • are difficult to demagnetise.

Both permanent and temporary magnets have useful applications in electrical engineering.

• **Permanent magnetic materials** are used in the manufacture of analogue electrical measuring instruments, loudspeakers, security sensors, etc.
• **Temporary magnetic materials** are used in the manufacture of transformers, relays, contactors, etc.

## Misconceptions

**A compass needle indicates the direction of True North.**

A compass needle lies along the earth's natural lines of magnetic flux, which link Magnetic North and Magnetic South, and does not indicate True North and True South.

**The term 'Magnetic North' refers to the magnetic polarity of that location.**

No. The terms 'Magnetic North' and 'Magnetic South' are used to distinguish their *locations* from True North and True South. They are not the magnetic polarities of those locations!

**The end of a compass needle which points to Magnetic North is really a south pole.**

No. Because the magnetic polarity of Magnetic North is a south pole, it attracts the north pole of a compass needle.

**Lines of magnetic flux exist.**

No they don't! They are simply a 'model' to help us visualise the pattern of a magnetic field.

## Review your learning

Now that we've completed this chapter, we need to examine the **objectives** listed at its start. Placing '*Can I…*' at the beginning, and a question mark at the end, of each objective turns that objective into a **test item**. If we can answer each of those test items, then we've met the objectives of this chapter.

**Online resources**
The companion website to this book contains further resources relating to this chapter. The website can be accessed via the following link:
www.routledge.com/cw/waygood

# Chapter 17

# Electromagnetism

On completion of this chapter, you should be able to

1 describe the evidence that confirms an electric current is always accompanied by a magnetic field.
2 determine the shape and direction of the magnetic field around a current-carrying straight conductor.
3 determine the shape and direction of the magnetic field around a current-carrying single loop conductor.
4 determine the shape and direction of the magnetic field around a current-carrying coil.
5 explain the term 'flux density', specifying its SI unit of measurement.
6 calculate the flux density of simple magnetic fields.
7 sketch the magnetic field resulting from a current-carrying conductor placed in a permanent magnetic field, and use this to explain why the conductor is subject to a force.
8 apply Fleming's Left Hand Rule to determine the direction of force on a current-carrying conductor placed in a magnetic field.
9 calculate the value of the force acting on a current-carrying conductor placed in a magnetic field.
10 describe the principle of operation of a simple d.c. motor.

**Important**: *conventional* current direction (positive to negative) is used throughout this unit.

## Introduction

Until the early nineteenth century, **electricity** and **magnetism** were generally considered to be *separate* phenomena. Even those scientists who instinctively believed in a connection between the two had no experimental evidence with which to support their views.

However, one evening in 1820, while preparing an experiment for his students, the Danish physicist, Hans Christian Ørsted (1777–1851), noticed that whenever he passed an electric current through a wire, the needle of a nearby compass deflected from its normal North–South direction.

Ørsted was already one of those scientists who believed that there must be a connection between electricity and magnetism, so his discovery may not have been quite as accidental as it has since been made out to be. Nevertheless, this observation spurred him into performing further experiments which convinced him that an electric current must create a magnetic field – indicating that *there was indeed a direct relationship between electricity and magnetism*.

Ørsted's observation triggered a great deal of research into this relationship, not only by Ørsted himself, but by others such as Ampère, all of whom realised that the magnetism produced by an electric current could produce *forces* which, in turn, might be harnessed to produce useful *motion*.

Furthermore, if the relationship between magnetism and electricity should prove to be *reversible*, then it might then also be possible to convert motion into electricity!

We call this relationship between electricity and magnetism '**electromagnetism**'. More specifically, electromagnetism is that branch of science concerned with those forces that result between electrically charged particles in motion.

## Electric currents and magnetic fields

Ørsted's experiments showed that a magnetic field *surrounds a current-carrying conductor*, and that the *direction of that field* depends upon the *direction of that current*.

An Introduction to Electrical Science, Waygood, ISBN 9780815391821, 2019. © Taylor & Francis

Figure 17.1

One of his experiments is very easy to reproduce, using simple equipment, as illustrated in Figure 17.1.

A conductor is passed vertically through a horizontal card, on which a number of plotting compasses have been placed around the conductor.

In Figure 17.1a, the switch supplying the vertical conductor is open, so there is no current, and all the compasses are seen to be pointing in their usual North–South direction.

In Figure 17.1b, the switch is closed and current moves through the conductor. In this case, the direction of *conventional*-current drift (i.e. positive to negative) is downward. When viewed from above, it will be seen the individual compass needles have now assumed a *clockwise* direction around the conductor.

Finally, in Figure 17.1c, the battery connection is reversed, so that the conventional current drift through the conductor is now upwards. This time, it will be seen that the individual compass needles have reversed direction, and have now assumed a counterclockwise direction around the conductor.

From this simple experiment, we can conclude that:

> 1   An electric current is always surrounded by a magnetic field.
> 2   The 'direction' of the magnetic field depends upon the direction of the current.

When we describe a magnetic field as 'surrounding a current', it applies whether or not there is a conductor present. For example, a magnetic field will also surround an **arc**, which is a current passing through air or a vacuum.

Remember, as we learned in the chapter on *magnetism*, a magnetic field consists of **magnetic flux** the direction of which, by common agreement, is determined by the direction in which a compass needle will point, when placed within that magnetic field – i.e from north to south.

## Electromagnetic fields

Before proceeding further, we need to learn how the *direction* of an electric current is represented, graphically, in this and in all other texts.

Imagine releasing an arrow in the same direction as the current (Figure 17.2). With the arrow moving *away* from you, you will see its *flight feathers* represented by a cross (×) but, if the arrow is coming *towards* you, then you will see its *point*, represented by a dot (•). This is called the '**cross/dot convention**'.

So, in a cross-sectional view of a conductor, it is conventional to show a cross (×) to indicate that the current is flowing *away* from you (i.e. *into* the page), and to show a 'dot' (•) to indicate that the current is flowing *towards* you (i.e. *out* of the page).

> Some textbooks represent current flow away from you as a plus (+) sign, rather than as a cross, but we will avoid this as it can be confused for polarity.

This convention applies to *both* conventional flow *and* to electron flow – so it is *very* important to know

a **cross** indicates
current flowing **away** from you

a **dot** indicates
current flowing **towards** you

Figure 17.2

*which* current direction is being used whenever electromagnetism is being discussed. Most textbooks assume **conventional flow** (i.e. positive to negative) and, for this reason, we will also use conventional flow throughout this and all other chapters, *except when otherwise specified.*

For electron flow, *all* the rules that you will be learning in this chapter are simply *reversed.*

## Magnetic field surrounding a straight conductor

Now, let's return to our experiment. By using small, transparent 'plotting compasses', it is possible to plot *the shape and direction* of the magnetic field surrounding a current, as follows (the black part of the compass needle represents its 'North-seeking' pole), by tracing out that field.

As you can see in Figure 17.3, when viewed from either end of the conductor, the magnetic field assumes the shape of a series of concentric circles which surrounds the conductor (or, more accurately, the *current*). As you will also see, the field is more concentrated immediately surrounding the conductor, than it is further away from the conductor.

So what is going on? Well, *the field actually starts expanding from within the conductor itself*, as illustrated in Figure 17.4.

When the current is initially switched on, it starts to rapidly increase from zero to a maximum value, limited by the resistance of the circuit.

Initially, a circular magnetic field, of infinitesimal radius, forms at the centre of the conductor. As the current increases, another tiny circular magnetic field is

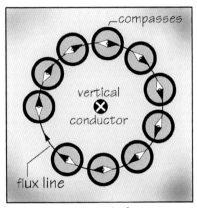

current drift
**away** from reader

compasses indicate flux forms
**clockwise** concentric circles
around current

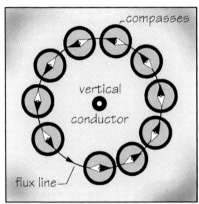

current drift
**towards** reader

compasses indicate flux forms
**counterclockwise** concentric circles
around current

magnetic field
around conductor

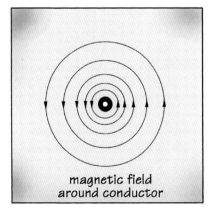

magnetic field
around conductor

Figure 17.3

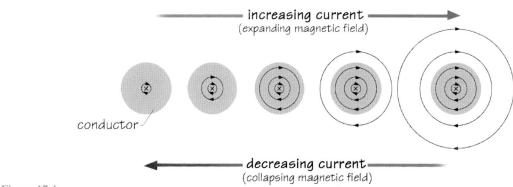

**Figure 17.4**

formed at the centre, which forces the existing magnetic field to expand outwards, rather like the ripples that radiate away from the point where a pebble is dropped into a pool. As the current increases further, yet more tiny magnetic fields are formed at the centre, forcing the others to expand even further outward. This process repeats itself, until the expanding circular magnetic fields start to surround the conductor externally, and it continues until the current has reached its maximum value, at which point the field becomes constant.

If you refer back to Faraday's model of a magnetic field, described in the chapter on *magnetism*, you will recall that the field's lines of force behave somewhat like rubber bands, and the tension within the circular flux lines causes them to be closer together towards the centre of the conductor than they are further away.

When the current is switched off, this process reverses itself, as the tension within each circular magnetic field causes them all to collapse inwards.

It's the action of these expanding and collapsing concentric magnetic field rings, 'cutting' the conductor internally, which is partially responsible for the '**skin effect**' described in the chapter on *resistance*. As we shall learn later, whenever a magnetic field passes through a conductor, it induces a voltage into the conductor which opposes the current at the centre of the conductor, forcing it to travel closer to the surface.

Although the magnetic field surrounding a current-carrying conductor doesn't have a north or a south pole, we can still allocate *direction* to the concentric circles. This is, of course, *the direction a compass needle would point if placed within the field* – from north to south.

Although these concentric circles of flux are shown in *two* dimensions, they are, of course, actually *three-dimensional* and enclose the current along its entire path throughout a circuit. It would, therefore, be more accurate to think of them as *concentric tubes*, rather than as concentric circles, as illustrated in Figure 17.5.

flux surrounds current in concentric tubes

**Figure 17.5**

However, concentric tubes are rather difficult to represent on a diagram, so it is standard practice to show the flux as concentric circles.

As we have seen from our explanation of Øersted's experiment, the *direction* of the magnetic field of a current is **clockwise** for current drifting *away* from you, and **counterclockwise** for current *drifting* towards you. To help you remember this, there are alternative rules we can use.

The first is called the '**corkscrew rule**', as illustrated in Figure 17.6. With this rule, the direction of the field is the same direction in which a corkscrew is turned, when pointing in the same direction as conventional current.

Figure 17.6

Figure 17.8

An alternative rule for determining the direction of flux around a current is the '**right-hand grip rule**' (Figure 17.7). If you grip a conductor, with your thumb pointing in the direction of conventional current, then the 'curl' of your fingers will represent the direction of the field.

Figure 17.7

Figure 17.9

## Magnetic field surrounding a single loop

If a length of current-carrying conductor is twisted to form a **single loop**, then the resulting magnetic flux lines would appear to leave the loop from one of its faces, and re-enter the loop via its opposite face. In the example, in Figure 17.8, the loop behaves as though its *front* face was a north pole (the face from which the flux leaves) and its opposite face would behave as a south pole (the face into which the flux re-enters).

A quick way of determining which face of the loop is north and which is south is to imagine the letter's **N** and **S**, pivoted so that they are free to 'spin' in the same direction as the arrow heads placed as shown in Figure 17.9.

The letter 'N' (for 'north') will spin counterclockwise, whereas the letter 'S' (for 'south') will spin clockwise. So, in the left-hand illustration, in Figure 17.9, the

current is flowing around the loop in the same direction as the letter 'N' would spin and, so, the front face of the loop is a *north* pole. In the right-hand illustration, in Figure 17.9, the current is flowing around the loop in the same direction as the letter 'S' would spin and, so, the front face of the loop is a *south* pole.

If an insulated conductor is wound to form *several loops*, then it will form a **coil**. And when a current passes through a coil, the magnetic field surrounding each individual loop acts to *reinforce* those around adjacent loops to form a single, strong, magnetic field – as shown in the coil section in Figure 17.10.

As you can see, the shape of the magnetic field surrounding a current-carrying coil closely resembles the field surrounding a bar magnet, with its poles concentrated towards opposite ends of the coil. In the example in Figure 17.10, the magnetic flux 'emerges' from the left-hand end – indicating that end is its north pole – and 'enters' the right-hand end – indicating that end is its south pole.

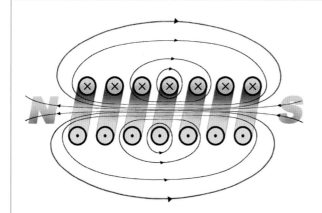

Figure 17.10

The polarity of each end of a coil may be determined using a variation of the '**right-hand grip rule**' which we learnt about earlier. In this case, simply grasp the coil with your *right* hand (Figure 17.11), such that the curl of the fingers represents the direction of (conventional) current around the coil, and the thumb will then point towards the **north pole** end of the coil. And the opposite end, of course, will be its **south pole**.

## Magnetic flux density

Magnetic fields are represented by lines of **magnetic flux** (symbol: $\Phi$, pronounced 'phi'). As we learnt in the chapter on *magnetism*, these lines are *imaginary* and simply represent a model for showing the shape or pattern of a magnetic field. Although they are imaginary, we can still quantify them, and the SI unit of measurement we use to do this is the **weber** (symbol: **Wb**) – pronounced *'vay-ber'* – after the German physicist, Wilhelm Eduard Weber (1804–1891).

In simple terms, we can think of each, individual, magnetic flux line within a magnetic field as representing one weber (we won't worry about its actual definition at this stage, but will return to it later).

The 'closeness' of a field's lines of magnetic flux (or number of webers) indicates the *intensity* of that field at any given point. This is termed the **magnetic flux density** (symbol: $B$) of the field at that particular point. Just as with a permanent magnet, the flux density is always greatest in the areas nearest the poles of the coil – as can be seen in the example in Figure 17.12.

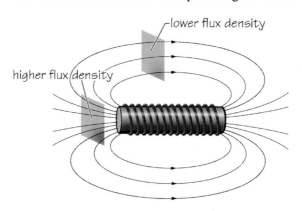

Figure 17.12

From Figure 17.12 it can be seen that more lines of flux pass through the shaded rectangle near the poles, than through the rectangle of identical cross-sectional area placed midway along the coil. So we can define **flux density** as the '*flux per unit area*', expressed as follows:

Figure 17.11

$$B = \frac{\phi}{A}$$

where:

$B$ = flux density, in teslas (T)

$\phi$ = flux, in webers (Wb)

$A$ = area, in square metres (m²)

The unit of measurement for flux density is the **tesla** (symbol: **T**), named in honour of the Serbian-American engineer, Nikola Tesla (1856–1943), and is equivalent to a 'weber per square metre'.

**Worked example 1**   The pole face of a magnet measures 30 mm by 30 mm. The flux leaving the pole is 0.25 mWb. Calculate the flux density at the pole face.

**Solution   Note!** You must change millimetres to metres, and milliwebers to webers.

$$\text{area of poleface} = (30 \times 10^{-3}) \times (30 \times 10^{-3})$$
$$= 900 \times 10^{-6} \text{ m}^2$$

$$B = \frac{\phi}{A} = \frac{0.25 \times 10^{-3}}{900 \times 10^{-6-3}} \approx 0.28 \text{ T (Answer)}$$

**Worked example 2**   A current-carrying coil, of inside diameter 50 mm, generates 1500 mWb of magnetic flux. Calculate the flux density within the coil.

**Solution   Note!** The millimetres must be changed to metres, and milliwebers to webers.

$$B = \frac{\phi}{A_{coil}} = \frac{\phi}{\pi r_{coil}^2}$$

$$= \frac{1500 \times 10^{-3}}{\pi \left(\dfrac{50}{2} \times 10^{-3}\right)^2} = \frac{1500 \times 10^{-3}}{\pi \times 25^2 \times 10^{-6-3}}$$

$$= \frac{1500}{1963 \times 10^{-3}} \approx 764 \text{ T (Answer)}$$

## Electromagnets

The flux density of the magnetic field produced by a coil can be *significantly* (*hundreds* or, even *thousands*, of times!) increased by winding the coil around a **ferromagnetic** (iron alloy) core, rather than around a hollow tube – as shown in Figure 17.13.

The reason for this was explained by the Scottish engineer and academic Sir James Ewing (1855–1935) as follows: when the core is magnetised, the individual magnetic field of each of its millions of domains align with and, therefore, *reinforce*, the relatively weak magnetic field due to the current alone, thus creating a very strong combined field.

Ferromagnetic materials, such as silicon steel, are used as **cores** for electromagnets, and for the **magnetic circuits** of electrical machines. We will learn what is meant by a 'magnetic circuit' in the next chapter.

Electromagnets are very much stronger than the strongest permanent magnet, and are used whenever we want to use electric current to create movement, such as in circuit breakers, motor starters, relays, etc.

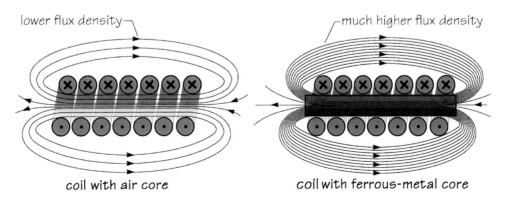

coil with air core            coil with ferrous-metal core

Figure 17.13

## Force on conductors

In this chapter, we've learnt that whenever a current drifts through a conductor, a magnetic field surrounds that current. It follows, therefore, that if currents move through each of two, parallel conductors, then the resulting magnetic fields will react with each other, and produce a force between the two conductors. This force may be one of **attraction** or of **repulsion**, *depending upon the relative directions of the electric currents in the conductors.*

If we apply Faraday's properties of lines of magnetic flux, which we learnt from the chapter on *magnetism* (specifically, 'parallel flux lines acting in the *same* direction *repel* other', while 'parallel flux lines acting in *opposite* directions *cancel*') then, as illustrated in Figure 17.14, if the currents are drifting *in opposite directions*, the resulting force between them will be one of *repulsion*; whereas if the currents are drifting *in the same direction as each other*, then the resulting force will be one of *attraction*.

You will recall that the SI unit of electric current, the **ampere**, has been defined in terms of *the force between two straight, parallel, current-carrying conductors*. The above description explains the reason for this force.

And these forces can be considerable. The exceptionally high fault currents that result from short-circuits in high-voltage power systems can actually cause severe distortion to any parallel conductors, such as busbars, carrying such currents.

The forces between magnetic fields are *very* important in electrical engineering because, as we shall see, the operation of **electric motors** is entirely dependent upon these forces.

Now let's turn our attention to the behaviour of a current-carrying conductor, placed within a permanent magnetic field – such as that illustrated in Figure 17.15. Again, the two fields will react with each other, causing a **force** to act upon the conductor.

Since (according to Faraday's rules on their behaviour) lines of magnetic flux never cross, they will either crowd together and act to reinforce each other whenever they act in the *same direction* or, when they act in *opposite directions*, they will act to *weaken* each other. The resulting **force** will then attempt to push the conductor out of the permanent field.

The force, which acts to push the conductor out of the magnetic field, is known as the '**Lorentz Force**', in honour of the Dutch physicist, Hendrik Antoon Lorentz (1853–1928).

This behaviour is called '**motor action**' because, as we shall see, this is the basis of how electric motors, as well as other devices, such as dynamic loudspeakers, work.

The above principle is also used to extinguish electric **arcs** in certain types of circuit breaker. By placing strong magnets (or electromagnets) either side of the circuit breaker's main contacts, the arc formed when the contacts break will be forced sideways, causing it to stretch. Because a stretched arc is unstable, it will extinguish relatively easily.

The **magnitude** of the resulting force on the conductor depends upon the flux density of the permanent magnetic field, the current in the conductor, and the length of conductor within the field, as expressed below:

$$F = BIl$$

where:

$F$ = force, in newtons (N)

$B$ = flux-density, in teslas (T)

$l$ = length of conductor in field, in metres (m)

conductors repelled

conductors attracted

Figure 17.14

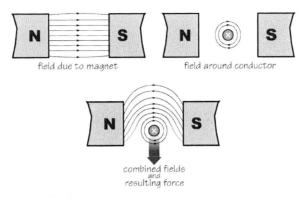

field due to magnet          field around conductor

combined fields
and
resulting force

Figure 17.15

**Worked example 3**   A current of 5 A drifts through a conductor placed between the poles of a magnet. If 50 mm of the conductor is within the magnetic field, and the flux density of the field is 500 mT, calculate the resulting force on the conductor.

**Solution**

$$F = BIl = (500 \times 10^{-3}) \times 5 \times (50 \times 10^{-3})$$
$$= 125 \times 10^{-3} \, \text{N} \quad \text{or} \quad 125 \, \text{mN (Answer)}$$

### Fleming's Left-Hand Rule

The **direction** of the force on a conductor may be determined from first principles, by drawing the individual magnetic fields, separately – as illustrated in Figure 17.15 – or, instead, by using **Fleming's Left-Hand Rule\*** for *conventional* current. This rule was created by the British engineer and academic, Sir John Ambrose Fleming (1849–1945), to help his students determine the direction of rotation of an electric motor's armature.

> \*Fleming, in fact, devised *two* such rules. His 'Left-Hand Rule' (for conventional flow) applies to motors, whereas his 'Right-Hand Rule' (for conventional flow) applies, as we shall learn, to generators. To avoid confusing the two, you might want to remember that the letters 'L' and 'M' follow each other and, so: **L** – **M** = **L**eft-Hand Rule – **M**otors. .

To apply Fleming's Left-Hand Rule, extend the first finger (index finger), the second finger, and thumb of the left hand at right angles to each other, as illustrated in Figure 17.16:

- the **f**irst finger indicates the direction of the permanent **f**ield
- the **s**e**c**ond finger indicates the direction of the **c**urrent
- the thu**m**b will then indicate the direction of the resulting **m**otion.

### Basic motor action

When scientists realised that it was possible to use an electric current to produce motion, a great deal of research followed to try and design motors capable of doing useful work.

A major step forward was the invention of the **electromagnet** because these were capable of producing far greater flux densities than permanent magnets and, therefore, producing a far greater force on a current-carrying conductor than was possible with permanent magnets.

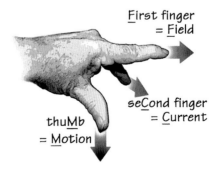

First finger
= Field

seCond finger
= Current

thuMb
= Motion

Figure 17.16

Up until this time, steam engines were the main means of driving mechanical loads and early electric motors were based on similar mechanisms – i.e. based on **linear motion** (equivalent to a steam engine's piston and cylinder) which could then be converted to rotary motion by driving a flywheel. Eventually, this approach was abandoned in favour of direct **rotational movement**, which led to the motors we are familiar with, today.

In a later chapter, we will examine how *linear* motor action is used to drive loudspeakers and how *rotational* motor action is achieved in a simple d.c. motor.

## Review your learning

Now that we've completed this chapter, we need to examine the **objectives** listed at its start. Placing '*Can I...*' at the beginning, and a question mark at the end, of each objective turns that objective into a **test item**. If we can answer each of those test items, then we've met the objectives of this chapter.

**Online resources**

The companion website to this book contains further resources relating to this chapter. The website can be accessed via the following link:
www.routledge.com/cw/waygood

# Chapter 18

# Magnetic circuits

An Introduction to Electrical Science, Waygood, ISBN 9780815391821, 2019. © Taylor & Francis

## Objectives

On completion of this chapter, you should be able to

1 explain the term 'magnetic circuit'.
2 compare a magnetic circuit with an electric circuit.
3 explain the factors that determine:
   • magnetomotive force
   • reluctance
   • magnetic field strength.
4 specify the SI units of measurement for
   • magnetomotive force
   • flux
   • flux density
   • reluctance
   • magnetic field strength.
5 explain the equivalent of 'Ohm's Law' for magnetic circuits.
6 explain the equivalent of 'Kirchhoff's Laws' for magnetic circuits.
7 briefly explain the relationship between absolute permeability ($\mu$), the permeability of free space ($\mu_o$) and relative permeability ($\mu_r$).
8 briefly explain the relationship between absolute permeability ($\mu$) and the flux density and magnetic field strength of a magnetic circuit.
9 explain the reason for the shape of a typical **B-H curve**.
10 describe the main features of a **hysteresis loop**, in particular:
   • residual flux density
   • coercive force
   • saturation points
   • cross-sectional area of the loop.
11 explain '**magnetic leakage**' and '**fringing**'.
12 solve problems on magnetic circuits.

## Introduction

In this chapter, we are going to explore the behaviour of **magnetic circuits**. A magnetic circuit is a key part of all electrical machines, for example generators, transformers and motors, as well as other electrical devices such as relays and contactors, and a great deal of effort goes into its design.

There are similarities between *magnetic* circuits and *electric* circuits, so it's usual to use an electric circuit as an analogy for explaining how magnetic circuits behave. And that's what we'll be doing throughout this chapter.

## Electric circuit analogy

motor's magnetic circuit

Figure 18.1

A **magnetic circuit** may be defined as one, or more, closed paths within which *lines of magnetic flux are confined*.

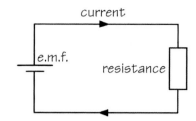

Figure 18.2

Magnetic circuits are manufactured from magnetically 'soft' ferromagnetic metals, such as iron or silicon steel. In the case of a transformer, the magnetic circuit is its 'core'. In the case of a generator or a motor, the magnetic circuit is more complex, and is a combination of the machine's 'stator', 'rotor' and the airgaps between the two, as illustrated in Figure 18.1 (where the broken lines respresent the flux paths set up by the field windings).

As already pointed out, there are certain similarities between magnetic and electric circuits, which will help us understand the behaviour of magnetic circuits, as shown in Figure 18.2.

The lines of **magnetic flux** (symbol: $\Phi$, pronounced 'phi'), confined within a magnetic circuit, are equivalent to the *electric current* in an electric circuit.

In order to establish this magnetic flux, we need a **magnetomotive force** (symbol: $F$), which is provided by means of a current-carrying winding or coil, and this is equivalent to the *electromotive force* applied across an electric circuit.

The material which forms a magnetic circuit always offers some degree of opposition to the formation of magnetic flux, and we call this opposition **reluctance** (symbol: $R_m$). This is equivalent to the *resistance* of an electric circuit.

Finally, **magnetic field strength** (symbol: $H$) of a magnetic circuit, is equivalent to a voltage gradient, or 'voltage per unit length'.

The original (but a, by far, more descriptive) term for magnetic field strength was '**m.m.f. gradient**', and it's a pity the scientific community hasn't retained that term instead of changing it to '**magnetising force**' and, finally, to '**magnetic field strength**', both of which are rather non-descriptive!

Table 18.1 summarises this.

Table 18.1

| For a magnetic circuit | | For an electric circuit |
|---|---|---|
| **magnetomotive force *(F)*** | *is equivalent to* | **electromotive force *(E)*** |
| **magnetic flux ($\Phi$)** | *is equivalent to* | **electric current *(I)*** |
| **reluctance ($R_M$)** | *is equivalent to* | **resistance *(R)*** |
| **magnetic field strength *(H)*** | *is equivalent to* | **voltage gradient *(E/l)*** |

## Series and series-parallel equivalent circuits

In the same way that we can have **series, parallel** or **series-parallel** *electric* circuits, we can also have equivalent series and series-parallel (but *not* 'parallel'!) *magnetic* circuits – examples of which are illustrated in Figure 18.3.

Figures 18.3 (a), (b) and (c) are examples of **series magnetic circuits**, together with their electric circuit equivalents. Figures 18.3 (d) and (e) are examples of **series-parallel magnetic circuits**, together with their electric circuit equivalents.

In **Figure 18.3 (a)**, $R_m$ represents the reluctance of the complete magnetic circuit.

In **Figure 18.3 (b)**, $R_{m1}$ represents the reluctance of the metal part of the magnetic circuit, while $R_{m2}$ represents the reluctance of the air gap.

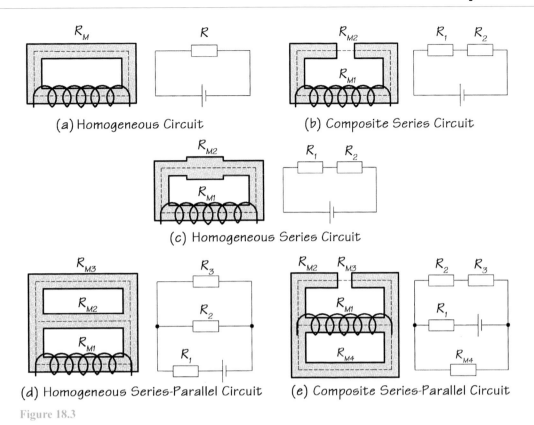

(a) Homogeneous Circuit          (b) Composite Series Circuit

(c) Homogeneous Series Circuit

(d) Homogeneous Series-Parallel Circuit     (e) Composite Series-Parallel Circuit

Figure 18.3

In **Figure 18.3 (c)**, $R_{m1}$ represents the reluctance of the metal part with the smaller cross-sectional area, while $R_{m2}$ represents the reluctance of the metal part with the large cross-sectional area.

In **Figure 18.3 (d)**, $R_{m1}$ represents the reluctance of the lower part of the magnetic circuit, while $R_{m2}$ represents the reluctance of the centre part, and $R_{m3}$ represents the reluctance of the upper part of the magnetic circuit.

In **Figure 18.3 (e)**, $R_{m1}$ represents the reluctance of the centre metal part of the magnetic circuit, while $R_{m2}$ represents the total upper metal part, and $R_{m3}$ represents the reluctance of the airgap in the upper branch, and $R_{m4}$ represents the reluctance of the lower part of the magnetic circuit.

We can further classify magnetic circuits as being either '**homogeneous**' or '**composite**', where:

- **homogeneous** (meaning 'the same throughout') describes a magnetic circuit in which the magnetic flux passes through entirely the same material. Figure 18.3 (a), (c) and (d) are examples of homogeneous magnetic circuits, because the flux is contained entirely within a ferrous-metal core. An example of an homogeneous magnetic circuit is a transformer core.
- **composite** describes a magnetic circuit in which the magnetic flux passes through two or more *different* materials. For example, in Figure 18.3 (b) and (e), the flux not only passes through ferrous metal but also through airgaps. Composite magnetic circuits are used in generators and motors, where there are airgaps between the stationary and rotating parts of the machines.

Now, let's move on and look in more detail at the *quantities* we listed in our comparison table of magnetic and electric circuits.

## Magnetomotive force (equivalent to 'electromotive force')

**Magnetomotive force** (symbol: $F$), which is often abbreviated to **m.m.f.**, is the term used to describe *the means by which magnetic flux is set up within a magnetic circuit*. In all practical magnetic circuits, this is provided using a current-carrying winding (coil), and is the product of the *current* and the *number of turns in the winding*. Its SI unit of measurement is the **ampere**

(symbol: A) which, in this context, is usually spoken as 'ampere turn' to avoid any confusion with the unit for electric current:

$$F = IN$$

where:

$F$ = m.m.f., in amperes (A)

$I$ = current, in amperes (A)

$N$ = number of turns (no units)

**Note!** Since the m.m.f is the product of amperes and a simple number (of turns), its SI unit of measurement is the 'ampere' but, to avoid confusion with the unit for current, it is usually spoken or pronounced as *'ampere turn'*. Despite its unit of measurement, it's important to realise that m.m.f. is equivalent to e.m.f. and *not* current!

## Magnetic flux (equivalent to 'current')

**Magnetic flux** (symbol: $\Phi$, pronounced 'phi') describes *the quantity of magnetic flux established within a magnetic circuit*. Its unit of measurement is the **weber** (symbol: Wb) – pronounced *'vay-ber'*. In practice, a weber is a *very* large unit and you are more likely to see flux expressed in milliwebers (mWb) or even microwebers (μWb).

## Reluctance (equivalent to 'resistance')

**Reluctance** (symbol: $R_M$), is *the opposition a magnetic circuit offers to the formation of magnetic flux*. As we shall explain shortly, its unit of measurement is the **ampere per weber** (symbol: A/Wb) – usually spoken as 'ampere-turn per weber', for the reason already explained.

Just like resistance, reluctance is *directly proportional* to the **length** (symbol: $l$) of a magnetic circuit, and is *inversely proportional* to its **cross-sectional area** (symbol: $A$):

$$R_M \propto \frac{l}{A}$$

To change the 'proportional' sign into an 'equals' sign we must, of course, introduce a *constant*. This constant is called '**magnetic reluctivity**', and is directly equivalent to 'resistivity' in an electric circuit. However, it's far more common to use its *reciprocal* instead, which we call '**absolute permeability**' (symbol: $\mu$,

pronounced *'mu'*), which is equivalent to 'conductivity' (the reciprocal of 'resistivity') in an electric circuit.

$$R_M = \left( \frac{1}{\mu} \right) \times \frac{l}{A}$$

$$R_M = \frac{l}{\mu A}$$

We will learn more about absolute permeability a little later in this chapter but, for now, you can think of it as a measure of the *ease* with which a particular material allows the formation of magnetic flux, making it directly equivalent to the *'conductivity'* (the reciprocal of 'resistivity') of an electrical conductor (see Table 18.2).

Table 18.2

| For a magnetic circuit | | For an electric circuit |
|---|---|---|
| $R_M = \dfrac{l}{\mu A}$ | *is equivalent to* | $R = \rho \dfrac{l}{A}$ |
| permeability (μ) | *is equivalent to* | conductivity |

## 'Hopkinson's Law' for magnetic circuits

The relationship between **magnetomotive force**, magnetic **flux** and **reluctance** is *exactly* equivalent to the relationship between potential difference, current and resistance, as established in the earlier chapter on *Ohm's Law*.

Although this relationship is, for obvious reasons, widely known as the *'Ohm's Law for Magnetic Circuits'*, strictly speaking, it should be termed '**Hopkinson's Law**', as the relationship was originally determined by the British electrical engineer, John Hopkinson (1849–1898).

We'll discuss whether Ohm's Law is, in fact, a good analogy or not, towards the end of this chapter. However, regardless of what we call it, the relationship is as follows:

$$\Phi = \frac{F}{R_M}$$

where:

$\Phi$ = magnetic flux, in webers (Wb)

$F$ = m.m.f., in amperes (A)

$R_M$ = reluctance, in amperes per weber (A/Wb)

**Note!** By rearranging this equation, and substituting the units of measurement, we can now confirm the unit of measurement of **reluctance**:

$$R_M = \frac{F}{\Phi} = \frac{\text{ampere}}{\text{weber}} = \text{ampere per weber}$$

Table 18.3

| For a magnetic circuit | | For an electric circuit |
|---|---|---|
| $R_M = \dfrac{F}{\Phi}$ | *is equivalent to* | $R = \dfrac{E}{I}$ |

**Worked example 1**   A mild steel ring has a coil of 200 turns, carrying a current of 10 A, wound uniformly around it. If a resulting flux of 750 μWb is established within the ring, calculate

a   the m.m.f.
b   the reluctance.

**Solution**

a   $F = IN = 10 \times 200 = 2000$ A (ampere turns)
    Answer (a.)

b   Using 'Hopkinson's Law' for magnetic circuits:

$$R_M = \frac{F}{\Phi} = \frac{2000}{750 \times 10^{-6}}$$

$$= 2.67 \times 10^6 \,\text{A/Wb} \ \text{ or } \ 2.67 \,\text{MA/Wb (Answer b.)}$$

## 'Kirchhoff's Laws' for magnetic circuits

You will recall that '**Kirchhoff's Voltage Law**' specifies that, '*in any closed loop, the applied voltage is equal to the sum of the voltage drops around that loop*'. Well, a similar principle applies to magnetic circuits, which is sometimes called '**Kirchhoff's Law for Magnetomotive Force**'; that is, *for any 'closed path' around a magnetic circuit, the applied magnetomotive force is equal to the sum of the m.m.f. drops around that same path*.

Since magnetic field strength is defined as 'm.m.f. per unit length', then an 'm.m.f. drop' must be the product of the magnetic field strength and the length *(Hl)* of the relevant part of the circuit; that is,

$$\text{since } H = \frac{F}{l} \quad \text{then} \quad F = Hl$$

... as shown in Figure 18.4 and the equations that follow.

$$F = (H_{\text{iron}} \, l_{\text{iron}}) + (H_{\text{air}} \, l_{\text{air}}) \qquad E = U_1 + U_2$$

You will also recall that '**Kirchhoff's Current Law**' specifies that *the sum of the currents approaching a junction must equal the sum of the currents leaving that junction*. Again a similar principle applies to magnetic circuits, which is sometimes called '**Kirchhoff's Law for Flux**'; that is '*the amount of flux approaching a junction must equal the sum of the fluxes leaving*', as shown in Figure 18.5 and the equations that follow.

$$\Phi = \Phi_1 + \Phi_2 \quad I = I_1 + I_2$$

It should be emphasised that these two 'laws' are actually the unnamed 'equivalents' of Kirchhoff's Laws for electric circuits, as Kirchhoff did *not* intend for his laws to apply to magnetic circuits.

Figure 18.4

**Figure 18.5**

## Understanding magnetisation (B-H) curves

'**Magnetisation curves**' are graphs that show the relationship between a magnetic material's **flux density** *(B)* and its **magnetic field strength** *(H)*. For that reason, they are also commonly called '*B-H* **curves**'.

### Flux density *(B)*

We have already briefly explained magnetic **flux density** in the chapters on *magnetism* and on *electromagnetism*, but it is sufficiently important for us to remind ourselves what it means here.

**Flux density** is a measure of the 'intensity' of the magnetic flux at any point within a magnetic circuit, and is defined as '*the flux per unit area*' at that point. Its unit of measurement is, therefore, the weber per square metre which, in SI, is given the special name **tesla** (symbol: **T**) in honour of the remarkable Serbian-American engineer, Nikola Tesla (1856–1943).

So, we can express flux density as. . .

$$B = \frac{\Phi}{A}$$

where:

$B$ = flux density (T)

$\Phi$ = flux (Wb)

$A$ = cross-sectional area of magnetic circuit (m²)

> **Worked example 2** A steel magnetic circuit has a rectangular section measuring 20 mm by 25 mm, and supports a flux of 750 μWb. What is its flux density in the magnetic circuit?

> **Solution** The cross-sectional area of the magnetic circuit is:
>
> $$A = (25 \times 10^{-3}) \times (20 \times 10^{-3}) = 500 \times 10^{-6} \, \text{m}^2$$
>
> . . . therefore, the flux density:
>
> $$B = \frac{\Phi}{A} = \frac{750 \times 10^{-6}}{500 \times 10^{-6}} = 1.5 \, \text{T (Answer)}$$

### Magnetic field strength *(H)*

At the beginning of this chapter, we compared magnetic circuits with electric circuits, and described the term **magnetic field strength** as being equivalent to a *voltage gradient* around an electric circuit. While this analogy is accurate, it's a little less obvious, as we don't usually talk in terms of voltage gradients around circuits but it is, nevertheless, a useful analogy!

> We do consider 'voltage gradient' in insulators or dielectrics, where it is known as 'dielectric strength', expressed in volts per metre.

As we have already briefly mentioned, 'magnetic field strength' was originally known as '**m.m.f. gradient**', which is by far the most descriptive and meaningful term for this quantity. It was more recently known as '**magnetising force**', which is just as undescriptive as its present term!

**Magnetic field strength** (symbol: $H$) is defined as '*the magnetomotive force per unit length of a magnetic circuit*':

$$H = \frac{F}{l}$$

where:

$H$ = magnetic field strength (A/m)

$F$ = magnetomotive force (A)

$l$ = length of magnetic circuit (m)

**Worked example 3**   A cast iron ring has a mean diameter of 50 mm, and is uniformly wound with 200 turns of insulated wire. If the current flowing in the wire is 2.5 A, what is the ring's magnetic field strength?

**Solution**   We start by calculating the magnetomotive force:

$$F = IN = 2.5 \times 200 = 500 \text{ A (ampere-turns)}$$

Next, we determine the mean length of the magnetic circuit:

length = circumference = $\pi D$

$\qquad = \pi \times (50 \times 10^{-3}) = 157 \times 10^{-3} \text{ m}$

Now, we can determine the magnetic field strength:

$$H = \frac{F}{l} = \frac{500}{157 \times 10^{-3}} = 3185 \text{ A/m (Answer)}.$$

Now that we've learnt a little more about **flux density (B)** and **magnetic field strength (H)**, let's move on and examine '**magnetisation curves**' or '**B-H curves**'.

By means of a relatively simple experiment, it is possible to produce a graph which shows the relationship between the two for a sample of ferrous metal. The experiment is illustrated in Figure 18.6.

The sample, in this case a toroid (ring) of iron, is wound with an insulated conductor and supplied, via a variable resistor, from a direct-current supply.

The magnetic field strength is proportional to the current so, by incrementally increasing the current, we are also able to incrementally increase the magnetic field strength. For each incremental increase in magnetic field strength, we can then measure the corresponding value of flux density within the iron ring, using an instrument, called a 'teslameter' (formerly called a 'gaussmeter') to do so.

If we were then to plot the values of magnetic field strength *(H)*, horizontally, and the resulting **flux density (B)**, vertically, on a graph, the result would look something like the curve shown to the right of the schematic diagram in Figure 18.6, or the one illustrated in Figure 18.7.

You may ask why should the resulting curve follow this particular shape instead of being, say, a straight line. Well, to understand this, we must remind ourselves of the '**domain theory of magnetism**' that we learnt about in the chapter on *magnetism*.

A '**domain**', you will recall, is a molecule which behaves just like a tiny permanent magnet. In a sample of *unmagnetised* ferromagnetic material, the domains form closed 'chains', with each north pole 'chasing' the south pole of an adjacent domain. In a sample of *magnetised* ferromagnetic material, the domain 'chains' have been broken, and all or most of the domains realigned to form rows.

So let's examine the magnetisation curve, and see what is happening to the domains within the iron sample as the magnetic field strength is gradually increased. This is illustrated in Figure 18.7.

**magnetising curve experiment**

**resulting B-H curve**

Figure 18.6

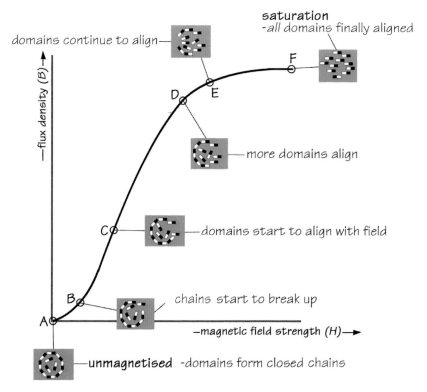

Figure 18.7

- At point **A**, before we actually start to increase the magnetic field strength, the sample is *unmagnetised*, with its domains forming closed 'chains'.
- At point **B**, the increasing magnetic field strength starts to cause the domain chains to break up and realign, creating a weak magnetic field (low flux density).
- At points **C** and **D**, the magnetic field strength continues to increase, causing more and more domains to align with the field and, so, the flux density increases yet further.
- At point **E**, the majority of domains have now realigned, and the sample of iron is starting to approach 'saturation'. 'Saturation' is the point of maximum flux density.
- At point **F**, *all* the domains have now realigned, and the sample has finally reached '**saturation**' – in other words, *there are no more domains to realign*, and the sample's flux density has reached its maximum possible value.

Once **saturation** has been reached, any further increase in magnetic field strength will have absolutely no effect whatsoever on the sample's flux density.

From the shape of the magnetisation curve, it should now be obvious that the ratio of *B:H* can only be (approximately) constant over the straight part of the magnetisation curve. Below point **B** on the above curve, and beyond point **D**, the ratio will change considerably. These points are called the 'lower knee' and 'upper knee', respectively (not all *B-H* curves have a noticeable 'lower knee', but they *all* have an 'upper knee').

## Absolute permeability: the ratio of flux density to magnetic field strength

The ratio of *flux density to magnetic field strength (B:H)* is important enough to be given its own name, and is called '**absolute permeability**' (symbol: $\mu$, pronounced *'mu'*). Earlier, we described absolute permeability as being equivalent to 'conductance' in an electric circuit.

For air, the value of absolute permeability is a *constant*, and is equal to a figure of $4\pi \times 10^{-7}$ H/m (henry per metre) or, if you prefer, $1.257 \times 10^{-6}$ H/m. You can think of the absolute permeability for air as being a sort of 'reference' permeability, and it is important enough to be given its own name: the **absolute permeability of free space** (symbol: $\mu_o$).

Strictly speaking 'free space' means a 'vacuum' but the permeability for air and for a vacuum are generally taken as being the same.

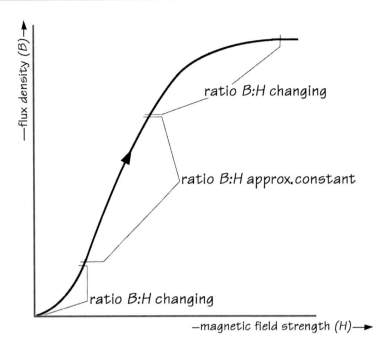

- flux density (B) →

ratio B:H changing

ratio B:H approx.constant

ratio B:H changing

- magnetic field strength (H) →

Figure 18.8

But for ferrous metals, the absolute permeability (unlike conductance) is a *variable*, and depends entirely on *where along the metal's magnetisation curve it is measured*. This is because the ratio of *B:H* changes along that curve (Figure 18.8).

Just like the figure for absolute permeability of free space, figures representing absolute permeability are quite awkward and clumsy. So, in an attempt to simplify them, permeabilities of ferrous metals are more usually expressed as a '*relative* permeability' ($\mu_r$), which is a simple ratio. For example, a metal having a relative permeability of, say, 1000, has a permeability that is 1000 times greater than that of air. This is equivalent to an absolute permeability of $1257 \times 10^{-6}$ H/m, but presented as a much more convenient figure. And this is the way in which you will normally see permeabilities expressed in data tables.

So the relationship between these three 'different' permeabilities is:

$$\mu_r = \frac{\mu}{\mu_o}$$

where:

$\mu_r$ = relative permeability (no units)

$\mu$ = absolute permeability (H/m)

$\mu_o$ = absolute permeability of freespace (H/m)

To summarise this section, we can say that the ratio *B:H* can *either* be expressed in terms of a material's **absolute permeability** *(μ)*:

$$\frac{B}{H} = \mu$$

... *or* in terms of the **absolute permeability of free space** *($\mu_o$)* and **relative permeability** *($\mu_r$)*:

$$\frac{B}{H} = \mu_o \mu_r$$

**Worked example 4**    What is the relative permeability, at a magnetic field strength of 1500 A/m, of a sample of ferromagnetic material whose *B-H* curve is reproduced in Figure 18.9?

**Solution**    For a magnetic field strength of 1500 A/m, the corresponding flux density (obtained from the graph) is 1.45 T, so:

$$\mu_r = \frac{B}{\mu_o H} = \frac{1.45}{4\pi \times 10^{-7} \times 1500} \simeq 769 \text{ (Answer)}$$

So, in the above worked example, the permeability of the sample at a magnetic field strength of 1500 A/m is **769** times higher than that for air.

Although the magnetisation curves for *all* ferromagnetic materials follow roughly the same shape, the positions of their lower and upper 'knees' and the gradient of their linear part will vary considerably – *depending on how easily the material is magnetised.*

**Figure 18.9**

## Solving problems on magnetic circuits

Most problems on magnetic circuits involve trying to determine the amount of magnetomotive force required to establish a specified amount of flux (or, perhaps, flux density) within a particular part, usually within an airgap, of a magnetic circuit. This has practical relevance to the design of magnetic circuits for rotating machines.

Most problems on magnetic circuits, then, can be worked out methodically, through the use of what is known as the '**results-table method**', in which Table 18.4 is gradually completed, starting with the information supplied in the question.

So, in the following worked example, we will use the '**results-table method**' to solve the problem, showing how the table is completed, step-by-step. Of course, when you actually use this method, you only need to complete the one table; we are repeating the image of the table, here, to demonstrate how each column is completed.

Table 18.4

| Part: | Length/m | $A$/m$^2$ | $\Phi$/Wb | $B$/T | $H$/A/m | $F$/A |
|-------|----------|-----------|-----------|-------|---------|-------|
|       |          |           |           | $B = \dfrac{\phi}{A}$ | $H = \dfrac{B}{\mu}$ | $F = Hl$ |

**Worked example 5**    (using the 'results-table method')

**Figure 18.10**

Determine how much current must flow in a 400-turn winding, in order to establish a flux of 1 mWb within the airgap of the magnetic circuit shown in Figure 18.10. The relationship between the magnetic field strength *(H)* and flux density *(B)* for the material from which the magnetic circuit is manufactured is as follows:

| H/A/m | 200 | 350 | 650 | 800 | 1250 | 2000 | 3500 |
|-------|-----|-----|-----|-----|------|------|------|
| B/T | 0.8 | 1.0 | 1.3 | 1.3 | 1.4 | 1.5 | 1.6 |

**Solution** The first step is to fill in as much of the results table as we can from the information supplied in the question:

| Part: | Length/m | A/m² | Φ/Wb | B/T | H/A/m | F/A |
|-------|----------|------|------|-----|-------|-----|
| Centre | 150×10⁻³ | 800×10⁻⁶ | *1×10⁻³ | | | |
| Outside | 340×10⁻³ | 500×10⁻⁶ | **0.5×10⁻³ | | | |
| Airgap | 1×10⁻³ | 800×10⁻⁶ | 1×10⁻³ | | | |

*As we are told fringing can be ignored, the flux in the centre limb will be equal to the flux in the airgap.
**As both outside limbs are identical, the amount of flux in each of these outer limbs must be *half* that of the inside limb/airgap.

Next, we can complete the flux density *(B)* column, by *dividing the flux by the cross-sectional area, i.e.*

$$B = \frac{\phi}{A}$$

| Part: | Length/m | A/m² | Φ/Wb | B/T | H/A/m | F/A |
|-------|----------|------|------|-----|-------|-----|
| Centre | 150×10⁻³ | 800×10⁻⁶ | 1×10⁻³ | **1.25** | | |
| Outside | 340×10⁻³ | 500×10⁻⁶ | 0.5×10⁻³ | **1.00** | | |
| Airgap | 1×10⁻³ | 800×10⁻⁶ | 1×10⁻³ | **1.25** | | |

To complete the magnetic field strength for the **airgap** is straightforward, as it is the flux density divided by the absolute permeability of free space:

$$H = \frac{B}{\mu_o} = \frac{1.25}{4\pi \times 10^{-7}} = 995 \times 10^3 \text{ A/m}$$

| Part: | Length/m | A/m² | Φ/Wb | B/T | H/A/m | F/A |
|-------|----------|------|------|-----|-------|-----|
| Centre | 150×10⁻³ | 800×10⁻⁶ | 1×10⁻³ | 1.25 | | |
| Outside | 340×10⁻³ | 500×10⁻⁶ | 0.5×10⁻³ | 1.00 | | |
| Airgap | 1×10⁻³ | 800×10⁻⁶ | 1×10⁻³ | 1.25 | **995×10³** | |

But to determine the magnetic field strength for the centre and outside limbs, we would normally first have to construct the B/H graph, from the data supplied. However, as you can see, from the supplied data, a flux density of 1 T corresponds to a magnetic field strength of 350 A/m.

For the centre limb, a flux density of 1.25 T corresponds to a magnetic field strength of **720 A/m**; for the outer limbs, a flux density of 1.00 T corresponds to a magnetic field strength of **350 A/m**.

| Part: | Length/m | A/m² | Φ/Wb | B/T | H/A/m | F/A |
|---|---|---|---|---|---|---|
| Centre | $150 \times 10^{-3}$ | $800 \times 10^{-6}$ | $1 \times 10^{-3}$ | 1.25 | **720** | |
| Outside | $340 \times 10^{-3}$ | $500 \times 10^{-6}$ | $0.5 \times 10^{-3}$ | 1.00 | **350** | |
| Airgap | $1 \times 10^{-3}$ | $800 \times 10^{-6}$ | $1 \times 10^{-3}$ | 1.25 | $995 \times 10^{3}$ | |

To complete the final column, the 'magnetomotive force drop' *(F)* is *the product of the magnetomotive force and the length of that part of the magnetic circuit:*

For the centre limb:

$$F = Hl = 720 \times (150 \times 10^{-3}) = 108 \text{ A}$$

For the outer limbs:

$$F = Hl = 350 \times (340 \times 10^{-3}) = 119 \text{ A}$$

For the airgap:

$$F = Hl = 995 \times (1 \times 10^{-3}) = 995 \text{ A}$$

| Part: | Length/m | A/m² | Φ/Wb | B/T | H/A/m | F/A |
|---|---|---|---|---|---|---|
| Centre | $150 \times 10^{-3}$ | $800 \times 10^{-6}$ | $1 \times 10^{-3}$ | 1.25 | 720 | **108** |
| Outside | $340 \times 10^{-3}$ | $500 \times 10^{-6}$ | $0.5 \times 10^{-3}$ | 1.00 | 350 | **119** |
| Airgap | $1 \times 10^{-3}$ | $800 \times 10^{-6}$ | $1 \times 10^{-3}$ | 1.25 | $995 \times 10^{3}$ | **995** |

To determine the total magnetomotive force required, we simply *add up the m.m.f. drops in the right-hand column:*

| Part: | Length/m | A/m² | Φ/Wb | B/T | H/A/m | F/A |
|---|---|---|---|---|---|---|
| Centre | $150 \times 10^{-3}$ | $800 \times 10^{-6}$ | $1 \times 10^{-3}$ | 1.25 | 720 | 108 |
| Outside | $340 \times 10^{-3}$ | $500 \times 10^{-6}$ | $0.5 \times 10^{-3}$ | 1.00 | 350 | 119 |
| Airgap | $1 \times 10^{-3}$ | $800 \times 10^{-6}$ | $1 \times 10^{-3}$ | 1.25 | $995 \times 10^{3}$ | 995 |
| | | | | | total mmf: | **1222 A** |

Finally, since m.m.f. is the product of the current through the winding, and its number of turns:

$$I = \frac{F}{N} = \frac{1222}{400} = 3.06 \text{ A (Answer)}$$

## Magnetic materials

Magnetic materials are classified according to their values of relative permeability, as shown in Figures 18.11 and 18.12.

Figure 18.11

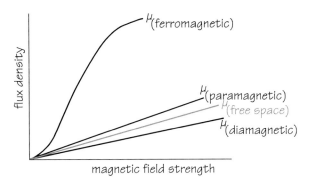

Figure 18.12

**Paramagnetic materials** (Figure 18.13, left) have values of relative permeability that are *slightly greater than unity*, which cause a very slight *increase* in the flux density within a sample of that type of material, when placed within a uniform magnetic field. Paramagnetic materials include magnesium and lithium.

**Diamagnetic materials** (Figure 18.13, right) have values of relative permeability that are *slightly less than unity*, which actually cause a slight *reduction* in the flux density within a sample of that material, when placed within a uniform magnetic field. Examples of diamagnetic materials include mercury, silver and copper.

**Ferromagnetic materials** are in a completely different category to the others, and have values of relative permeability that lie in the *hundreds or thousands*. These materials are further sub-classified as being either 'hard' or 'soft'. 'Hard' ferromagnetic materials (e.g. steel) are relatively difficult to magnetise and demagnetise, whereas 'soft' ferromagnetic materials (e.g. iron) are relatively easy to magnetise and demagnetise. Rather oddly, some ferromagnetic alloys (e.g. heusler alloy) are actually mixtures of entirely paramagnetic or diamagnetic materials!

Paramagnetic and diamagnetic materials are of little practical importance, in comparison with ferromagnetic materials – *all large-scale energy conversion, using generators, transformers and motors, depends entirely on the use of ferromagnetic materials.*

## Hysteresis loops

In the magnetisation curve experiment described earlier, we increased the magnetic field strength until the sample became saturated.

But what would happen if we continued the experiment further, by *reducing* the magnetic field strength *back to zero* and, then, *reversing* the direction of the magnetic field strength? Well, let's find out!

First of all, let's repeat the last experiment, so that the sample reaches saturation. So, in Figure 18.14, as we increase the current flowing through the coil, the magnetic field strength increases, and the flux density increases, following the curve until saturation is reached at point *a*.

In Figure 18.15, we gradually reduce the current back towards zero. As we do this, the magnetic field strength reduces to zero as well, reducing the flux density in the sample. However, the reduction in flux density *doesn't follow the original curve* but, instead, it

paramagnetic material

diamagnetic material

Figure 18.13

Figure 18.14

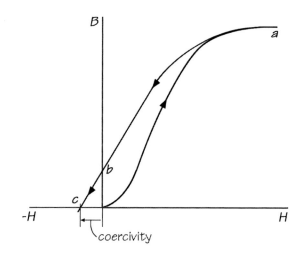

Figure 18.16

follows the *second* curve *(a–b)*. So when the magnetic field strength eventually reaches zero, the flux density has *lagged behind* and has *not* fallen to zero, but only to point *(b)* on the new curve! In other words, the sample hasn't completely demagnetised, but has *retained* a certain amount of its flux density. We call this remaining flux density '**residual flux density**', or the '**remanence**' of the sample.

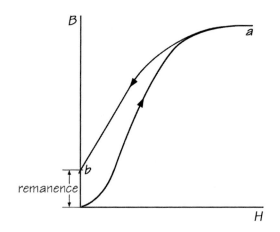

Figure 18.15

To *completely* demagnetise the sample (that is, to entirely remove the residual flux density), it's necessary to *reverse* the direction of the current in the coil, thereby *reversing the direction of the magnetic field strength* and, then, applying just enough to completely demagnetise the sample – i.e. dropping its flux density to zero (point *c*). The amount of 'negative' magnetic field strength required to achieve this is called '**coercivity**' or the '**coercive force**' – as illustrated in Figure 18.16.

On a humorous note, you may think of this property in the following terms: 'it takes more **effort** to get a bus-load of football fans *out* of a pub than the amount of **effort** it took to get them *into* that pub'! In this context, '**effort**' is equivalent to '**magnetic field strength**', and the 'bus-load of football fans' is equivalent to flux density!

In Figure 18.17, we have continued to increase the current in the *reverse* direction, increasing the magnetic field strength in the same direction. This now causes the flux density to increase in the negative direction (i.e. reversing the polarity of the sample) until it, once more, reaches saturation again at point *d*.

In Figure 18.18, we now reduce the current again, reducing the magnetic field strength and causing the flux density to fall, following the curve *d–e*. Again, when the magnetic field strength has fallen to zero, some flux density remains – so, again, a certain amount of residual magnetism remains within the sample. By *reversing* and *increasing* the current (back to its original direction) we can remove this residual magnetism *(e–f)*. As the current continues to increase, the flux density follows curve *(f–a)* back to saturation in the original direction.

From now on, no matter how often we continue to cycle the value of the current through the magnetising coil, the *outside* curve, (a–b–c–d–e–f–a) will *always* result – as illustrated in Figure 18.19.

This curve is called a '**hysteresis loop**', and is particularly important when we apply an alternating current (and, therefore, an alternating magnetic field

Figure 18.17

Figure 18.18

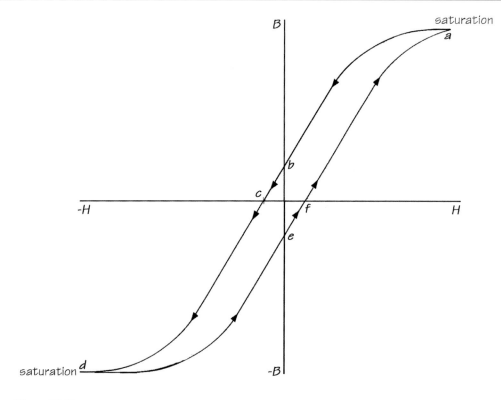

**Figure 18.19**

strength) to the sample – e.g. in transformers or a.c. motors.

The term '**hysteresis**' was coined by the Scottish engineer and academic Sir James Ewing (1855–1935), who conducted a great deal of research into the magnetic properties of metals in the 1920s, and who is credited with discovering this property. The word itself is derived from a Greek word, meaning *'to lag behind'* and, in this context, describes how a change in **flux density** 'lags behind' any corresponding change in **magnetic field strength**.

The cross-sectional area enclosed by the hysteresis loop represents the energy (expressed in 'joules per cubic metre per cycle') required to magnetise and demagnetise the sample – the *greater* this area, the *greater* the energy required to do so. Again, this is particularly important for alternating-current magnetic circuits (i.e. generators, transformers and motors) and accounts for part of the energy losses (called 'hysteresis losses') that occur within the magnetic circuits of a.c. machines.

Different ferromagnetic materials have hysteresis loops of different cross-sectional areas – e.g. 'soft'

iron and silicon steel have very *narrow* cross-sections, whereas 'hard' steels have very *wide* cross-sections, as shown in Figure 18.20.

The left-hand figure shows a relatively narrow hysteresis loop, with low remanence (residual flux density). This is typical for a 'soft' ferromagnetic material, which is easily magnetised and demagnetised, and which retains very little residual flux density when the magnetic field strength is removed. This represents a material that is ideal for making *temporary magnets* and magnetic circuits for electrical machines. The low cross-sectional area indicates that relatively little energy is required to magnetise and demagnetise the sample.

The right-hand figure shows a relatively wide hysteresis loop, with high remanence. This is typical for a 'hard' ferromagnetic material, which is difficult to magnetise, and just as difficult to demagnetise. This represents a material that is ideal for making *permanent magnets*, but would be no use whatsoever for making the magnetic circuits for electrical machines. The large cross-sectional area indicates a relatively large amount of energy is required to magnetise and demagnetise the sample.

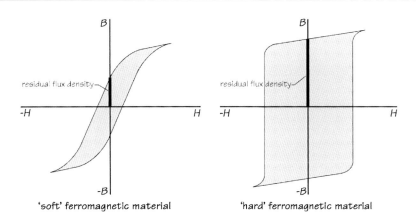

'soft' ferromagnetic material          'hard' ferromagnetic material

Figure 18.20

## Practical applications of magnetic circuits

As briefly mentioned at the beginning of this chapter, the main application for magnetic circuits is to provide a low-reluctance 'path' for the magnetic flux utilised in electrical machines.

In the case of a rotating machine, such as a motor or generator, we can think of the iron part of a magnetic circuit as providing the means by which flux can be 'guided' to the airgaps within which the machine's armature windings are located, in much the same way as conductors are used, in an electric circuit, to guide the current to the wherever its load is located.

Taking this analogy further, the low-reluctance iron part of a magnetic circuit is equivalent to the low-resistance conductors in an electric circuit. And the high-reluctance air gaps are equivalent to the higher-resistance loads in an electric circuit. In the case of a magnetic circuit, this ensures that maximum m.m.f.

appears across the airgaps in the same way that low-resistance conductors ensure that maximum e.m.f. appears across an electrical load.

### Transformers

In the case of a simple, two-winding **transformer**, its core is an example of a *homogeneous* magnetic circuit, the purpose of which is to efficiently link the magnetic flux created by its primary winding with its secondary winding. There are two different types of transformer core design: 'core type' (Figure 18.21, left) and 'shell type' (Figure 18.21, right). By using a low-reluctance ('soft') ferromagnetic material (typically, silicon steel), maximum flux density is assured, with minimum remenance, and with a minimum energy requirement – thereby maximising the efficiency of the transformer.

### Rotating machines

**Electrical rotating machines**, such as generators and motors, use a *composite* magnetic circuit in the

Figure 18.21

**Figure 18.22**

construction of their stators (the stationary part of the machine) – again, usually laminated silicon steel. These magnetic circuits (see Figure 18.1) are 'composite' because the magnetic path includes airgaps. The magnetic circuits of rotating machines comprise silicon steel and air (the airgap) and have *two* purposes. The first is to *link the field windings and armature winding*. The second is to *concentrate and maximise the flux density within the airgap* between the stator and rotor, through which the armature windings move, to ensure a maximum induced e.m.f. in those windings.

### Relays, contactors, etc.

A **relay** or **contactor** (a heavy-duty relay) is a remote-controlled switch, and consists of a *solenoid* and a spring-loaded, hinged, *armature* – as illustrated in Figure 18.22. The function of the armature is to open or close contacts which are used to control an external circuit. The solenoid and armature form part of a composite magnetic circuit (shown by the broken line, above), which includes an airgap between the face of the solenoid and the armature – as illustrated in Figure 18.22a.

When the solenoid is initially energised, the magnetic circuit includes the airgap and, so, the reluctance of the magnetic circuit is relatively high, although the flux is sufficient for the solenoid to attract the armature towards its face.

As soon as the armature makes contact with the face of the solenoid (Figure 18.22b), the airgap disappears and the reluctance of the magnetic circuit falls appreciably – so far less flux is required to 'hold' the relay closed, than is needed to cause it to close in the first place.

**Magnetic leakage and fringing**

Only in an 'ideal' magnetic circuit will *all* the magnetic flux be contained entirely within the medium of the circuit itself. In all *practical* magnetic circuits (transformers, motors, generators, etc.), part of the flux produced by the magnetomotive force 'bypasses' the magnetic circuit. This is called '**magnetic leakage**'.

Similarly, repulsion between adjacent lines of magnetic force within any airgaps (in the case of motors, generators and measuring instruments) cause the outermost lines of flux to bulge outwards, essentially reducing the flux density (flux per unit area) within the airgap. This is called '**magnetic fringing**' as shown in Figure 18.23.

**Figure 18.23**

Magnetic leakage and fringing are relatively minor within a well-designed magnetic circuit, but nevertheless act to reduce the theoretical flux density achievable within the magnetic circuit.

**Limitations of the 'electric circuit' analogy**

At the beginning of this chapter, we learnt that there are useful similarities between **magnetic circuits** and **electric circuits**, which we then used to help us understand the behaviour of magnetic circuits.

This analogy is a widely used approach for teaching students about magnetic circuits, but it is, nevertheless, an *analogy* (a likeness) and, like all analogies, *it has its limitations*.

So we will finish this chapter by briefly looking at some of the significant *differences* between magnetic circuits and electric circuits.

First of all, an electric current is a *movement*, or *drift*, of electric charges (free electrons, in the case of metal conductors) *around* a circuit. Magnetic flux is *not* a movement, or drift, of anything at all! It is simply the 'shape' assumed by lines of magnetic flux contained within the magnetic circuit.

As we learnt in the chapter on *resistance*, *'the consequence of resistance is heat'*. Whenever current passes through resistance, energy is dissipated through heat transfer away from the circuit. A magnetic circuit's equivalent of resistance is **reluctance**, but *absolutely no energy is dissipated by reluctance due to the formation of magnetic flux*.

Although we described absolute permeability as being 'equivalent' to conductance, unlike conductance its value is not constant as it depends on the ratio of flux density to magnetic field strength, which varies according to where the ratio is measured along its magnetisation curve.

With the exception of insignificantly small 'leakage currents' through a conductor's insulation, electric current is confined within its circuit. Magnetic flux, on the other hand, is subject to both '**leakage**' and '**fringing**'. 'Leakage' describes magnetic flux which is created by a magnetomotive force but which, then, partially or entirely 'bypasses' the magnetic circuit, and 'fringing' describes the reduction in flux density whenever flux passes through an airgap in the magnetic circuit. Because of the difficulty is coping with leakage and fringing, most magnetic-circuit calculations you will be expected to perform include the words, *'ignore any leakage or fringing'*.

Most, although not all, metal conductors are 'ohmic' or 'linear' and, therefore, obey Ohm's Law; in other words the ratio of voltage to current (resistance) is usually *constant* for wide variations in voltage. However, this is most definitely *not* the case for magnetic circuits, which are exclusively '**non-linear**' – that is their ratios of magnetomotive force to flux (reluctance) are *not* constant. As well as the length and cross-sectional area of a magnetic circuit, reluctance also depends upon the absolute permeability of that circuit's material. Relative permeability, in turn, depends upon the ratio of flux density to magnetic field strength *(B/H)*, which can vary considerably for different values of magnetising current.

Finally, when the potential difference is removed from an electric circuit, no current can continue to flow in that circuit. However, in a magnetic circuit, when the magnetomotive force is removed, some (residual) magnetic flux usually remains, due to the circuit's **remanence**.

To summarise, then. Continue to use the analogy between magnetic circuits and electric circuits to understand the behaviour of magnetic circuits. But be aware that it is *only* an analogy and, like most analogies, it tends to ignore any significant differences!

## Summary

A **magnetic circuit** is the path that encloses magnetic flux in air, or in a ferromagnetic material.

There are similarities between **magnetic circuits** and *electric circuits*, where:

- **magnetomotive force** *(F)* is equivalent to *electromotive force*.
- **magnetic flux** ($\Phi$) is equivalent to *electric current*.
- **reluctance** *($R_m$)* is equivalent to *resistance*.
- **magnetic field strength** *(H)* is equivalent to *voltage gradient*.

These quantities are related by the magnetic equivalent of Ohm's Law:

$$\Phi = \frac{F}{R_m}$$

A **magnetomotive force** is created when an electric current flows through a coil, and is the product of the current and number of turns:

$$F = IN \text{ amperes}$$

**Reluctance** is directly proportional to the *length* of the magnetic circuit, and inversely proportional to its *cross-sectional area:*

$$R_m = \frac{l}{\mu_o \mu_r A}$$

**Absolute permeability** *(µ)* is equivalent to the *conductivity* of an electric circuit, and is the product of the **absolute permeability of free space** *(µ$_o$)* and **relative permeability** *(µ$_r$)*, where:

- **absolute permeability of free space**, $\mu_o = 4\pi \times 10^{-7}$ H/m.
- **relative permeability**, $\mu_r$, varies from material to materal, and can be as high as 100 000 (no units).

**Flux density** *(B)* is defined as *the flux per unit area*, measured in teslas (T):

$$B = \frac{\Phi}{A}$$

**Magnetic field strength** *(H)* is defined as *the magnetomotive force per unit length of magnetic circuit*, measured in amperes per metre (A/m):

$$H = \frac{F}{l}$$

**Absolute permeability** is equal to the ratio of flux density to magnetic field strength:

$$\mu_o \mu_r = \frac{B}{H}$$

The **relative permeability** for any particular material is not constant because the ratio of *B:H* is not constant, because it varies according to the shape of the *B-H* curve (Figure 18.24).

A **hysteresis loop** shows the variation in *flux density* as the magnetic field strength is varied in a positive and negative sense. The resulting change in flux density 'lags' behind changes in magnetic field strength – which is what 'hysteresis' means (Figure 18.25).

'**Soft**' ferromagnetic materials (iron, silicon steel) have narrow hysteresis loops which exhibit little residual flux density when the magnetic field strength is removed, and require little energy to magnetise and demagnetise the materials.

'**Hard**' ferromagnetic materials have wide hysteresis loops which exhibit large residual flux densities when the magnetic field strength is removed, and require larger amounts of energy to magnetise and demagnetise the materials.

The magnetic circuits of electrical machines need to be manufactured from '**soft**' ferromagnetic materials.

'**Leakage**' describes flux that is not contained within the magnetic circuit.

'**Fringing**' describes flux within a motor's or generator's magnetic circuit airgap which does not link with the rotor.

**Leakage** and **fringing** both act to reduce the theoretical effective flux within a magnetic circuit.

**Figure 18.24**

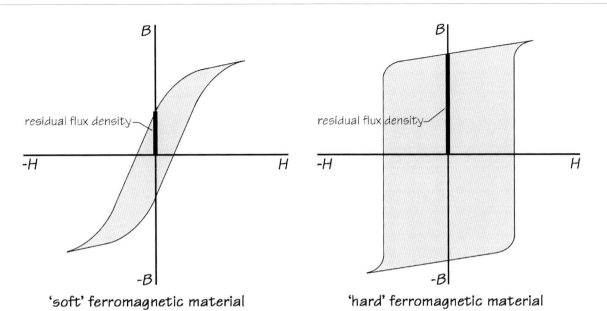

'soft' ferromagnetic material          'hard' ferromagnetic material

Figure 18.25

## Review your learning

Now that we've completed this chapter, we need to examine the **objectives** listed at its start. Placing '*Can I...*' at the beginning, and a question mark at the end, of each objective turns that objective into a **test item**. If we can answer each of those test items, then we've met the objectives of this chapter.

**Online resources**
The companion website to this book contains further resources relating to this chapter. The website can be accessed via the following link:
www.routledge.com/cw/waygood

# Chapter 19

# Electromagnetic induction

## Objectives

On completion of this chapter, you should be able to

1  explain the result of moving a permanent magnet towards, or away from, a coil.
2  describe the effect of moving a straight conductor perpendicularly through a magnetic field.
3  determine the magnitude and direction of the potential difference induced into a straight conductor moved perpendicularly through a magnetic field.
4  demonstrate the application of Fleming's Right-Hand Rule for generator action.
5  briefly explain the effects of self-induction.
6  explain Faraday's and Lenz's Laws for electromagnetic induction.
7  list the factors that affect the self-induction of a coil.
8  describe the effect of self-induction on the growth and decay of direct currents in inductive circuits.
9  solve simple problems on the growth and decay of direct currents in inductive circuits.
10  explain the terms 'mutual inductance' and 'coupled circuits'.
11  describe the behaviour of the mutual induction between two inductive circuits.
12  solve simple problems on mutual induction.
13  describe how energy is stored in magnetic fields.
14  solve simple problems on the energy stored in magnetic fields.
15  recognise and give simple examples of the functions of inductors.
16  solve simple problems on inductors in series and in parallel.

**Important!** Conventional current (positive to negative) is assumed throughout this chapter.

## Introduction

One of the most important advances in the science of electrical engineering was made in 1831, when **Sir Michael Faraday** (1791–1867) discovered that whenever there is relative motion between lines of magnetic flux and a conductor, a potential difference is 'induced' into the conductor.

So, as predicted by scientists such as Ørsted, it was indeed possible to use *movement* to produce electricity!

We call this behaviour **electromagnetic induction** and, within a year of Faraday's discovery, it was discovered quite independently by the American physicist **Joseph Henry** (1797–1878). Rightly, both men have been credited with this discovery.

This discovery directly led to the invention and development, by Faraday, of the **generator** and the **transformer** without which large-scale generation, transmission and distribution of electrical energy would be quite impossible.

### Faraday's experiments

Like so many other great scientific breakthroughs, this behaviour was apparently discovered quite by accident while Faraday was conducting a completely separate experiment, and it led him to develop more experiments with which to further investigate the behaviour. The basis of the experiments conducted by Faraday is explained as follows.

With the ends of an insulated coil connected to a sensitive measuring instrument, called a 'galvanometer', we have a closed circuit (Figure 19.1). Of course, as things stand, no current will drift, because there is no *electromotive force*.

However, if a permanent magnet is now suddenly plunged *towards* the coil, the galvanometer's needle

An Introduction to Electrical Science, Waygood, ISBN 9780815391821, 2019. © Taylor & Francis

Figure 19.1

will be seen to respond by giving a sharp 'kick', or sharp deflection, in one direction or the other (Figure 19.2).

Figure 19.2

If the magnet is then suddenly *withdrawn* from the coil, the galvanometer will again register a sharp 'kick' – but, this time, in the *opposite* direction (Figure 19.3).

Figure 19.3

The reaction of the galvanometer confirmed the presence of a potential difference across the ends of that coil, causing a current to drift through the coil and be registered by the galvanometer.

Faraday described these potential differences as being '**induced**' into the coil.

It's important to understand that it's a **potential difference** that's being induced into the coil,

*not* a current. A current only flows if the coil is connected to a load, such as the galvanometer. There is no such thing as an 'induced current'!

The experiment also demonstrated that the *direction* of the induced potential difference *reverses whenever the direction of movement of the magnet is reversed.*

Importantly, the experiment also confirmed that this potential difference is induced *only when there is movement by the magnet.* If the magnet is held stationary, *no* electromotive force is induced into the coil (Figure 19.4).

Figure 19.4

Actually, it doesn't really matter whether *the magnet is moved relative to the coil*, or *the coil is moved relative to the magnet.* Providing there is *relative movement between the two*, then a potential difference will be induced into the coil.

From this experiment Faraday was able to conclude that . . .

*Whenever there is relative motion between a conductor and a magnetic field, a potential difference will be induced into that conductor. The direction of the induced potential difference depends upon the direction of motion of the conductor relative to the lines of magnetic flux.*

Faraday repeated his experiments, while changing the various conditions involved, and discovered that the magnitude of the induced voltage varied according to

- the **velocity** (i.e. speed and direction) at which the magnet was moved
- the **flux density** of the bar magnet
- the **number of turns** wound on the coil.

Increasing *any* of these factors caused the magnitude of the induced voltage to increase. We shall return to the effect of these variations a little later in this chapter.

Faraday also found that replacing the magnet with an electromagnet would achieve the same effect, not by moving the electromagnet, but by *varying the current through it*. Increasing the current through the electromagnet produced exactly the same result as moving a magnet *towards* the coil, while decreasing the current produced exactly the same effect as moving a magnet *away* from the coil.

What Faraday had discovered was that it was possible to induce a voltage into a coil, through no physical movement whatsover! He had, in fact, invented a primitive **transformer**.

Figure 19.5 shows Faraday's experimental set-up with the electromagnet: two coils, wound around a common iron 'core' (Faraday actually used a wooden core!). One coil (the electromagnet), termed the 'primary winding', is connected to a battery, via a switch, while the other, termed the 'secondary winding', is connected to a galvanometer.

Figure 19.5

Whenever Faraday closed the switch in the primary circuit, he noticed a sharp deflection by the secondary's galvanometer. With a steady current drifting through the primary circuit, the galvanometer indicated zero. When he then re-opened the switch, he noticed another sharp deflection by the galvanometer but, this time, in the opposite direction.

Faraday went on to vary, rather than to switch, the current in the primary circuit, and noticed that as the primary current varied, the galvanometer deflected in one direction when the primary current was increased, but in the opposite direction when the primary current was reduced.

## Generator action

Let's now move on to examine the **magnitude** and **direction** of an induced potential difference – initially

using the simple case of a straight conductor that is forced to move perpendicularly (at right angles) through a permanent magnetic field set up between the poles of a magnet (Figure 19.6).

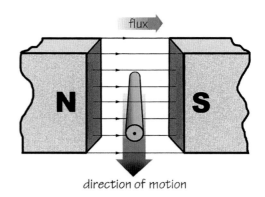

direction of motion

Figure 19.6

As the conductor is forced vertically downwards through the magnetic field, we say that it is *'cutting'* the lines of magnetic flux. The term 'cutting', in this context, isn't meant literally; it simply means that the conductor is *'moving across'* or *'passing through'* the magnetic field.

Because there is now relative motion between the conductor and the magnetic field, a potential difference is induced into that conductor.

For a conductor cutting the magnetic flux perpendicularly, the **magnitude** of this induced potential difference is given by the following equation:

$$E = Blv$$

where:

$E$ = potential difference (V)

$B$ = flux density (T)

$l$ = length of conductor (m)

$v$ = velocity of conductor (m/s)

**Important!** By 'length of conductor', what we *really* mean is that length of the conductor that lies *within the magnetic field – not* the entire length of the conductor!

Actually, this equation is *only* true when the conductor *cuts the flux at right angles*. For any other angle, the above equation needs to be modified as follows.

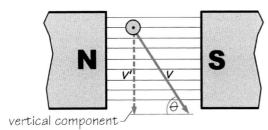

vertical component

conductor cutting flux at an angle θ

Figure 19.7

If the conductor cuts the flux at an angle, *θ* (pronounced *'theta'*) at a velocity represented by vector *v*, then we must find the perpendicular component of that velocity vector, *v'* – as represented by the broken line in Figure 19.7.

This can be determined from the sine ratio, as follows:

$$\sin \theta = \frac{\text{opposite}}{\text{hypotenuse}} = \frac{v'}{v}$$
$$v' = v \sin \theta$$

So we can now modify the original equation so that it can be used for conductors moving at *any* angle, *θ*, through the flux:

$$E = Blv \sin \theta$$

where:

$E$ = potential difference (V)

$B$ = flux density (T)

$l$ = length of conductor (m)

$v$ = velocity of conductor (m/s)

$\theta$ = angle cutting flux (°)

**Worked example 1**   A conductor moves through a permanent magnetic field of flux density 250 mT at a velocity of 20 m/s. If the length of conductor within the field is 175 mm, calculate the voltage induced into the conductor, if it cuts the flux (a) perpendicularly, and (b) at 60°.

**Solution**

$$E = Blv \sin \theta$$
$$= (250 \times 10^{-3}) \times (175 \times 10^{-3}) \times 20 \times \sin 90°$$
$$= 0.875 \text{ V (Answer a.)}$$

$$E = Blv \sin \theta$$
$$= (250 \times 10^{-3}) \times (175 \times 10^{-3}) \times 20 \times \sin 60°$$
$$= 0.875 \times 0.866$$
$$= 0.758 \text{ V (Answer b.)}$$

Of course, no potential difference will be induced in the conductor should that conductor be moved *parallel* to the lines of magnetic flux, because it will not be 'cutting' them.

## Fleming's Right-Hand Rule

The **direction** of this induced potential difference may be determined by using (for conventional flow) **Fleming's Right-Hand Rule** for 'generator action', which works as follows.

The *thumb, first finger (index finger)* and *second finger* of the right hand are held at right angles to each other, as shown in Figure 19.8.

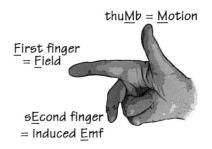

thu**M**b = **M**otion

**First** finger
= **F**ield

s**E**cond finger
= induced **E**mf

Figure 19.8

Then:

- the **thuMb** indicates the direction of **M**otion of the *conductor relative to the field*
- the **First finger** (index finger) indicates the direction of the magnetic **F**ield (i.e. north to south)
- the **sEcond finger** indicates the direction of the induced **E**.m.f. (Potential difference).

Note that, for Fleming's Right-Hand Rule to apply, the direction of motion *always* refers to the movement of the *conductor* relative to the magnetic field, *never* the other way around!

So if we apply **Fleming's Right-Hand Rule** to the downward-moving conductor, shown previously, you will find that the resulting induced potential difference will act *towards you* (i.e. *out* from the page). In other words, the nearer end of the conductor

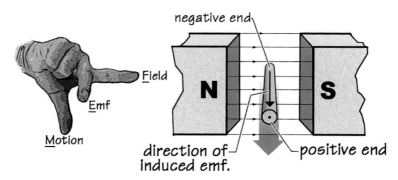

Figure 19.9

will be the positive end, and the far end will be the negative end.

Now, if this conductor is connected to a closed *external circuit*, the induced potential difference will cause a current to drift in the same direction as that induced potential difference – i.e. in this case, *towards* you as the reader, as indicated by the 'dot' convention shown in Figure 19.9.

> **Reminder**: For conventional current, a **dot** represents current drifting *towards* you, and a **cross** represents current drifting *away* from you.

As already pointed out in an earlier chapter, when we talk about the 'direction of current', we are *always* referring to the *direction of current through the load*; **never** *through the voltage source itself* – so, in this particular case, the current will be leaving the moving conductor from its positive (nearest) end, drifting through the load, and re-entering the conductor at its negative (far) end.

The *plan* (downward) view of this arrangement (shown in Figure 19.10) should help clarify this.

As you can see, with the downward-moving conductor acting as a 'voltage source' for the external load, the resulting load current drifts from positive to negative (while the current drift *within* that moving conductor is from negative to positive).

## Action and reaction: Lenz's Law

In the previous section, we learnt that if the conductor moving through a magnetic field is connected to a load, then the induced potential difference will cause a current to drift through any external circuit.

> This current is sometimes referred to as an *'induced current'*. This, however, is incorrect. It's the *potential difference* that is induced, *not* the resulting current! Current will *only* flow as a result of this induced potential difference, provided the moving conductor is connected to an external load.

This current, of course, is capable of expending energy in the external load – e.g. if the load were, say, a lamp,

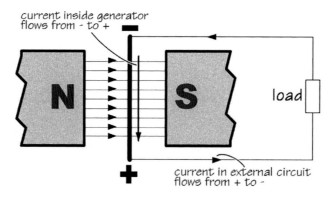

Figure 19.10

then the current would heat its filament. If the **Law of the Conservation of Energy** (*'energy can neither be created nor destroyed, but only changed from one form to another'*) is to be maintained, then this current can only expend energy *if work has been done to produce it in the first place!*

The *work expended* in moving the conductor through a magnetic field is the product of the **force** applied to that conductor and the **distance** through which it is moved. And, according to **Newton's Third Law**, *this force must be opposed by an equal and opposite force* called a '**reaction**' (*'for every force, there is an equal and opposite reaction'*).

And this reaction *can only come from a magnetic field which the current itself produces!* That is, the magnetic field set up around the conductor by the load current flowing through it.

This reasoning led the Russian physicist, **Emil Lenz** (1804–1865), to conclude that . . .

*The current resulting from the induced potential difference, due to the motion of a conductor through a magnetic field, must act in such a direction that its own magnetic field will then oppose the motion that is causing that induced potential difference.*

This statement is one interpretation of what is known as **Lenz's Law**, and we will meet other interpretations later.

You may find it necessary to re-read this statement through a couple of times, as it can be rather confusing the first time around!

In the meantime, let's see if we can make Lenz's Law more understandable, by considering it step by step.

Firstly, let's see if we can explain it in terms of **Fleming's Right-** and **Left-Hand Rules**.

Figure 19.11 shows the lower side of a continuous rectangular loop of wire being pushed downwards through a magnetic field.

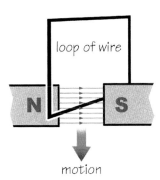

Figure 19.11

If you apply **Fleming's Right-Hand Rule** (for 'generator' action) to this conductor, you will see that the resulting induced potential difference acts towards you, causing a current to drift in the *same* direction – as illustrated in Figure 19.12.

If we now apply **Fleming's Left-Hand Rule** (for 'motor' action) to this current we see that the resulting force (i.e. the 'reaction') due to that current will act *upwards*. That is, it will react *against* the direction of downward motion (Figure 19.13).

In case you also found *this* explanation rather confusing, let's try yet another approach . . . This time, we'll ignore Fleming's Rules, and concentrate on the magnetic flux set up around the conductor due to the current drifting through it

In Figure 19.14, the conductor is forced downwards through the magnetic field, and a potential difference is induced into the conductor.

If the conductor is connected to an external load, then a current will flow in the direction shown (towards you).

Figure 19.12

Figure 19.13

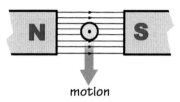

motion

Figure 19.14

In Figure 19.15, we see the magnetic field surrounding the conductor, caused by the current drifting through it. For clarity, we have *not* shown the permanent field.

field due to 'induced' current

Figure 19.15

In Figure 19.16, we show how the conductor's magnetic field *and* the permanent field combine, and react with each other to create an upward-acting *reaction* against the force responsible for the conductor's downward motion.

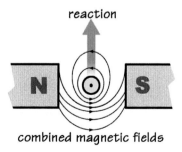

combined magnetic fields

Figure 19.16

If we extend the same logic (for every action there must be a reaction) to Faraday's magnet and coil experiment described earlier, then we must conclude that if we move the magnet *towards* the coil, then the direction of the induced potential difference must be such that the resulting current causes the polarity of the coil-end nearest the magnet to be the same as the nearest pole of the magnet – thus *opposing* (reacting against) the force moving the magnet *towards* the coil (like poles repel). And when we move the magnet *away* from the coil, the induced potential difference and resulting current reverses, causing the coil's polarity to *reverse* and oppose the force moving the magnet *away from* the coil (unlike poles attract) – as illustrated in Figure 19.17!

This **action-reaction** effect is a *very* important concept, and you will meet it again when you study both generators and motors. *So it is highly recommended that you re-read this section again – perhaps several times – so that you fully understand the important principle of Lenz's Law.*

## Self-inductance

In this section we are going to learn how, whenever the current drifting through a coil *changes in either magnitude or direction*, the resulting change in its magnetic field will induce a potential difference into that *same* coil, and the direction of this induced potential difference will *always* act *to oppose that change in current*.

As we now know, there are two *laws* which define the effect of changing the current drifting through a coil: **Faraday's Law** and **Lenz's Law**.

polarity of coil opposes
movement of magnet *towards* coil

polarity of coil opposes
movement of magnet *away* from coil

Figure 19.17

**Faraday's Law** tells us that *'the potential difference induced into a winding (coil)\* by a change in magnetic flux is proportional to the rate of change of that flux'.*

We can write this as:

$$u \propto \frac{\text{change in flux}}{\text{change in time}}, \quad \text{written as:} \quad u \propto \frac{\Delta \Phi}{\Delta t}$$

[\*Actually, this also applies to a single, straight, conductor. But its effect is practically insignificant compared with that of a coil.]

In the above expression, we use a lower-case '*u*' to represent a *changing*, as opposed to a continuous, value of induced voltage; and the Greek letter Δ ('delta') which mathematicians use to mean 'change in'. So the above expression simple means that *the value of the induced potential difference at any instant is proportional to the rate of change of flux at that same instant.*

**Lenz's Law**, tells us that *'the direction of an induced potential difference is such that it always opposes the change producing it'.* This is the 'action-reaction' process already described.

By *combining* these two laws, we can rewrite the above expression as:

$$u \propto -\frac{\Delta \Phi}{\Delta t}$$

The negative sign simply indicates the *sense*, or the *direction*, in which the induced potential difference is acting; it has nothing to do with the polarity of electric charge.

As always, to change a *proportional* sign into an *equals* sign, we must introduce a *constant*. For a coil, this constant is the **number of turns** *(N)* linked by the magnetic flux. So, the above expression may be rewritten as:

$$u = -N\frac{\Delta \Phi}{\Delta t} \qquad \text{—equation (1)}$$

where:

$u$ = potential difference (V)

$N$ = number of turns

$\Delta \phi$ = change in flux (Wb)

$\Delta t$ = change in time (s)

Now, if you refer back to the chapter on *magnetic circuits*, you will recall that

$$\text{flux density} = \frac{\text{flux}}{\text{area}} \quad i.e: \quad B = \frac{\Phi}{A} \quad \text{so} \quad \Phi = BA$$

For any given coil, its cross-sectional area, *A*, is fixed (a 'constant'), so if we substitute for Φ in equation (1), we have:

$$u = -NA\frac{\Delta B}{\Delta t} \qquad \text{—equation (2)}$$

If you again refer back to the chapter on *magnetic circuits*, you will recall that:

$$H = \frac{IN}{l} \text{ and } \mu_o \mu_r = \frac{B}{H} \quad \text{so} \quad B = \mu_o \mu_r H = \mu_o \mu_r \frac{IN}{l}$$

So, if we substitute for *B* in equation (2), we have:

$$u = -\left(\frac{NA\mu_o\mu_r N}{l}\right) \times \frac{\Delta I}{\Delta t} = -\left(\frac{N^2 A\mu_o\mu_r}{l}\right) \times \frac{\Delta I}{\Delta t}$$

—equation (3)

The entire expression appearing *inside the brackets* is known, simply, as '**self-inductance**' (symbol: L), and its SI unit of measurement is the **henry** (symbol: H). So, the final equation becomes:

$$u = -L\frac{\Delta I}{\Delta t}$$ —equation (4)

where:

$$u = \text{induced voltage (V)}$$
$$L = \text{inductance (H)}$$
$$\Delta I = \text{change in current (A)}$$
$$\Delta t = \text{change in time (s)}$$

From this equation, we should now be able to understand the significance of the definition of the **henry**:

> *A circuit has an inductance of one **henry** when a potential difference of **one volt** is induced into it when its current changes at a uniform rate of **one ampere per second**.*

Although you are not expected to remember this definition, you should be able to see how this relates to equation (4).

**Worked example 2**   The current drifting through a coil of self-inductance 0.2 H is increased from 0 A to 5 A in 0.01 s. What is the value of the potential difference induced into the coil?

**Solution**   In this example, the change in current can be determined from:

$$\Delta I = \left(I_{final} - I_{initial}\right) = (5-0) = 5\,\text{A}$$

Substituting for $\Delta I$ in equation (4):

$$u = -L\frac{\Delta I}{\Delta t} = -0.2 \times \frac{(5)}{0.01} = -100\,\text{V (Answer)}$$

Note that, in this worked example, the negative sign indicates that the induced potential difference is acting to **oppose** *the increase in current.*

**Worked example 3**   The current drifting through a coil of self-inductance 0.25 H is reduced from 5 A to 0 A in 0.01 s. What is the value of the potential difference induced into the coil?

**Solution**   In this example, the change in current can be determined from:

$$\Delta I = \left(I_{final} - I_{initial}\right) = (0-5) = -5\,A$$

Notice, here, that the current is getting smaller, hence the negative sign.

Substituting for $\delta I$ in equation (4):

$$u = -L\frac{\Delta I}{\Delta t} = -0.25 \times \frac{(-5)}{0.01} = -0.25 \times (-500)$$
$$= +125\,\text{V (Answer)}$$

This time, the positive sign indicates that the induced potential difference is acting in the *same* direction as the current – i.e. it is trying to *oppose its collapse* – or, to put in another way, it is acting to *maintain the current.*

## Factors affecting the self-inductance of a coil

The list of factors that affect the **self-inductance (L)** of a coil were developed in equation (3), above:

$$L = \frac{N^2 A \mu_o \mu_r}{l}$$ —equation (5)

where:

$$L = \text{self-inductance (H)}$$
$$N = \text{number of turns}$$
$$A = \text{area of coil (m}^2\text{)}$$
$$\mu_o = \text{permeability of free space}$$
$$\mu_r = \text{relative permeability}$$
$$l = \text{length of coil (m)}$$

Figure 19.18 summarises these relationships:

* *doubling* the **number of turns** will *quadruple* its inductance
* *doubling* its **cross-sectional area\*** will *double* its inductance
* *doubling* its **length** will *halve* its inductance
* inserting a 'soft' **ferromagnetic core** will increase its inductance by *hundreds* or even *thousands of times*

[\*this is the csa of the **coil**, *not* the conductor]

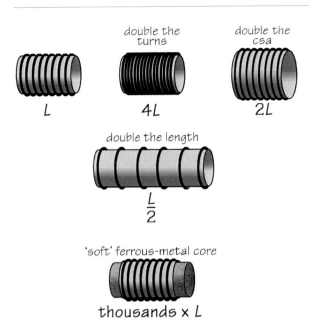

double the
turns

double the
csa

*L*

*4L*

*2L*

double the length

$\dfrac{L}{2}$

'soft' ferrous-metal core

**thousands × L**

Figure 19.18

You will recall, from the chapter on *magnetic circuits* that 'soft' ferrous metals (iron, silicon steel, etc.) will have a **relative permeability ($\mu_r$)** that is *thousands* of times greater than the **absolute permeability of free space** (i.e. that of a hollow coil), so inserting a ferromagnetic core inside the coil will have, *by far*, the greatest effect on increasing a coil's self-inductance.

In the previous chapter, we learnt that the absolute permeability *($\mu$)* of a magnetic material only remains (approximately) constant over the linear portion of its **B-H curve**. Accordingly, the inductance value of an iron-cored inductor applies only when that inductor is operating within the linear portion of its *B-H* curve, and the core of an inductor is always designed in such a way that its flux density never, for example, approaches its saturation level. Therefore, manufacturers of inductors normally specify the range of currents for which the specified inductance value applies.

Another way of expressing **self-inductance, *L*,** is as follows.

If you refer back to the chapter on *magnetic circuits*, you will recall that a magnetic circuit's reluctance is given by the equation:

$$R_m = \frac{l}{\mu_o \mu_r A}$$

(the equivalent of $R = \rho \dfrac{l}{A}$ for an electric circuit).

Most of the variables in the equation for reluctance also appear in the equation for inductance. In fact, if you separate $N^2$ from the rest of the equation, you will notice that the remaining part is equivalent to the *reciprocal* of reluctance:

$$L = \frac{N^2 A \mu_o \mu_r}{l}$$

which we can rewrite as

$$L = N^2 \left( \frac{A \mu_o \mu_r}{l} \right) = N^2 \times \frac{1}{R_m}$$

$$\boxed{L = \frac{N^2}{R_m}}$$

where:

$N$ = number of turns

$R_m$ = reluctance

**Worked example 4**   A ferromagnetic ring has a mean circumference of 3 m, a circular cross-sectional area of 2000 mm², and a relative permeability of 2000. If it is wound with a coil of 1000 turns, calculate (a) its reluctance, and (b) its inductance.

**Solution**

$$R_m = \frac{l}{\mu_o \mu_r A} = \frac{3}{(4\pi \times 10^{-7}) \times 2000 \times (2000 \times 10^{-6})}$$

$$= \frac{3}{5.03 \times 10^{-6}} = 597 \text{ kA/Wb (Answer a.)}$$

$$L = \frac{N^2}{R_m} = \frac{1000^2}{597 \times 10^3} = 1.675 \ H \text{ (Answer b.)}$$

## Behaviour of d.c. inductive circuits

Figure 19.19 represents a circuit of inductance ***L*** and resistance ***R***, supplied from a d.c. source via a two-way switch.

The circuit could represent an **inductor** of inductance, *L*, and negligible *resistance*, in series with a **resistor** of resistance, *R*. On the other hand, it could also be an equivalent circuit, representing an **inductor** having an *inductance, L,* and a *resistance R*.

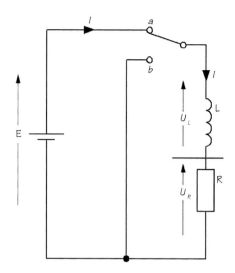

**Figure 19.19**

At the instant the switch is moved to position *(a)*, the external d.c. supply voltage, *E,* is suddenly applied to the inductive circuit, and a resulting load current *(i)* starts to increase in value. The maximum value *($I_{max}$)* this current can reach is limited by the resistance *(R)* of the circuit, where:

$$I_{max} = \frac{E}{R}$$

Now, if this were a purely resistive circuit, this current would reach its maximum value *instantaneously*. But this *cannot* happen in an inductive-resistive circuit, as explained by **Faraday's** and **Lenz's Laws**, which tell us that, due to its self-inductance, a changing current induces a potential difference *($u_L$)* into that circuit, the direction of which *will always oppose that change in current.*

So it's important to understand that, while the direction of the induced voltage is such that it always *opposes the growth in current*, it does **not** *prevent it from eventually reaching its maximum value.* As we shall see, the reverse is also true, whenever a current decays towards zero, the direction of the resulting induced voltage is always such that it *opposes that decay – in other words, it tries to sustain that current, but it does **not** prevent it from eventually reaching zero.*

> A circuit's **resistance** will determine the maximum current in a circuit; whereas its **self-inductance** will *reduce its rate of growth.*

The increasing load current initially follows the line **O-A**, as shown in Figure 19.20. If it *were* to continue to follow that line, then it would reach its maximum value (determined by the potential difference divided by resistance) in a period of time, called a **time constant** (symbol: *T*), measured in seconds, where:

$$T = \frac{L}{R}$$

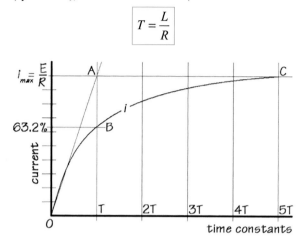

**Figure 19.20**

But, instead of continuing to follow line **O-A**, and reaching its maximum value in that time period it, instead, follows curve **O-C** and, so, actually only reaches approximately 63.2 per cent of its final value (position **B** on the graph).

The time it *actually* needs to reach its maximum value (point *y* on the curve) is almost exactly *five* time constants:

> time needed to reach maximum value $\simeq 5\dfrac{L}{R}$

Let's see where this curve comes from.

In Figure 19.21a., we have created a graph whose horizontal axis is divided into **five time constants**. As we learnt, above, if the initial current continued to rise at its intial rate *(a – b)*, it would reach its maximum value in one time constant *(T)*.

In Figure 19.21b, we start to construct the actual curve followed by the current. We start by choosing a random point, *c*, a short distance from a, along line *a – b*. Again, if the current were to continue to rise at its initial rate from point *c*, it would also reach its maximum value in one time constant. So, from point *c*, we construct a horizontal broken line, exactly one time constant *(T)* long and, from its end, a vertical broken line to point *d*. Next, we draw a straight line from point *c* to point *d*.

(Fig a.)

(Fig b.)

(Fig c.)

**Figure 19.21**

In Figure 19.21c. we repeat the process by choosing point *e*, a little way along line *c – d*, and constructing the line *e – f*. This process is then repeated, until the shape of the final curve emerges.

Although beyond the scope of this book, the shape of the curve can also be derived, mathematically, using calculus. The resulting equation will show us that the shape of the curve is '**exponential**' which, in this case, simply means that *the rate of change of current decreases as the value of current increases*.

To summarise, then. For an inductive-resistive circuit suddenly connected to a d.c. voltage, the current *cannot* rise to its maximum value instantaneously but, instead, *will always follow the curve shown above*. Once the current has eventually reached its maximum value, it becomes constant, so is no longer opposed by the inductive-effect of the circuit, and will remain at this value.

The *exact* value of current at *any* point along this graph line can be determined using calculus – which is beyond the scope of this text. So, you should bear in mind that the figures of '**63.2%**' and '**5 × time constants**' used in this chapter are very *good approximations* of the actual values derived by using calculus.

**Worked example 5**   A coil which has an inductance of 16 mH and a resistance of 4 Ω is suddenly connected across a 100-V d.c. supply. What will be the

a   maximum value that the current will eventually reach
b   coil's time constant.
c   current reached in that amount of time
d   time taken for the current to reach its maximum value?

**Solution**

a   $I_{max} = \dfrac{E}{R} = \dfrac{100}{4} = 25$ A (Answer a.)

b   $T = \dfrac{L}{R} = \dfrac{16 \times 10^{-3}}{4} = 4 \times 10^{-3}$ s or 4 ms (Answer b.)

c   $i \approx 0.632\, I_{max} \approx 0.632 \times 25 \approx 15.8$ A (Answer c.)

d   time to reach $I_{max} \approx 5T \approx 5 \times 4 \approx 20$ ms (Answer d.)

Now, *with a constant current flowing through the circuit*, Figure 19.22 shows what happens if we suddenly move the switch from position *(a)* to position *(b)*.

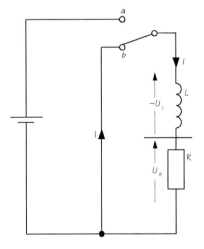

**Figure 19.22**

With the switch in position *(b)*, the external voltage supply is removed and the inductive circuit is short-circuited, so the current will start to collapse although its *direction* will remain the same.

If this were a purely resistive circuit, then the current would simply collapse to zero instantaneously. However, the collapsing current induces a voltage ($-u_L$) which acts to *oppose* that collapse of current or, to put it another way, *tries to maintain that current.*

Once again, it's important to understand that the induced voltage does not *prevent* the current from collapsing to zero but, rather, it acts to prevent it reaching zero *instantaneously,* by acting to *try to maintain the current!* So the collapse of current follows the curve shown in Figure 19.23. As you will notice, this curve has *exactly* the same shape as before, except that it is now inverted.

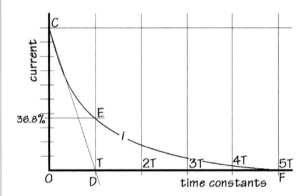

**Figure 19.23**

Initially, the current follows the line **C-D** and, if it continued to follow that line, it would reach zero in a period (**O-D**) called the time constant, **T**, measured in seconds, where:

$$T = \frac{L}{R}$$

But, instead of reaching zero in that time period, it *actually* follows the curve **C-F** and only falls by approximately 63.2 per cent of its initial value (position **E** on the graph), reaching **36.8 per cent** of $i_{max}$.

And the time it actually takes to reach zero (point **F** on the graph) is, again, approximately *five* time constants:

$$\text{time needed to reach zero} \approx 5\frac{L}{R}$$

### Inductance and arcing

For circuits, such as the one in Figure 19.24, when the switch is closed onto a highly inductive load, the resulting current will, of course, increase to its maximum value, following the curve described above.

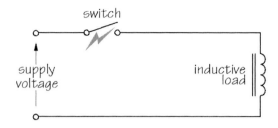

**Figure 19.24**

However, when the switch is opened, and the current starts to collapse, an **arc** can occur across the gap between the switch's contacts as the load's self-induced voltage tries to maintain the current across that gap.

The temperature of an electric arc is very high and may actually cause the switch contacts to partially melt. For this reason, *it can be hazardous to break the supply to any load that has a substantial amount of self-inductance* – such as the field winding of a d.c. machine.

There are various ways of preventing, or at least minimising, this arcing problem. One such method is to use a special type of 'make-before-break' switch which, when operated, temporarily connects a **discharge resistor** in parallel with the inductive load before it actually breaks the supply to that load. This discharge resistor then provides a route for the collapsing current to pass through, rather than by arcing across the switch contacts.

## Energy stored in a magnetic field

When a current drifts through a purely inductive circuit (i.e. a circuit without any resistance) increasing in magnitude, it does **work** by *expanding*, and storing energy in, the magnetic field. You will remember from the chapter on *magnetism* that lines of magnetic flux are considered to have an 'elastic' property, and work must be done to *stretch* them.

The energy, thus stored in the magnetic field, is later returned to the circuit when the magnetic field collapses as it attempts to maintain current drift.

An analogy can be made, here, to winding up the rubber-band 'motor' of a model aeroplane. The energy thus stored in the twisted rubber band, when released, will then drive the propellor until all that stored energy is released.

The expression for the work done in storing energy in an inductive circuit's magnetic field is:

$$W = \frac{1}{2} L I_{max}^2$$

where:

$W$ = work (J)

$L$ = inductance (H)

$I_{max}$ = steady-state current (A)

**Worked example 6**   How much energy will be stored in the magnetic field of a coil of self-inductance 0.05 H and resistance 3 Ω, when connected to a 12-V d.c. supply?

**Solution**   We first need to find out the steady-state current $(I_{max})$ that will drift in the coil:

$$I_{max} = \frac{E}{R} = \frac{12}{3} = 4\,A$$

Now, we can determine the energy stored:

$$W = \frac{1}{2} L I_{max}^2 = \frac{1}{2} \times 0.05 \times 5^2$$
$$= 0.625\,J\ or\ 625\,mJ\ (Answer)$$

## Mutual inductance

**Mutual inductance** (symbol: $M$) exists between *two* separate self-inductive circuits when a change in the flux set up by one circuit induces a potential difference into the second circuit – just like Faraday's experiment with the coil and electromagnet, explained towards the beginning of this chapter.

Although it is common to use the term 'inductance', when we strictly mean 'self-inductance', we should *never* use 'inductance' whenever we mean 'mutual inductance'.

In Figure 19.25, the self-inductive circuit connected, via a switch, to the battery is called the **primary** circuit, and the self-inductive circuit connected to the resistive load is called the **secondary** circuit.

In Figure 19.25(a), when the switch is closed, a current drifts through the primary winding in the direction shown. As we have already learnt, this current *cannot* rise *instantaneously* to its maximum value, due to the **self-inductance** of the primary winding. However, as the current increases towards its maximum value (determined by the primary circuit's resistance) it creates, in this example, a correspondingly expanding magnetic flux in a *clockwise* direction around the core. You can confirm the direction of this flux by applying the '**right-hand grip rule**' to the primary winding (coil).

Most of this changing flux will be contained within the core, and link with the secondary winding, inducing a secondary potential difference into that winding by **mutual induction**.

Since the secondary circuit is connected to a load, the secondary induced potential difference will cause a current to drift through the secondary circuit. In accordance with Lenz's Law, the direction of this secondary current must create a secondary flux which will act *counterclockwise* around the core – in order to *oppose the increase in the primary flux*.

In Figure 19.25(b), the switch is opened and the primary flux starts to collapse (remember, the primary current is collapsing, *not reversing direction*, so its flux still acts *clockwise* around the core). This collapse in primary flux again induces a potential difference into the secondary winding, by **mutual induction** – this time, in the *opposite* direction to the originally induced secondary potential difference. The resulting reverse in the secondary current now causes the secondary flux to reverse, which will now act clockwise around the core, trying *to sustain the primary flux*.

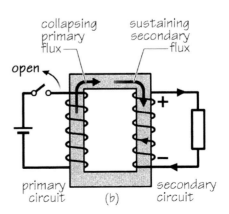

**Figure 19.25**

It's very important to understand that the behaviour we are describing *only* occurs for *changing* currents. When a current in the primary circuit becomes steady or constant, the associated magnetic flux also becomes steady, and cannot induce a potential difference into the secondary circuit.

We say that the primary and secondary circuits, in the above example, are 'coupled by **mutual induction**' (symbol: **M**) which, just like self-inductance, is measured in **henrys**.

The value of the potential difference induced into the secondary circuit is given by the expression:

$$e_s = -M \frac{\Delta i_p}{\Delta t}$$

. . . where $\frac{\Delta i_p}{\Delta t}$ means the *rate of change* in the primary current, and the *negative* sign indicating that, by **Lenz's Law**, the potential difference induced into the secondary circuit acts in a direction that will *oppose* the change in primary current.

From the above expression, we can say that '*two circuits will have a mutual inductance of one henry if a uniform change in current of one ampere per second in one circuit causes a potential difference of one volt to be induced into the other*'.

**Worked example 7** Two self-inductive circuits, *A* and *B*, have a mutual inductance of 0.05 H. If the current in circuit *A* increases from zero to 10 A in 0.2 s, what will be the value and direction of the potential difference *(u_B)* induced into circuit *B*?

**Solution** In this example the change in current in circuit *A* may be found from:

$$\Delta I = i_{initial} - i_{final} = 10 - 0 = 10 \text{ A}$$

so, $u_B = -M \dfrac{\Delta I}{\Delta t}$

$$= -0.05 \times \frac{10}{0.2} = -0.05 \times 50 = -2.5 \text{ V (Answer)}$$

Once again, the negative sign indicates that the direction of the induced voltage is acting to oppose the change in current in the circuit *A*.

## Expression for mutual inductance

For two circuits, having self-inductances of $L_1$ and $L_2$ respectively, and which are linked by *all* of the flux created by the current in the primary circuit, it can be shown that the **mutual inductance**, *M*, is given by:

$$M = \sqrt{L_1 L_2}$$

## Coupled circuits

In practice, not *all* the flux created by the primary current will actually link with the secondary circuit. That flux which does *not* link with the secondary circuit is called **leakage flux**, or is simply referred to as '*leakage*'.

If two self-inductive circuits (such as a pair of coils), are placed so that there is mutual induction between them, then we say that the two circuits are magnetically **coupled**.

The magnitude of the circuits' mutual inductance depends upon their 'magnetic closeness'. By this, we *don't* mean how physically close the two coils are but, rather, the degree by which the magnetic flux is able to link the two circuits, or *how closely the circuits are magnetically 'coupled'*:

- If the circuits are linked by *only part* of the flux, then they are said to be **loosely coupled**.
- If the circuits are linked by *most* of the flux, then they are said to be **closely coupled**.

In Figure 19.26, the pair of coils to the left are 'loosely coupled' whereas the pair of coils to the right, while being the same *physical* distance apart, are 'closely coupled' – thanks to the iron core that links them.

The degree of **looseness** or **closeness** of coupling is represented by a number called the '**coupling coefficient**' or '**coefficient of coupling**' (symbol: *k*), the value of which depends upon the *degree of coupling*.

For two circuits that are *perfectly coupled* (no leakage), the coupling coefficient is **unity** (1); for two circuits that are so separated that there *is no mutual inductance between them whatsoever*, the coupling coefficient is **zero**.

So the **mutual induction** between the two self-inductive circuits, expressed above, must be modified, as follows:

$$M = k\sqrt{L_1 L_2}$$

where:

$M$ = mutual inductance (H)

$k$ = coupling coefficient

$L_1$ = primary circuit self-inductance (H)

$L_2$ = secondary circuit self-inductance (H)

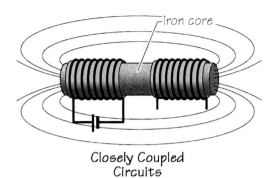

Loosely Coupled Circuits                   Closely Coupled Circuits

Figure 19.26

**Worked example 8**   Two circuits, of 8 mH and 16 mH respectively, are closely coupled with a coupling coefficient, $k = 0.75$. What is the mutual inductance between the two circuits?

**Solution**

$$M = k\sqrt{L_1 L_2}$$
$$= 0.75\sqrt{8 \times 16}$$
$$= 0.75\sqrt{128} = 0.75 \times 11.31 = 8.48 \text{ mH (Answer)}$$

**Worked example 9**   What is the coupling coefficient between two circuits, having self-inductances of 240 μH and 360 μH, if their mutual inductance is 250 μH?

**Solution**

$$M = k\sqrt{L_1 L_2}$$
$$\text{so, } k = \frac{M}{\sqrt{L_1 L_2}}$$
$$= \frac{250}{\sqrt{240 \times 360}} = \frac{250}{\sqrt{86\,400}} = \frac{250}{294} = 0.85 \text{ (Answer)}$$

## Practical examples of mutually inductive circuits

Two practical examples of mutually-inductive circuits are a motor vehicle's **ignition coil**, and the **transformer**.

The **ignition coil** is the term applied to an autotransformer used to step-up the 12-V d.c. battery supply to the thousands of volts required by the vehicle's spark plugs. In order to provide the changing-current in the primary coil, necessary to induce the high-voltage into the secondary coil (which has many more turns than the primary coil), the primary circuit is continually switched on and off by a rotary switch, mechanically-driven by the vehicle's engine or, these days, by electronic switching.

**Transformers** are used to step up, or to step down, voltages. However, unlike the ignition coil, there is no requirement to continuously switch the primary circuit on and off, because a transformer is an **alternating-current** device, whose primary current is continually changing in value.

## Inductors

*All* circuits have some degree of *natural* self-inductance – even a single conductor has some self-inductance, albeit extremely low compared to that of a coil.

However, we can *modify* a circuit's natural self-inductance using circuit components called **inductors** – in just the same way in which we can modify a circuit's resistance by using resistors.

Inductors have a great many applications. At one end of the scale, they are used in low-power electronic circuits while, at the other end of the scale, they are used with the 400-kV supergrid transmission system. So they come in an enormous range of shapes, and in sizes measuring from a few millimetres to several metres.

The terms '**reactor**', '**choke**' and '**ballast**' are often used instead of 'inductor' but, strictly speaking, these terms describe the *application* of inductors, rather than the devices themselves.

- '**Reactors**' are inductors used, for example, in electrical transmission systems to limit the magnitude of switching and fault currents.

- The term 'ballast' is applicable to resistors as well as to inductors, and describes their use in limiting the rise of alternating currents – e.g. to limit the current flow through a fluorescent lamp once its gas has been ionised.
- 'Chokes' are inductors used to limit the flow of alternating currents while allowing the passage of direct currents, or to block ('filter out') high frequency currents (e.g. telecommunication signals) from low frequency currents (e.g. mains current).

Despite their differences in size, inductors are essentially *all* constructed in very much the same way, consisting of a **coil** of insulated wire wound around a 'soft' ferromagnetic laminated **core**, such as silicon steel.

Like resistors, inductors can be connected in series, parallel, series-parallel and complex. In this section, we will only examine inductors that are connected in **series** and in **parallel**. You will notice that inductors are handled in exactly the same way as resistors, as we shall describe below.

## Inductors in series

Figure 19.27

Ignoring any *mutual* inductance that may exist, the **equivalent self-inductance** for series self-inductance is given by:

$$L = L_1 + L_2 + L_3 + etc.$$

**Worked example 10** What is the total self-inductance if four inductors, of self-inductance 2 mH, 3 mH, 4 mH and 5 mH, are connected in series.

**Solution**

Figure 19.28

$$L = L_1 + L_2 + L_3 + L_4 = 2+3+4+5 = 14 \text{ mH (Answer)}$$

## Inductors in parallel

Figure 19.29

Ignoring any *mutual* inductance that may exist, **the equivalent self-inductance** for parallel self-inductance is given by:

$$\frac{1}{L} = \frac{1}{L_1} + \frac{1}{L_2} + \frac{1}{L_3} + etc.$$

**Worked example 11** What is the total self-inductance if three inductors of 6 mH, 18 mH and 36 mH, are connected in parallel (Figure 19.30)?

Figure 19.30

**Solution**

$$\frac{1}{L} = \frac{1}{L_1} + \frac{1}{L_2} + \frac{1}{L_3}$$

$$\frac{1}{L} = \frac{1}{6} + \frac{1}{18} + \frac{1}{36} = \frac{6+2+1}{36} = \frac{9}{36}$$

$$L = \frac{36}{9} = 4 \text{ mH (Answer)}$$

## Unintentional coupling between adjacent inductors

In each of the examples above, we specified 'ignoring any mutual induction that might exist'. But what happens if we can't ignore it?

If adjacent inductors are located too close together in the same circuit, then the magnetic field of one inductor can affect the other and *vice versa*. In other words, mutual inductance can occur *unintentionally* between pairs of inductors if they are located too close together.

We call this effect '**cumulative**' or '**differential**' coupling.

For a pair of inductors connected in series, the effect of this mutual inductance can be to either increase or to decrease the total inductance of the circuit – depending on how they are wound relative to each other.

For example, suppose two inductors, $L_1$ and $L_2$, are connected in series close enough, and in such a way that their fields *reinforce* each other ('cumulatively coupled'), as shown in Figure 19.31, then their total inductance will be given by the equation:

$$L = L_1 + L_2 + 2M$$

If, on the other hand, the two inductors are connected in series close enough, and in such a way that their fields *oppose* each other ('differentially coupled'), as shown in Figure 19.32, then their total inductance will be given by the equation:

$$L = L_1 + L_2 - 2M$$

For each of these equations, the mutual inductance, $M$, is given by:

$$M = k\sqrt{L_1 L_2}$$

. . . where $k$ is the coefficient of coupling.

**Worked example 12**   Two inductors, of inductance 20 mH and 30 mH are connected in series and close enough to each other such that they are cumulatively coupled. If the coefficient of coupling is 0.25, what is the total inductance of the circuit?

**Solution**

$$L = L_1 + L_2 + 2M = L_1 + L_2 + 2k\sqrt{L_1 L_2}$$
$$= 20 + 30 + 2 \times 0.25\sqrt{20 \times 30}$$
$$= 50 + 0.5\sqrt{600}$$
$$\simeq 50 + 12.25$$
$$\simeq 62.25 \text{ mH (Answer)}$$

Figure 19.31

Figure 19.32

**Worked example 13**   The same two inductors are connected in series with each other such that they are differentially coupled, with the same coefficient of coupling. What will be the total inductance of the circuit?

**Solution**

$$L = L_1 + L_2 - 2M = L_1 + L_2 - 2k\sqrt{L_1 L_2}$$
$$= 20 + 30 - 2 \times 0.25\sqrt{20 \times 30}$$
$$= 50 - 0.5\sqrt{600}$$
$$\simeq 50 - 12.25$$
$$\simeq 37.75 \text{ mH (Answer)}$$

The worst case scenario for this condition is for the mutual induction between a pair of inductors to either *double* the series inductance of the inductors or to completely *cancel* the series inductance of the two inductors!

## Review your learning

Now that we've completed this chapter, we need to examine the **objectives** listed at its start. Placing '*Can I...*' at the beginning, and a question mark at the end, of each objective turns that objective into a test item. If we can answer each of those **test items**, then we've met the objectives of this chapter.

**Online resources**

The companion website to this book contains further resources relating to this chapter. The website can be accessed via the following link:

www.routledge.com/cw/waygood

# Chapter 20

# Motor principle

## Objectives

On completion of this chapter, you should be able to

1. apply Fleming's Left- and Right-Hand Rules.
2. apply the 'Corkscrew Rule' to determine the direction of a magnetic field surrounding a conductor.
3. specify the factors that affect the magnitude of the force acting on a conductor placed within a magnetic field.
4. specify the factors that affect the voltage induced into a conductor as it moves within a magnetic field.
5. identify the main components of an electrodynamic loudspeaker, and specify the function of each.
6. explain the principle of operation of an electrodynamic loudspeaker.
7. identify the main components of a d.c. motor's stator and rotor.
8. explain the principle of operation of a simple, single-loop, d.c. motor.
9. explain the need for a d.c. motor's commutator system.
10. outline the principle of action/reaction of a d.c. motor under load.
11. identify a shunt-wound and a series-wound d.c. motor.
12. describe the difference between the output torque of a shunt- and series-wound d.c. motor.
13. explain why the starting current of a d.c. motor can be high, and how it may be reduced.

## First, a reminder

### Fleming's Hand Rules

Whenever we use *conventional* current flow, **Fleming's Left-Hand Rule** (Figure 20.1) is used for *'motor action'*, while **Fleming's Right-Hand Rule** (Figure 20.2) is used for *'generator action'*. Each rule relates the relationships between the direction of a magnetic field, the direction of motion of a conductor within that field and the *conventional* direction of the current (and corresponding voltage).

Fleming's Left-Hand Rule

**Figure 20.1**

Fleming's Right-Hand Rule

**Figure 20.2**

When a current-carrying conductor is placed in a magnetic field, the conductor is subject to a **force** that acts to move the conductor out of the field.

The *direction of the magnetic field* (north to south) is represented by the *first finger*.

The direction of the conventional *current* is represented by the *second finger*.

When a conductor is moved through a magnetic field, a **voltage** is induced into the conductor. If this conductor forms part of a closed loop, then a current will flow in the same direction as the induced voltage.

The *direction of the magnetic field* (north to south) is represented by the first finger.

An Introduction to Electrical Science, Waygood, ISBN 9780815391821, 2019. © Taylor & Francis

| The direction of the *motion*, caused by the resulting force, is then represented by the *thumb*. | The direction of the *motion* of the conductor through the field is then represented by the *thumb*.<br><br>The direction of the *induced e.m.f.* (and any resulting *current*) is represented by the *second finger*. |
|---|---|

## The Corkscrew Rule

Figure 20.3

The direction of the magnetic flux surrounding a current-carrying conductor can be determined using the '**Corkscrew Rule**' – as illustrated in Figure 20.3.

## D.C. motor fundamentals

Before proceeding, it is necessary to remind ourselves that, throughout this chapter, we will be assuming 'conventional current flow' – i.e. *current direction, through the load, from positive to negative*. So, **Fleming's Left-** and **Right-Hand Rules**, together with the **Corkscrew Rule**, assume conventional flow.

## Motor action

In the chapter on *Electromagnetism*, we learnt that a current-carrying conductor placed between the poles of a magnet will *experience a force at right-angles to the field's magnetic flux*. This force, which is known as the '*Lorentz Force*', acts to push the conductor out of the magnetic field.

The **magnitude** of the Lorentz Force is given by the following equation:

$$F = BIl$$

where:
$F$ = force (in newtons)
$B$ = flux density (in newtons)
$I$ = current (in amperes)
$l$ = length of conductor in field (in meters)

For example, let's assume for a flux density of 0.25 T, a current of 10 A, and a conductor length, within the field, of 150 mm. The resulting force will be:

$$F = BIl = 0.25 \times 10 \times \left(150 \times 10^{-3}\right) = 0.375\,\text{N (Answer)}$$

To put this amount of force in terms of something we can understand, it's worth realising that a typical apple weighs between 1.0–1.5 N.

The **direction** of this force is due to the interaction of the permanent magnetic field, and the magnetic field surrounding the conductor, and may be determined by applying **Fleming's Left-Hand Rule** (for conventional-current flow). For example, in Figure 20.4, the field direction is north to south, and the current is flowing away from us, so the direction of resulting force will be *downwards*.

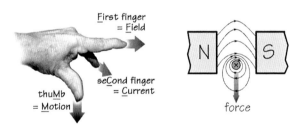

Fleming's Left-Hand Rule determines Direction of Force

Figure 20.4

## Generator action

The English physicist Michael Faraday (1791–1867) showed that the *opposite* of 'motor action' also occurs. That is, moving a conductor through a magnetic field (or moving the magnetic field relative to the conductor) will induce a voltage into the conductor.

The direction in which this voltage acts can be determined by applying **Fleming's Right-Hand Rule** – as illustrated in Figure 20.5. If the conductor forms part of a closed circuit, then this induced voltage will cause a current to flow in the same direction.

The **magnitude** of this induced voltage, when the conductor is moved perpendicularly through a magnetic field, is given by the following equation:

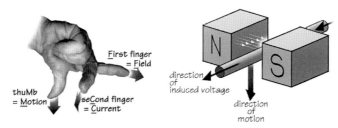

**Fleming's Right-Hand Rule determines Direction of Induced E.M.F.**

Figure 20.5

$$U = Blv$$

where:
$U$ = induced voltage (in volts)
$B$ = flux density (in teslas)
$l$ = length of conductor in field (in meters)
$v$ = velocity of conductor (in meters per second)

For example, let's assume a conductor of length 150 mm within a magnetic field of flux density of 0.25 T, is moved perpendicularly at a velocity of, say, 5 m/s. The value of the induced voltage will be:

$$U = Blv = 0.25 \times \left(150 \times 10^{-3}\right) \times 5 = 0.19\,\text{V (Answer)}$$

### Comparing 'generator action' with 'motor action'

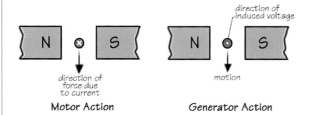

Motor Action          Generator Action

Figure 20.6

If we now compare 'generator action' with 'motor action' (Figure 20.6), we will notice that the voltage *(U)* induced into the conductor by 'generator action' acts in the *opposite* direction to the voltage *(E)* that drives the current that causes 'motor action'. For this reason, we call the voltage induced into the conductor by generator action a '**back-e.m.f.**'.

To summarise, then, we can say that **whenever there is 'motor action', there is** *always* **'generator action' too**, and **the generated voltage** (i.e. the **back-e.m.f.**) *always opposes* **the voltage that causes the motor action to take place in the first place**.

It is very important to understand this relationship, because this is the reason for the '**action–reaction**' effect that takes place in a motor, which we will describe later in this chapter.

## Motor action

**Electric motors** are so common that we hardly give them a second's thought; well, unless they stop working, that is! But if we were to carry out an inventory of the various electric motors we have, say, in our car, we would probably be very surprised to discover how many there are. For even the most basic vehicle, we rely on electric motors to start the engine, cool the radiator, run the ventilation system, operate the windscreen wipers, wind the windows up and down, and so on. The more luxurious the vehicle, the more electric motors there tend to be – for example, it may have motors which adjust the position of its seating. And, of course, we are now entering the era of electrically powered vehicles, in which the electric motor is actually replacing the internal-combustion engine!

Until the early nineteenth century, **electricity** and **magnetism** had been considered to be two completely separate phenomena. But, in 1820, while preparing an experiment for his students, the Danish physicist Hans Christian Ørsted (1777–1851) noticed that whenever he passed an electric current through a wire, it caused the needle of a nearby compass to deflect – proving that there was, indeed, a link between the two.

The publication of Ørsted's discovery triggered a great deal of further research and development by numerous other physicists, engineers and inventors, worldwide, all of whom realised that *if an electric current could produce a force* then maybe, just maybe, *this force could then be used to produce useful motion.*

One of these physicists was Michael Faraday (1791–1867), and he was able to create a laboratory demonstration which, for the first time, demonstrated the extremely important concept of **electromagnetic rotation**.

An illustration of Faraday's experimental apparatus is shown in Figure 20.7. It consists of a squat glass tube, containing some liquid mercury surrounding a vertical bar magnet. The free end of a straight wire or rod, suspended from a hook in the cork stopper, is immersed in the mercury and, when a current was passed along the rod from the hook into the mercury (which, of course, is a conductor), the resulting electromagnetic field set up around that rod reacted with that of the permanent magnet, causing it to continuously rotate around the magnet.

(You might care to figure out, using the 'Corkscrew Rule', *why* the rod should rotate around the magnet.)

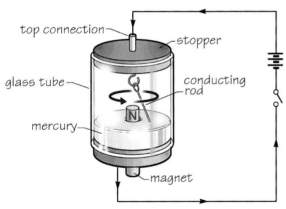

**Faraday's Demonstration of Motor Action**

Figure 20.7

Although this experiment certainly demonstrated that rotational motion was possible, Faraday's device was clearly of no practical use whatsoever. It was simply a laboratory demonstration of the results of Faraday's theoretical research, and any torque it might have produced would have been quite incapable of driving any type of mechanical load. Nevertheless, it was a very important milestone in the quest to create a machine capable of converting electrical energy into useful, rotational motion.

During the years that followed the publication of the discoveries made by Ørsted, Faraday and others, inventors from all over the world began working quite

independently from each other to produce a practical '**electromagnetic engine**' (as 'electric motors' were known in those days) capable of producing sufficient force or torque to do useful work by driving a mechanical load. It might come as a surprise to learn that many of these early electromagnetic engines worked on the principle of *linear* rather than of *rotational* movement, and were obviously designed to emulate the mechanisms of reciprocating steam engines, which they were hoping to replace. Steam engines, of course, used a cylinder and piston to produce linear motion in order to drive a flywheel which, in turn, converted that linear motion into rotational motion. It was some time, however, before successful directly rotating electric motors were designed.

Whether linear or rotational, the biggest problem was to design a machine capable of producing sufficient output force or torque to drive mechanical loads and, probably, the most important step forward in this direction was the invention of the **electromagnet** by William Sturgeon (1783–1850) in 1825. Prior to this, only relatively weak permanent magnets were available to provide the magnetic field necessary to produce motion. Electromagnets were capable of producing a significantly higher level of flux density than any permanent magnet then available. This accelerated the development of practical motors capable of producing sufficient torque to drive substantial mechanical loads.

## Electrodynamic loudspeakers

As already mentioned, '**motor action**' doesn't necessarily describe *rotational* movement. One very familiar device that uses *linear* motor action is the '**moving-coil**' or '**electrodynamic**' loudspeaker – the most-common type of loudspeaker.

**Electrodynamic Loudspeaker**

Figure 20.8

Figure 20.8 shows the main components of an electrodynamic loudspeaker. The 'heart' of the loudspeaker is a powerful permanent '**E-section**' (see Figure 20.9) or '**pot magnet**', consisting of a cylindrical, or pot-shaped, magnet surrounding a solid core. The cylinder and solid core are magnetised in such a way as to create opposite polarities, thus establishing a radial magnetic field across the air gap within the pot magnet.

The moving coil, or '**voice coil**' as it is known, consists of a very light tube-shaped former, around which is wound a coil of fine insulated wire. This voice coil is centred, and free to move, within the radial magnetic field inside the pot magnet.

This combination of pot magnet and voice coil is what creates the 'linear electric motor', which drives the loudspeaker.

The voice coil is bonded to a '**diaphragm**': a light cone, manufactured from stiffened cardboard, or some other rigid proprietary material such as carbon fibre. The loudspeaker's components are all supported by means of a non-magnetic, cone-shaped, metal 'exoskeleton' (not shown), with the diaphragm suspended in place by means of a 'circular concertina' arrangement around its outer circumference which allows it to freely 'pump' forward and backwards.

The loudspeaker's principle of operation is simple, as illustrated in Figure 20.9.

A signal voltage is fed from the output of an audio amplifier on to the voice coil. Let's assume, as shown in Figure 20.9, that at a particular moment, the resulting (conventional) current is flowing around the coil such that it is travelling *away* from us over the top of the voice coil and, therefore, *towards* us from under the bottom. If we now apply **Fleming's Left-Hand Rule** to both the upper and lower turns of the voice coil, we will see that the resulting force acts to push the voice coil towards the right. If the current direction then reverses direction, the voice coil would, instead, be drawn towards the left.

The *direction of movement* of the voice coil depends, then, upon the *direction of the current*, and the *amount of thrust* applied to the voice coil depends upon the *magnitude of the current*.

This linear, or 'pumping', movement of the voice coil, therefore, mechanically replicates the direction/magnitude of the audio signal supplied from the amplifier, and this is transferred into the air by means of the attached diaphragm. The resulting sound then reaches our ears due to the resulting compression/rarefaction of the air caused by the moving diaphragm.

Electrodynamic loudspeakers are widely used in many audio applications, including high-fidelity sound systems, and they vary in diameter according to the volume of air they are required to move. High-frequency loudspeakers, called '**tweeters**', have relatively small diameters and short linear movements, whereas low-frequency loudspeakers, called '**woofers**', tend to have relatively large diameters and longer linear movements. Tweeters and woofers are normally mounted adjacent to each other within the same cabinet, but are individually supplied from an electronic 'cross-over network', located inside the speaker cabinet, which separates the amplifier's output audio signal into higher and lower frequency signals.

Not *all* loudspeakers work on the principle described above. Some types use completely different operating principles entirely, including those based on **piezoelectricity** (usually tweeters) or **electrostatics** (normally full-range speakers).

**Piezoelectric tweeters** work on the principle that certain crystal materials vibrate in sympathy with signal voltages applied across opposite faces.

The **electrostatic loudspeaker**, which first appeared as a consumer product in the 1950s, was designed by Peter Walker, founder of the British hi-fi manufacturer, *Quad Electroacoustics*, and is a high-end type that works on the principle of attraction/repulsion forces applied to large, flat, conductive-coated plastic diaphragms due to high-voltage electrostatic charges. Electrostatic loudspeakers, which require a power-supply to create the required high voltages, tend to be significantly larger than electrodynamic types, are far more expensive and are, it is argued, capable of far more accurate sound reproduction than electrodynamic loudspeakers.

## D.C. electric motor

Now let's turn our attention to the production of *rotational motion* which, of course, is what we usually

Figure 20.10

Figure 20.12

think of whenever we hear the term '**motor**' or '**motor action**'.

In this section, we'll learn how a d.c. motor works. A.C. motors are beyond the scope of this book, but are covered in detail in the companion book *Electrical Science for Technicians*.

Before we consider its principle of operation, let's take a look at an exploded view of a typical d.c. motor (Figure 20.10).

In common with *all* types of motor and generator, the various components of a d.c. motor can be collected together under *two* main headings: the **stator** and the **rotor**, as illustrated in Figure 20.11.

Figure 20.11

The **stator**, as the name suggests, comprises all the *stationary* components of the motor, including: the **yoke** (the cylindrical housing), the **pole pieces** and their **field windings**, **bearings** and the **brushes**. In the example illustrated, there are four pole pieces, but they can vary in number from two, four, six, etc.

The **rotor** comprises all the *rotating* parts of the motor, including: the **drive shaft**, **armature**, **armature windings**, **commutator** and **impeller fan**. We won't be discussing the impeller fan in this chapter, but its purpose is to draw air through the motor in order to prevent it from overheating.

The **yoke**, **pole pieces**, **field windings** and **armature** (a laminated-iron cylinder which supports the armature windings) form the motor's

**magnetic circuit**, as illustrated in Figure 20.12. D.C. motors can have two, four or more (even numbers) of poles. The function of the low-reluctance magnetic circuit is to guide the magnetic flux, established by the field windings, to where it is needed: across the radial **airgap** between the poles pieces and the armature.

Now that we are familiar with what a 'real' d.c. motor looks like, and recognise its main parts, let's move on to learn *how it works*.

## Principle of operation

In Figure 20.13, a single loop of conductor is placed between the poles of a permanent magnet, and is pivoted at its near and far ends in such a way that it is free to rotate.

Figure 20.13

With the current ($I_a$) flowing around the loop in the direction shown, if we apply **Fleming's Left-Hand Rule** to the left-hand side of the loop, we would find that it is subject to a *downward-acting force*; applying the same rule to the right-hand side of the loop would confirm that it is subject to an *upward-acting force*. These forces couple to create a rotating force, or **torque**, which will act to rotate the loop in a counter-clockwise direction.

The total **torque** acting on this single loop, obviously, will be *twice* the torque acting on either side of the loop:

torque = 2 (force × perpendicular distance from centre of rotation)

torque = 2 (force × radius of the loop)

$$T = 2Fr = 2(BIl)r$$

From this, we can see that for a motor of given dimensions and flux density, the *torque is proportional to the armature current*.

At the same time, *a back-e.m.f. will be induced into the loop* (we can confirm this by applying Fleming's Right-Hand Rule to Figure 20.13), the value of which will be:

$$E = 2Bl(r\,\omega)$$

… where ($r\,\omega$) – radius × angular velocity – represents the velocity ($v$) of the conductor.

From this, we can see that for a motor of given dimensions and flux density, the *induced e.m.f. is proportional to the angular velocity of the armature winding*.

Of course, a single-loop motor will produce very little torque so, in practice, a *coil*, called an **armature winding** (Figure 20.14), is used instead. If the armature winding has $z$ loops, then the torque (and corresponding back-e.m.f.) will be increased by $z$.

**Figure 20.14**

## Importance of back e.m.f. – action and reaction

As we have already learnt, whenever the armature winding is rotating, it behaves both as a motor *and* as a generator *simultaneously*! Passing a current through the winding causes it to move through the magnetic field ('motor action'), but the movement of that winding through the field causes it to generate a voltage ('generator action') which always *opposes* the applied voltage that produces the current in the first place!

So, the *actual* current flowing through the armature winding will be:

$$I_a = \frac{(E-U)}{R_a}$$

where:
$I_a$ = current in loop (in amperes)
$E$ = supply voltage (in volts)
$U$ = back e.m.f. (in volts)
$R_a$ = resistance of armature loop (in ohms)

The 'action' and 'reaction' effect, described above, plays a **very important** part in causing d.c. machines (both motors and generators) to *automatically react to changes in load*.

For example, in the case of a d.c. motor, if its mechanical load should *increase*, then the armature winding will tend to slow down. As it slows down, the back-e.m.f. *decreases*, allowing the current through the armature winding to *increase*, creating more torque to match the increasing mechanical load.

Similarly, if the motor's mechanical load should *decrease*, then the armature winding's speed will *increase*. As its speed increases, its back-e.m.f. also *increases*, reducing the load current and reducing the torque to match the reduced load.

In the earlier chapter on **Magnetism**, we learnt that magnetic lines of force don't really exist. They are simply a 'model' to help us understand what is going on in the area surrounding a magnet. Anyone giving serious thought to the motor/generator behaviour described here will soon begin to realise the weakness of this 'lines of force model', as it requires us to believe 'motor action' is due to the *distortion* of the magnetic field which pushes the conductor out of that field while, at the same time, asking us to believe that the simultaneous 'generator action' is due to the conductor *cutting* that field. Clearly *both* cannot be happening at the same time! So, the time has come to accept that there is no perfect model for what is actually going on, and some things simply happen because they happen!

## Maintaining rotation

Returning, now, to the rotation of the armature winding: once the armature winding has rotated through 90° counterclockwise, and is in the vertical plane, the two perpendicular forces then exactly oppose each other, so the torque completely disappears, and the armature winding will immediately stop rotating. Even if the momentum of the armature winding is such as to cause it to rotate a little way past the vertical, there would be an immediate reversal of the two forces, resulting in a corresponding reversal in torque which would quickly return the loop back to the perpendicular position.

Of course, we *don't* want this to happen! We want the armature winding to continue rotating in the same direction..

So, how can we encourage it to continue rotating in its original counterclockwise direction?

Well if, somehow, we could (1) *increase the loop's momentum*, to force it several degrees *past* the vertical position and, then, (2) *immediately reverse the direction of the current* around the loop, then the forces acting on each coil would also immediately reverse and the torque would continue to act in a counterclockwise direction, and the loop would continuously rotate.

Figure 20.15 shows the sequence of events described above. For the sake of clarity, a single loop is shown, rather than an armature winding.

To *increase the momentum* of the armature winding it is wound around a laminated iron cylinder, called an '**armature**', which, due to its mass, introduces a 'flywheel' effect (Figure 20.16). At the same time, the low-reluctance of the armature will help increase the flux density of the field, as it will act to concentrate the flux within the airgap between itself and the poles of the magnet. Remember, the reluctance of iron is *thousands* of times lower than that of air.

We can *reverse the direction of the current* every time the coil passes its vertical position by supplying current via a simple rotary switch, called a '**commutator**' or, more accurately, a '**split-ring commutator**' (Figure 20.16).

One end of the armature winding is soldered or brazed to one half of the 'split ring', and its other end is soldered to the opposite half of the 'split ring'. The

Forces couple to create counterclockwise torque.

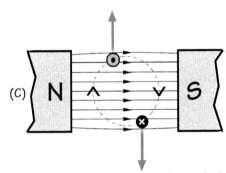

Loop approaches the vertical position.

Momentum causes loop to pass the vertical position and torque reverses.

So, to maintain rotation in the same direction, the current in the loop must be reversed.

Figure 20.15

commutator and the armature winding share the same shaft, and rotate together.

Current is fed on to, and away from, the commutator by means of a stationary pair of spring-loaded **carbon brushes**, which press against opposite sides of the commutator and 'ride' its surface as it rotates between them. As well as being a good conductor, carbon is

Figure 20.16

*self-lubricating* – making it an ideal material from which to manufacture brushes.

In simple terms, we can say that the function of the split-ring commutator is to ensure that, as the armature winding of our motor rotates, current **always** flows *away* from us in whichever side of the winding happens to be left of vertical, and the current always flows *towards* us in whichever side of the winding happens to the right of vertical.

This ensures that, immediately the armature winding passes its vertical position, the current around the winding reverses direction and the armature winding will be subject to a continuous counterclockwise torque, causing it to continue to rotate in a counterclockwise direction without hesitating as it passes through its vertical position.

The combination of split-ring commutator and its carbon brushes is known as the motor's '**commutation system**'.

The split-ring commutator also conveniently solves the problem of how to continuously supply current to and from the coil while it is rotating.

There are plenty of animations available on the internet which demonstrate the action of a motor's commutation system far better than it can be described in a textbook.

### Increasing torque

Now that we have achieved our goal of getting the armature coil to continuously rotate, we can turn our attention to how to *maximise its torque*, so that it can provide sufficient work to drive a mechanical load. The obvious method is to maximise the flux density in the airgap.

The flux density of even the best permanent magnets is not very great. In fact, motors that use permanent magnets to provide their field are generally limited to low-torque applications, such as toys. To provide enough torque to drive larger mechanical loads, we need to use **electromagnets**, which will provide a substantially greater flux density than permanent magnets. These are provided by **field windings** which are wound around the pole pieces, and form the motor's '**excitation system**'. The term 'excitation' simply describes the generation of magnetic flux.

At the same time, and in order to establish maximum flux density within the airgaps in which the armature rotates, those airgaps need to be *as narrow as possible*. This is achieved by using shaped **pole pieces**, which partially enclose the armature, as shown in Figure 20.17.

Figure 20.17

Finally, we need a low-reluctance **magnetic circuit** whose function is to guide the flux created by the excitation system to the airgaps. This magnetic circuit, then, comprises the excitation windings, the yoke, the pole pieces and the armature – as shown in Figure 20.18.

Magnetic Circuit for Two-Pole Motor

Figure 20.18

The motor's **magnetic circuit** can be compared with an **electric circuit**, in which we use low-resistance copper wires to guide the current to where it is needed: a lamp, or some other load. Because the resistance of the wire is so low, practically all the electromotive force applied to the circuit will appear across the load. In a similar way, thanks to the low-reluctance of the magnetic circuit, practically all the magnetomotive force (m.m.f.) created by the field windings will appear across the airgaps.

## Practical d.c. motors

As can be seen from Figure 20.10, a *practical* d.c. motor bears little resemblance to the simple model we've used to describe its principle of operation.

The two most-noticeable differences are that (1) most d.c. motors have *four or more poles*, rather than just two, (2) the armature windings are numerous, far more complex and occupy longitudinal slots that are distributed around the entire circumference of the armature, and (3) the commutator has numerous segments, not just two. These features ensure a constant torque is applied throughout the complete 360° rotation of the armature.

A less obvious difference is the way in which the field windings are supplied. In our basic model, we have rather assumed that they are supplied from a completely separate voltage source from that supplying the armature winding.

Although it is perfectly feasible to supply the excitation windings from a completely independent source of voltage, this is *not* usually the case. In practice, the *same* voltage supply is used to energise *both* the field windings *and* the armature windings.

The two most obvious methods of achieving this are (1) by connecting the field windings *in parallel* with the armature windings, or (2) by connecting the field windings *in series* with the armature windings – as illustrated in Figure 20.19. Motors connected in this way are termed **'shunt motors'** and **'series motors'**, respectively. The term 'shunt', of course, is an alternative name for 'in parallel'.

Shunt Motor    Series Motor

Figure 20.19

Another method is a combination of these two connections, termed a '**compound motor**', but this connection is beyond the scope of this chapter.

In the case of a **shunt motor**, the current drawn by the field windings is determined simply by the resistance of those windings and, so, is *constant* – providing a constant magnetic field in the airgap. The torque developed by the motor, therefore, will be proportional only to the *armature* current.

In the case of a **series motor**, however, the current through the armature windings also passes through the field windings. Since the flux density produced by the field windings is proportional to the field current (well, until saturation is reached) and the torque on the armature winding is proportional to the armature current *and* the flux density of the field then, for a series motor, the torque must be proportional to the *square* of the armature current. The **current/torque characteristic curves** for shunt and series motors are shown in Figure 20.20.

Current/Torque Characteristics of D.C. Motors

Figure 20.20

A series d.c. motor, therefore, has a very much higher starting torque, compared to an equivalent shunt motor so, in practice, series motors are used to drive very heavy mechanical loads, such as drag lines (these are very large excavators used in open-cast coal mining).

The '**compound motor**', described briefly earlier, produces a combination of the two characteristics, and can be designed to a match specific load requirement.

### Starting current

When we first switch on a d.c. motor, the armature, of course, is stationary, so there is no back-e.m.f. Accordingly, the initial load current can be very high. But, as the motor runs up to speed, the back-e.m.f. starts to increase, and the load current reduces significantly.

Once the machine is up to speed, the load current varies with its mechanical load. If the mechanical load should *increase*, the motor will tend to *slow down*, its back-e.m.f. will *decrease*, its load current will *increase*, and its torque will *rise* to match the demand of the higher mechanical load.

If the mechanical load should *decrease*, the motor will tend to *speed up*, its back-e.m.f. will *increase*, its load current will *decrease*, and its torque will *fall* to match the demand of the lower mechanical load.

So we can describe motors as being 'self-regulating': constantly adjusting their torque to match any variations in their mechanical load.

A motor's **starting current** is *always* very much higher than its normal operating load current, because the armature is simply not yet rotating fast enough to generate sufficient back-e.m.f. to oppose that starting current. For larger motors, high starting currents will inevitably result in excessively high temperatures which may break down the insulation. So larger d.c. motors usually require *some means of reducing their high starting currents*. One method of achieving this is by inserting a variable resistor in series with the motor, and gradually lower the value of that resistance as the motor picks up speed. Such a device is called a '**motor starter**', and is beyond the scope of this chapter.

Before finishing this chapter, it should be pointed out that, in the past, d.c. motors have always had a major advantage over a.c. motors. And that is, it has always been very easy to *control the speed of d.c. motors*, whereas the speed of a.c. motors has, to a large extent, been governed by the frequency of the a.c. supply voltage. This is no longer the case, thanks to developments in industrial electronics, and to a large extent, a.c. motors (which are far easier to construct) have replaced d.c. motors in many applications.

For a far more detailed examination of the construction, operation and performance of practical d.c. motors, and their motor starters, we can refer to the companion book *Electrical Science for Technicians*.

## Review your learning

Now that we've completed this chapter, we need to examine the **objectives** listed at its start. Placing *'Can I …'* at the beginning, and a question mark at the end, of each objective turns that objective into a **test item**. If we can answer each of those test items, then we've met the objectives of this chapter.

# Chapter 21

# Transformers

On completion of this chapter, you should be able to

1 state the main function of a transformer.
2 explain the major difference between a 'mutual transformer' and an 'autotransformer'.
3 explain the function and properties of a transformer's core.
4 explain what is meant by a transformer's
   a primary winding
   b secondary winding.
5 outline the basic principle of operation of a mutual transformer.
6 explain the terms 'turns ratio', 'voltage ratio' and 'current ratio', and the relationship between them.
7 solve basic problems relating to a transformer's turns, voltage and current ratios.
8 describe the differences between a transformer's 'copper losses' and its 'iron losses'.
9 explain the significance of transformers in electricity transmission/distribution systems.
10 explain why high voltages are so important in the transmission and distribution of electrical energy.

## Introduction

In the previous chapter, we learned how the physicist Michael Faraday discovered that by constantly *varying the current* in one coil, he was able to *induce a voltage* into an adjacent coil, through a process called **'mutual induction'**.

What Faraday had discovered was a primitive **transformer**, although he didn't use that term and it's very unlikely he fully realised the importance of his discovery. Why 'important'? Well, as we shall learn, without transformers, the transmission and distribution of electrical energy over any distance would be quite impossible.

A **transformer** is classified as an 'electrical machine' – one which transfers electrical energy from one circuit (the 'primary circuit') to another (the 'secondary circuit'), through inductively coupled coils.

Faraday's 'transformer' had a long way to go before it became the transformer we would recognise today.

The first practical 'transformer' was, in fact, an **induction coil** – a direct-current device, invented in 1836, by an Irish clergyman-cum-scientist, the Reverend Nicholas Callan (1799–1864). In Callan's time, the *only* practical source of electrical energy was the battery, which, of course, provides a steady direct current. So he arranged to vary the primary current by continuously interrupting it with an electromagnetic switching-mechanism similar to an electric bell or buzzer. This rapidly interrupted primary current was then able to induce a near-continuous high voltage into the secondary coil.

A **transformer** differs from an induction coil in that it uses the continuously varying *alternating current* in the primary coil to induce a voltage into the secondary coil – doing away with the need for an interrupting mechanism in the primary circuit.

Callan was also one of the first researchers to realise that the magnitude of the voltage induced into the secondary coil depended on the **turns ratio** between the primary and secondary coils. For example, if there were *twice* as many turns on the secondary coil than on the primary coil, then the secondary voltage would be *twice as high* as the voltage applied to the primary winding.

We still use induction coils today. They are used in petrol-driven motor vehicles to step up the low d.c. voltage supplied by the vehicle's battery or generator to the *thousands* of volts necessary to operate the spark plugs. In older vehicles, the primary current was interrupted by a switch incorporated into the distributor head, driven by the engine; in modern vehicles, this is done using an electronic switch.

The transformer with which we are familiar, today, however, is credited to the American electrical engineer William Stanley (1858–1916), but his design was really the culmination of numerous incremental improvements made by various other engineers worldwide, dating all the way back to Callan's induction coil.

There are basically *two* types of transformer: '**mutual transformers**' and '**autotransformers**'.

- **Mutual transformers** are, by far, the more common and have completely *separate* primary and secondary coils, electrically insulated from one another, but magnetically coupled through mutual induction. When anyone uses the word 'transformer', they are *usually* referring to a mutual transformer. A schematic diagram for a mutual transformer is shown in Figure 21.1; the two horizontal lines represent the transformer's iron core.

**Figure 21.1**

- **Autotransformers**, on the other hand, use a single coil, in which a voltage applied to *part* of that coil induces a voltage into the *remainder* of the coil. A schematic diagram for a mutual transformer is shown in Figure 21.2; the lower voltage is obtained by tapping (connecting) one terminal part way along the coil.

**Figure 21.2**

A common application for a 'step up' autotransformer is a petrol-engined vehicle's **ignition coil**. The d.c. supply voltage of 12 V, provided by the vehicle's battery, is chopped up into a series of pulses using a switching transistor (triggered by the engine) and fed to the low-voltage terminals, and a very high pulsed voltage then appears across the high-voltage terminals' windings.

The prefix '**auto**', in '**autotransformer**', is not short for 'automatic', as some students believe. In fact, it is derived from a Greek word, meaning 'self'. So an 'autotransformer' is one in which a voltage applied to part of a coil induces a voltage *into 'itself'* – meaning *into the same coil*.

In this chapter, we are going to learn the basic operation of a **mutual transformer**. We will *not* be going into the subject in great depth, because transformers are *alternating current* machines, which means we require an understanding of alternating current to study it in any depth. And we are not quite ready for that … yet! For an in-depth study of transformers, including autotransformers, you are referred to the companion book, *Electrical Science for Technicians*.

Mutual **power transformers** are also sometimes referred to as '**constant voltage transformers**', by which we mean there is no significant difference

Circuit-Board Transformer     Transmission Grid Transformer

**Figure 21.3**

between their no-load secondary voltage and their full-load secondary voltage. We may need to refer back to the chapter on *internal resistance* (Chapter 14) to remind ourselves of the significance of this.

**Mutual transformers** vary enormously in their physical size: from tiny transformers mounted on electronic circuit boards, measuring just a few cubic millimetres, right up to huge transformers used in high-voltage electricity transmission systems (Figure 21.3). But, regardless of their physical size, *they are all basically the same in terms of their construction and operation.*

A basic mutual **transformer** comprises *three* parts: a **magnetic circuit**, called the '**core**', around which are placed two coils – the '**primary winding**' and the '**secondary winding**'. Some transformers may have more than one primary or secondary winding, but this is beyond the scope of this chapter.

The **core** is constructed from a stack of **laminations** (thin sheets), manufactured from an alloy of iron and silicon: a 'soft' ferromagnetic material which means that it can be very easily magnetised and demagnetised and, then, remagnetised in the opposite direction. Although the core is manufactured from an *iron*-silicon alloy, it is more commonly referred to, somewhat inaccurately, as 'transformer steel' or 'silicon-steel', or by a manufacturer's trade name, such as 'stalloy'.

During operation, circulating currents – called '**eddy currents**' – are induced into the core. These are undesirable, as they would cause the core to overheat resulting in unnecessary energy losses. However, the silicon content of the core increases its resistivity, which acts to reduce the magnitude of these eddy currents, and laminating the core restricts their path around the core. So, the eddy currents themselves, and the losses they cause, are always very low

The horizontal parts of a transformer's core are referred to as 'yokes', the vertical components are called 'limbs' and the spaces containing the windings are called 'windows' – as illustrated in Figure 21.4.

*The top 'yoke' of this transformer core is removable, allowing the windings to be placed around the centre limb.*

**Transformer 'Shell-Type' Core**

Figure 21.4

Electrical engineers always refer to a transformer's coils as '**windings**'. A winding is normally manufactured from numerous turns of **insulated copper wire**, although aluminium is sometimes used. Copper is preferred because, for any given current, a copper conductor (being a better conductor) will have a *smaller* cross-sectional area than an equivalent aluminium conductor, so a copper-wound transformer will have a smaller overall volume, or 'footprint', compared with an aluminium-wound transformer of equivalent rating.

By definition, the '**primary winding**' is the one connected to the *supply*, while the '**secondary winding**' is the one connected to the *load*. Note, the terms 'primary' and 'secondary' do *not* relate to the levels of voltage applied to them.

If the secondary winding operates at a *higher* voltage than the primary winding, we describe the transformer as being a '**step-up transformer**'.

If the secondary winding operates at a *lower* voltage than the primary winding, we describe the transformer as being a '**step-down' transformer**.

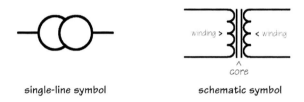

single-line symbol          schematic symbol

Figure 21.5

For the purpose of clarity, schematic diagrams (Figure 21.5) *always* show the primary and secondary windings mounted *separately* on different limbs of the core. But, in reality, they are *always* wound, concentrically, *around each other* before being placed over the core's limbs. This ensures maximum magnetic coupling between the primary and secondary windings.

core-type core          shell-type core

Figure 21.6

Figure 21.6 shows the two types of single-phase transformer core design: '**core-type**' and '**shell-type**'. They are easily recognisable when you examine a transformer: for the 'core-type', *the windings surround the core* while, for the 'shell-type', *the core surrounds the windings*.

With the core-type, half the lower-voltage windings and half the higher-voltage windings are distributed between the outer limbs, with the lower-voltage windings located inside the higher-voltage windings.

With the shell-type, both windings are placed around the centre limb.

In each case, the low-voltage windings are located between the core and the high-voltage windings.

## Principle of operation

### Relationship between voltage ratio and turns ratio

When a transformer's primary winding is connected to an alternating current supply (Figure 21.7), a primary current flows through the primary winding and establishes an alternating magnetic flux within the core. Providing all this flux is contained within the core of the transformer, this magnetic flux is common to both the primary *and* secondary windings, and induces an identical voltage into *each* of their individual turns.

**Ideal Transformer**

Figure 21.7

On the primary side, the 'volts per turn' is simply the primary-voltage ($E_p$) divided by the number of primary turns ($N_p$):

$$(\text{volts per turn})_{\text{primary}} = \frac{Ep}{Np}$$

The 'volts per turn', on the secondary side, is exactly the same as the 'volts per turn' on the primary side, so the *total* secondary voltage, $E_s$, must be the product of the volts per turn and the number of secondary turns ($N_s$):

$$E_s = \left(\frac{E_p}{Np}\right) \times N_s$$

… rearranging this equation:

$$\frac{E_p}{E_s} = \frac{N_p}{N_s} \quad \text{— equation 1}$$

Putting this into words, we can say:

An ideal transformer's **voltage ratio** is *equal* to its **turns ratio**.

A transformer's 'turns ratio' is also known as its '**transformation ratio**'.

> **Worked example 1** A transformer has a 200-turn primary winding, and a 1000-turn secondary winding. If the transformer is connected to a supply voltage of 230 V, what will be its secondary voltage?
>
> **Solution**
>
> $$\frac{E_p}{E_s} = \frac{N_p}{N_s}$$
>
> $$E_s = \frac{N_s}{N_p} E_p = \frac{1000}{200} \times 230 = 1150 \text{ V (Answer)}$$

### Relationship between current ratio and voltage ratio

Partly because transformers have no moving parts, *transformers are highly efficient*. In fact, those very large transformers used in transmission/distribution systems can be between 96–98% efficient when operating at full load.

So, in order to understand the basic operation of a transformer, we can't go far wrong by assuming that we are dealing with an '**ideal transformer**' – that is, one which has no losses whatsoever.

The '**power rating**' of a transformer is the product of its rated *secondary* voltage and its rated *secondary* current. For transformers, this is *always* expressed in 'volt amperes' (**V·A**), *never* in 'watts'. The reason for this (i.e. the difference between 'volt amperes' and 'watts') will become apparent when we study alternating current but, for now, we simply need to accept that this is so, because measuring power in a.c. circuits is somewhat more complicated than it is with d.c. circuits!

> A transformer's power rating is *always* expressed in **volt amperes (V·A)**, *never* in watts.

If we assume, then, that an ideal transformer is 100% efficient, then we can say that its **input** (primary) **power** must equal its **output** (secondary) **power**:

$$P_p = P_s$$
$$E_p I_p = E_s I_s$$

… rearranging:

$$\frac{E_p}{E_s} = \frac{I_s}{I_p} \quad \text{— equation 2}$$

Putting this into words, we can say:

> An ideal transformer's **current ratio** is the *reciprocal* of its **voltage ratio**.

As an analogy, we can think of a transformer as being the *electrical equivalent* of a **balanced beam** in mechanical science. No doubt we can all recall from our science lessons that, for a beam to remain balanced,

the sum *of the clockwise moments* (i.e. the horizontal distance from the fulcrum multiplied by vertical force) *must equal the sum of the counterclockwise moments*.

Figure 21.8 shows a balanced beam. For the beam to remain balanced, the product of the horizontal distance, *x*, and the vertical force, $F_1$, must *exactly equal* the product of the horizontal distance, *y*, and the vertical force $F_2$.

If force $F_2$ should, say, *increase*, then to maintain balance, force $F_1$ must also be *increased* in proportion. So if, for example, we were to double $F_2$, then we must also double $F_1$.

In the case of a transformer, the product of the primary voltage, $E_p$, and the primary current, $I_p$, must *exactly equal* the product of the secondary voltage, $E_s$, and the secondary current, $I_s$.

So, if the secondary (load) current, $I_s$, were to *increase*, then, to maintain balance or remain in 'electrical equilibrium', the transformer's primary current, $I_p$, would *also* have to *increase* in proportion.

In fact, *a transformer's **primary** current is entirely dependent upon its **secondary**, or load, current*. If the secondary (load) current were to, say, *double*, then the primary current would have to *double*, too. If the secondary (load) current were to, say, *halve*, then the primary current would have to *halve*, too.

However, if there is *no* secondary current, because the secondary winding isn't connected to a load, then the primary current does *not* fall to zero but it does fall to a *very* low value – just sufficient to maintain the magnetic flux around the core. This very small primary current is called the '**magnetising current**'.

Balanced Beam:
x·F1 = y·F2

Transformer Equivalent:
Ip·Ep = Is·Es

Figure 21.8

**Worked example 2**  A step-up transformer has a primary voltage of 100 V and a secondary voltage of 500 V. If the current drawn by a load (i.e. the secondary current) is 5 A, calculate the resulting primary current.

**Solution**

$$\frac{E_p}{E_s} = \frac{I_s}{I_p}$$

$$I_p = I_s \frac{E_s}{E_p} = 5 \times \frac{500}{100} = 25\,\text{A} \ (\text{Answer})$$

## Relationship between current ratio and turns ratio

Since a transformer's voltage ratio is the *same* as its turns ratio, we can modify equation (2) as follows:

$$\boxed{\frac{I_s}{I_p} = \frac{N_p}{N_s}} \ — \ (\text{equation 3})$$

Putting this into words, we can say:

An ideal transformer's **current ratio** is the *reciprocal* of its **turns ratio**.

**Worked example 3**  A step-down transformer has a primary winding with 250 turns and a secondary winding with 750 turns. If a supply voltage of 120 V is applied to the primary winding, determine: (a) the secondary voltage, (b) the secondary rated load current if the transformer is rated at 5000 V·A, and (c) the resulting primary current.

**Solution**

(a)  To find the secondary voltage:

$$\frac{E_s}{E_p} = \frac{N_s}{N_p}$$

$$E_s = \frac{N_s}{N_p} E_p = \frac{750}{250} \times 120 = 360\,\text{V} \ (\text{Answer a.})$$

(b)  To find the secondary rated load current, we simply divide its rated power output by the rated secondary voltage:

$$I_{s(rated)} = \frac{P_{(rated)}}{E_{s(rated)}} = \frac{5000}{360} = 13.89\,\text{A} \ (\text{Answer b.})$$

(c)  To find the primary current, when the secondary rated load current flows:

$$\frac{I_p}{I_s} = \frac{N_s}{N_p} \ — \ (\text{from equation 3})$$

$$I_p = \frac{N_s}{N_p} I_s = \frac{750}{250} \times 13.89 = 3 \times 13.89$$

$$= 41.67\,\text{A} \ (\text{Answer c.})$$

Students often believe that a transformer will 'step down' or 'step up' the *primary* current. This is, strictly, incorrect.

As worked example 2 shows, the primary current is *completely dependent upon the secondary, or load, current* and the *reciprocal* of the turns ratio.

So, it's more accurate to say that a transformer steps up or steps down the **secondary** current NOT the primary current.

## Transformer energy losses

Up to this point, we have assumed that transformers are 'ideal' – i.e. they are 100% efficient. 'Real' transformers are not, of course, 100% efficient although large ones come pretty close, at around 96–98% at full load.

One of the main reasons that transformers are so efficient is that they contain no moving parts. So, just what are the causes of the small energy losses that do occur in transformers? Well, the losses can be summed up under two categories, termed '**copper losses**' and '**iron losses**'.

- **Copper losses** are energy losses that occur in the copper *windings*. These are the $I^2R$ (heating) losses due to the currents in the transformer's primary *and* secondary windings.
- **Iron losses** are energy losses that occur within the iron *core*. These are due to:
  - **eddy-current losses**: these, too, are $I^2R$ losses caused by the small eddy currents that

circulate within the iron core. These are very small, thanks to the silicon content of the iron, and the fact that the core is laminated – both of which act to minimise the eddy currents.

○ **hysteresis losses**, resulting from the energy required to continually reverse the direction of the magnetic field. If we refer back to the chapter on *magnetic circuits* (Chapter 18), we will recall that these are proportional to the area of the iron's *B-H* loop.

Copper losses vary as the square of the current. Iron losses, on the other hand, remain approximately constant from no load to full load.

The **efficiency** of a transformer is the ratio of its output power to its input power, and usually expressed as a percentage:

$$\eta = \left( \frac{P_{output}}{P_{input}} \right) \times 100\%$$

This can be rewritten in terms of losses, as follows:

$$\eta = \left( \frac{P_{output}}{P_{output} + P_{copper\,loss} + P_{iron\,loss}} \right) \times 100\%$$

In Chapter 16 on *magnetism*, we learnt that magnetic lines of force don't really exist. They are simply a 'model' to help us understand what is going on in the area surrounding a magnet. Anyone giving serious thought to transformer action, described here, will begin to realise the weakness of this 'lines of force model'. The magnetic field set up by the primary winding is practically all confined within the core of the transformer and, so, has no opportunity to 'cut' either the primary or secondary windings. So something is going on, here, that we don't fully understand and which the 'lines of force' model can't really explain.

## Advantages of a.c. in electricity transmission/distribution systems

Electricity transmission and distribution is only possible thanks to the invention of the **transformer**.

During the nineteenth century, a great controversy arose in the United States, between two industrial giants, over whether *direct current* or *alternating current* was better suited for electricity transmission and distribution. The famous inventor Thomas Edison (1847–1931) had invested heavily in direct current machines and, so, was an advocate of using direct current as a means of distribution. At the same time, the industrialist George Westinghouse (1846–1914), who was greatly influenced by what he had seen in Europe and by one of his brilliant young engineers, Nikola Tesla (1856–1934), had invested heavily in alternating-current machines and, so, was an advocate of alternating current.

This led to what has since become known as '*The War of the Currents*'.

Westinghouse argued that a.c. generators should be located where fuel for their prime movers was readily available and, using step-up transformers, electricity should then be transmitted at high voltage to the load centres where it was needed and, using step-down transformers, the voltage reduced to match the needs of the consumer. As we shall see shortly, by using high voltage, the resulting transmission current would be low, resulting in negligible voltage drops and energy losses along the transmission lines.

Edison, on the other hand, argued that d.c. generators should be located near where the electricity was needed, and the generated voltage should match the needs of the consumer. He had little choice, of course, because transformers are a.c. machines, and d.c. voltages could not, at that time, be easily changed. So a d.c. distribution system would, of course, require numerous local power stations, requiring a means by which fuel could be shipped in to them.

Whichever system triumphed would obviously have a huge effect upon the fortunes of either Edison or Westinghouse.

Edison, who was certainly no stranger to self-promotion, embarked upon a major public relations exercise arguing that, because a.c. was transmitted at high voltage, it was highly dangerous. There was some truth in his arguments, because the safety regulations at that time hadn't kept up with the technology, and were inadequate regarding the locations of transmission poles and the clearance of the high-voltage conductors. Even at low-levels of voltage, a.c. is more dangerous than d.c. So the dangers were real, but seriously exaggerated by Edison's publicity machine.

So, out of self-interest, Edison concentrated his arguments against alternating current, and he became famous for the lengths to which then he went to demonstrate 'how dangerous' it was by using it to electrocute stray dogs and cats, horses and, on one occasion, an elephant, at public gatherings. He also advocated the use of a.c. as a means of execution and supported the introduction of the electric chair.

Finally, he manipulated the English language to belittle alternating current – for example, after the introduction of the electric chair, he introduced the term *'to Westinghouse'*, meaning *'to electrocute'*, condemned prisoners! (Indeed, the term *'electrocution'*, which we use today, is derived from the expression *'execution by electricity'*!) Another term he introduced was *'to rectify'* when describing the process of changing a.c. into d.c., as though a.c. had something inherently wrong with it and needed to be rectified or corrected! That particular term stuck, of course, and we still use 'rectify' today!

In the end, however, the clear engineering and financial advantages of Westinghouse's alternating current transmission system overcame Edison's opposition, as it made sense to locate power stations near coal fields or in mountainous areas where hydro-electric dams could be constructed and, then, to transmit the electricity to wherever it was needed.

Having lost *'The War of the Currents'*, Edison was forced to accept that he had backed the wrong horse and, at great personal expense, had to go into the business of designing and manufacturing a.c. machines.

Although *'The War of the Currents'* took place in the United States, it perfectly illustrates *why* alternating current, rather than direct current, has became universally accepted for the transmission and distribution of electricity. A system that is *only* possible thanks to the **transformer** which allows alternating voltages to be easily, and with very little energy loss, stepped up or down.

Today, in the United Kingdom, electricity is generated in power stations at 11 kV or 25 kV and, using transformers, stepped up to 400 kV for transmission via the national grid. Towards load centres, such as towns and large industrial consumers, the voltages are gradually stepped down, using other transformers, through 275 kV, 132 kV, 33 kV, 11 kV and, eventually, to the 400/230 V required by commercial and domestic consumers.

> A general 'rule of thumb' used in the electricity supply industry specifies a transmission distance of *one kilometre per kilovolt*, which means that a 400-kV transmission line can be up to around 400 km long.

But *why* is it so important to transmit electricity at such high voltages? Well, let's look at a very simple example that demonstrates why.

**Worked example 4** Determine the 'receiving end' voltage, and the energy loss in a cable having a total resistance of 0.5 $\Omega$, when transmitting energy at the rate of 10 kW at a supply voltage of (a) 100 V and (b) 100 000 V.

**Solution**

(a) At 100 V, the resulting load current can be found as follows:

$$\text{since } P = EI$$

$$\text{then } I_{load} = \frac{P}{E} = \frac{10000}{100} = 100 \, \text{A}$$

we can now determine the voltage drop along the cable:

$$U_{load} = I_{load} R_{cable} = 100 \times 0.5 = 50 \, \text{V}$$

... so the terminal voltage will be:

$$U_{terminal} = E_{supply} - U_{drop} = 100 - 50 = 50 \, \text{V (Answer)}$$

... and the energy loss in the line is:

$$P_{loss} = I_{load}{}^2 R_{cable} = 100^2 \times 0.5 = 5000 \, \text{W or 5 kW (Answer)}$$

(b) At 100 000 V, the resulting load current can be found as follows:

$$\text{Since } P = EI$$

$$\text{then } I_{load} = \frac{P}{E} = \frac{10000}{100000} = 0.1 \, \text{A}$$

we can now determine the voltage drop along the cable:

$$U_{drop} = I_{load} R_{cable} = 0.1 \times 0.5 = 0.05 \, \text{V}$$

... so the terminal voltage will be:

$$U_{terminal} = E_{supply} - U_{drop} = 100000 - 0.05$$
$$= 99999.5 \, \text{V (Answer)}$$

... and the energy loss in the line is:

$$P_{loss} = I_{load}{}^2 R_{cable} = 0.05^2 \times 0.5$$
$$= 1.25 \times 10^{-3} \, \text{W or 1.25 mW (Answer)}$$

In the above example, we can see that supplying energy at 100 V results in a voltage drop along the line of 50 V, which means the *half the voltage is lost during transmission!* At the same time, the power loss along the line is 5 kW accounting for *half of the power transmitted to the load!*

On the other hand, by supplying energy at 100 000 V, the voltage drop along the same line is a *negligible* 0.05 V while the power loss along the line is a *negligible* 1.25 mW!

Granted, this example is a gross oversimplification of a real-life transmission system, because it doesn't take into account the effects of a transmission line's inductance and capacitance and other important factors, but it *does* nevertheless demonstrate why high voltages are *essential* to electricity transmission and distribution and which *is only made possible thanks to the transformer.*

Before finishing with this chapter, it should be mentioned that direct current *is* sometimes used for high-voltage transmission in particular situations where it offers important advantages over alternating current. However, an explanation of these advantages is well beyond the scope of this chapter.

## Review your learning

Now that we've completed this chapter, we need to examine the **objectives** listed at its start. Placing *'Can I ... '* at the beginning, and a question mark at the end, of each objective turns that objective into a **test item**. If we can answer each of those test items, then we've met the objectives of this chapter.

# Chapter 22

# Capacitance

## Introduction

In 1745, barely three years before his death, a German cleric and physicist Ewald Georg von Kleist (1700–1748) invented a device for 'storing static electricity' for the purpose of his experiments. It consisted of a glass jar, coated both inside and out with metal foil (often, gold leaf), and with its inner coating connected, via a metal chain, to a brass rod that passed through an insulated wooden stopper – as illustrated in Figure 22.1.

A year later, in 1746, before details of von Kleist's invention had even been published, a Dutch physicist working at the University of Leiden, Pieter van Musschenbroek (1692–1761), independently invented an almost identical device which became known as a 'Leiden Jar' or **'Leyden Jar'**, the forerunner of what we know, today, as a **capacitor**.

Leyden Jar

Figure 22.1

Although invented well over 250 years ago, the Leyden and Kleistien Jars each have *exactly* the same components as a modern capacitor: two conducting **'plates'** (its separate inner and outer metal foils), separated by an insulator or **dielectric** (glass).

Prior to the invention of these devices, physicists studying electricity were only able to store separated electrical charges on physically large, electrically isolated conductors; the Leyden Jar offered a more compact, convenient and portable alternative.

Until around the 1950s, in the English-speaking world, capacitors were called '**condensers**' from the Italian word, '*condensatore*', coined by Alessandro Volta, meaning to 'squeeze' or to 'compress', because of their ability to store a *greater density* of separated electrical charge than the larger normally isolated conductors used at that time. Incidentally, this is the word by which capacitors are *still* known by engineers and electricians who live, for example, in France ('*condensateur*') and in Germany ('*kondensator*').

In fact, in the English-speaking automotive industry, the capacitors used in ignition systems are still widely known as 'condensers'!

## Capacitors

In this chapter we will learn about the behaviour of **capacitors**, and about a naturally occuring electrical quantity, called **capacitance**, which exists in most circuits.

Throughout this chapter, we will be using **electron flow**, as the use of conventional flow will, otherwise, make this particular topic unreasonably confusing.

Let's begin by looking at what a capacitor *does*.

> A **capacitor** is defined as '*a passive two-terminal electrical component that stores electrical energy in an electric field*'.

We should *not* interpret this statement as meaning that most capacitors offer a practical means of storing useful amounts of energy for later use (although this is indeed true for so-called, '**supercapacitors**', which we will discuss later), rather in the same way a cell or battery does! Rather, this ability gives a capacitor some very useful properties which, as we will learn later, are used by both d.c. and a.c. applications.

As we shall learn, whereas inductors oppose any change in *current, capacitors oppose any change in voltage*.

> A **capacitor** acts to '*oppose any change in voltage*'.

A common misconception that we need to dispel right from the very beginning is that 'capacitors store *charge*'! Unfortunately, a great many textbooks state this to be the case but, as we shall learn, this is both misleading and prevents us from understanding their behaviour.

So it's worth repeating that capacitors store **energy**, *not* charge.

In their simplest form, *all* capacitors consist of two thin, parallel metal sheets (or foils), called '**plates**', placed very close together, but separated by an insulating material which we call a '**dielectric**'. In Figure 22.2, the dielectric is simply air, but another dielectric material could be used instead, including mica, plastics, oil-impregnated paper, etc. In fact, many, *but not all*, capacitors are named according to their dielectric, hence: 'mica capacitor', 'paper capacitor', etc.

As already explained, these components are exactly equivalent to the interior and exterior metal foil coatings and the glass dielectric of the original Leyden Jar.

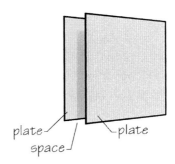

Figure 22.2

If the capacitor's plates are connected to an external source of d.c. potential difference, such as that provided by a battery, electrons will be transferred from one plate to the other. The resulting deficiency of electrons on the first plate will cause that plate to acquire a positive potential, while the excess of electrons on the other plate will cause that plate to acquire a negative potential – creating a potential difference across the plates, and an electric field between them. And *it is within this electric field that a capacitor stores energy*.

We call this process of charge transfer between its plates 'charging the capacitor' (although a more accurate expression would be 'energising the capacitor'), and it is explained as follows.

### Before charging takes place

With the switch open, each plate is electrically *neutral* (Figure 22.3) – i.e. they each contain equal quantities of protons and electrons (represented in the illustrations by the small positive and negative signs).

Figure 22.3

Remember, only the negatively charged *free electrons* are able to move, as their corresponding positively charged protons are bound within the nucleii of their immovable atoms.

## Charging action

When the switch is closed, the positive terminal of the battery will immediately attract electrons away from the left-hand plate, while its negative terminal will drive electrons onto the right-hand plate (Figure 22.4).

Figure 22.4

We call this movement of free electrons through the capacitor's external circuit, a '**charging current**'.

As more and more free electrons arrive at the right-hand plate, the amount of negative charge on that plate continues to increase while the loss of these electrons from the left-hand plate means that the amount of positive charge on that plate increases by the same amount, causing the potential difference (U) between the plates to rapidly increase.

The direction of this increasing potential difference, of course, is *opposite* that of the battery (E), and eventually:

$$U = E$$

During the process, the charging voltage follows the curve shown in Figure 22.5. So, as we can see, when the battery is first connected across the capacitor, the resulting current is high. But as the opposing potential difference (U) builds up across the plates of the capacitor, the resulting charging current falls towards zero.

Charging Voltage          Charging Current

Figure 22.5

As we shall learn later, if the resistance of the charging circuit is negligible, then the charging current will fall to zero in just microseconds.

As we can see from the charging-voltage curve in Figure 22.5, as the capacitor charges, it *opposes*, but does not prevent, the increase in voltage appearing across its plates.

The charging voltage and current curves always follow the specific shapes shown, as they are defined by a particular mathematical equation, and they are the same shapes as those for d.c. inductive-resistive circuits. This means that, if we wish, we can determine the *exact* potential difference or discharge current at any point along the curves.

However, the derivation and application of this equation is beyond the scope of this book.

> The potential difference across the plates of a capacitor *will always rise to equal the potential difference of the external source*, but will act in the opposite sense – that is, the polarities of a capacitor's plates will always match those of the external potentials to which they are connected.

It's important to understand that, during the 'charging' process, all that is happening is that *electric charge is being transferred from one plate to the other*. There is no change to the *net amount* of charge on those plates – i.e. the battery is *not* forcing any additional charges onto the plates of the capacitor.

So, when we refer to the amount of 'charge' on a capacitor, we are actually referring *either* to the amount of positive charge on the positive plate, *or* to the identical amount of negative charge on the negative plate – *not to the sum of these two charges!* However, in keeping with the electron theory, it is convenient to express the 'charge on a capacitor' in terms of *the amount of negative charge, expressed in coulombs, accumulated on the negative plate of the capacitor.*

> **Important!**
> By convention, the 'charge on a capacitor' is the amount of negative charge, expressed in coulombs, that has accumulated on the negative plate of that capacitor. It is *not* the sum of the positive and negative charges accumulated on *both* plates.

### With the switch open

Figure 22.6

If the switch is now opened (Figure 22.6), the capacitor will *retain the potential difference that has built up across its plates.* This is because the open switch has broken the only direct path by which those electrons which have accumulated on the right-hand plate can return to the left-hand plate – other than through the dielectric.

In practice, there is no such thing as a *perfect* dielectric. So, *some* of the excess electrons on the right-hand plate do, in fact, manage to find their way back to the left-hand plate through the dielectric. This is termed **leakage current**, and results in *a very gradual reduction in the potential difference across the capacitor's plates.*

### Discharge action

If the battery is now removed from the circuit (Figure 22.7), and replaced by a load (in this example, a resistor), the excess electrons crowded onto the right-hand plate now have a direct route back around the circuit to the left-hand plate, and the fully charged capacitor will begin to **discharge.** As shown in Figure 22.8, the resulting discharge current will follow exactly the same-shaped curve as the charging current but, of course, *in the opposite direction.* As the electrons return to the opposite plate, the potential difference collapses, too.

Figure 22.7

Discharge Voltage

Discharge Current

Figure 22.8

As we can see from the discharging voltage curve in Figure 22.8, the capacitor acts to *oppose* the collapse or, to put it another way, to sustain that voltage.

From the charging and discharging voltage curves, we can state that…

> A capacitor acts to oppose any change in voltage.

Again, as we shall learn later, if the resistance of the load is negligible, then the capacitor will completely discharge in milliseconds or microseconds.

### Charge distribution on capacitors in series

We earlier described the charge on a capacitor as the 'amount of negative charge accumulated on its negative plate'. So, what happens if a number of capacitors are connected in *series?*

If several capacitors are connected in series across a d.c. supply, as illustrated in Figure 22.9, free electrons will be transferred from the left-hand plate of capacitor $C_1$, counterclockwise, through the circuit to the right-hand plate of capacitor $C_4$.

Figure 22.9

The negative charge acquired by the right-hand plate of capacitor $C_4$ will then repel electrons from its left-hand plate, driving them onto the right-hand plate of capacitor $C_3$. This action is repeated for capacitor $C_2$, and the resulting charge distribution will be as indicated by the positive and negative signs shown in Figure 22.9.

However, the **total charge** *on all four capacitors* will simply be the original amount of negative electric charge transferred from the left-hand plate of capacitor $C_1$ onto the right-hand plate of capacitor $C_4$.

In other words, the total charge will remain the same, *regardless of how many capacitors are connected in series between the outer plates of the outermost two capacitors!* We will return to this topic, for further explanation, a little later.

### Water-flow analogy of a capacitor

Imagine a hollow, water filled, chamber divided in half by a flexible rubber diaphragm, and connected to a water pump via water-filled pipework (Figure 22.10).

Figure 22.10

The volume of water on either side of the diaphragm is equivalent to the balanced charge on the plates of

an uncharged capacitor, and the diaphragm itself is equivalent to the electric field that will appear between the two plates.

When the pump starts, water will flow (in this case) counterclockwise around the system and start to push against the diaphragm (see Figure 22.11).

Figure 22.11

Initially, there will be a sudden rush of water around the system. But, as the diaphragm stretches, it will increasingly oppose the flow of water until, eventually, it becomes fully stretched, its opposition will be equal and opposite to the pump's pressure and the flow will stop altogether (Figure 22.12). At the same time, the work done in stretching the diaphragm means that the diaphragm has stored energy. The **diaphragm** *is the equivalent of a capacitor's electric field*, and the action described is equivalent of *a capacitor being charged.*

The increased volume of water to the right of the diaphragm represents the negative charge that has built up on the negative plate of the capacitor. But, although the volume of water to the right of the diaphragm has increased, the volume to the left has decreased by exactly the same amount, so there has been no change in the overall volume of water within the chamber, yet energy has been stored in the stretched diaphragm.

Figure 22.12

Imagine, now, that the pump is allowed to freewheel (Figure 22.13). The stretched diaphragm will now relax and return to its normal shape, forcing water to flow in the *reverse* direction (clockwise) around the system. At the same time, the energy stored in the fully stretched diaphragm is allowed to dissipate. Initially, there will be a sudden rush of water, but this flow will reduce as the diaphragm returns to its normal shape. *This action is equivalent to a capacitor being discharged.*

Figure 22.13

If we think of the volume of water to the right of the diaphragm, within the chamber, as being equivalent to the 'charge transferred' to the negative plate of a capacitor, it should become apparent that, as no overall change in the volume of water has taken place, then it should be equally clear that no 'additional electric charge' has been 'stored' on the plates of a capacitor during the 'charging' process. What *has* been 'stored' in the capacitor is **energy**, and this energy has been stored in the its **electric field**.

So we can say that **a capacitor is a circuit component that stores** *energy*.

Perhaps we should replace the terms, 'charging' and 'discharging' with 'energising' and 'de-energising'? Well, unfortunately, the terms 'charging' and 'discharging' are too well established for us to change them at this stage!

Despite this, many textbooks continue to insist that a capacitor 'stores charge', but this is misleading because, as we have learnt, there is no increase in the overall charge on the plates whenever a capacitor is charged. All that has happened is that charge has been *separated*, and has been moved from one plate across to the other; *no additional charge has been introduced or stored.*

Actually, these textbooks aren't really 'wrong' because what they *mean* (as opposed to what they *say*!) is that a capacitor is a circuit component which 'stores *separated* charges' which, of course, is quite correct!

## Circuit symbols for capacitors and capacitance

Capacitors may be **fixed value** or **variable value**. A special type of variable capacitor is the **trimmer** type; this is a variable capacitor that has been pre-set to a particular value of capacitance and is not intended to be adjusted further.

Figure 22.14

'Electrolytic' capacitors (Figure 22.15) are fixed-value capacitors which are 'polarised', which means that it is important which way around a potential difference is applied to their plates according to their markings (for the circuit symbol, shown in Figure 22.14, the solid black plate represents its negative plate).

Figure 22.15

We should be aware that the US circuit symbol of a capacitor is quite different from the European version, and can be confusing for the following reason.

What appears to Europeans as being the symbol for a capacitor is, in fact, the US circuit symbol for a set of normally open electrical **contacts** (normally *closed* contacts use the same symbol, but with a diagonal line through it). The American symbol for a capacitor uses a *curved* line to represent one of its plates (the negative plate, in the case of a polarised capacitor).

Figure 22.16

So care must be taken when reading US-sourced schematic diagrams, to ensure that you do not confuse the two symbols.

## Capacitance

During the charging process, as free electrons accumulate on the negative plate of a capacitor, this causes the potential of that plate, measured with respect to the positive plate, to increase.

The resulting potential difference *(U)* between the two plates is directly proportional to the amount of charge *(Q)* accumulating on the negative plate, and can be expressed as follows:

$$U \propto Q$$

To change the proportional sign to an equals sign, we need to introduce a constant:

$$U = \text{constant} \times Q$$

Expressed another way, we can say:

$$\text{constant} = \frac{Q}{U}$$

We call this constant the **capacitance** *(C)* of the capacitor, which is defined as *'the ratio between the amount of electric charge on the negative plate, and the potential difference between the two plates'*.

$$\boxed{C = \frac{Q}{U}} \qquad \text{—(equation 1)}$$

where:

C = capacitance, in farads

Q = charge, in coulombs

U = potential difference, in volts

The SI unit of measurement for capacitance is equivalent to a 'coulomb per volt' which, in SI, is given the special name: the **farad** (symbol: F). While the ratio of charge to voltage tells us what the capacitance happens to be for that particular ratio, the capacitance itself is a constant, as we are about to learn, determined by the physical characteristics of the capacitor and the nature of its dielectric.

> The **farad** (symbol; **F**) is defined as *'the capacitance of a capacitor, between the plates of which there appears a difference in potential of 1 volt, when it is charged to 1 coulomb'*.

The farad, in practical terms, is absolutely enormous so, in practice, capacitance is usually expressed in **microfarads** (μF), in **picofarads** (pF) or in **nanofarads** (nF).

> Interestingly, one of the earliest units for capacitance was the '**jar**' (named after the 'Leyden Jar'). This unit was used well into the 1930s, where a capacitance of one microfarad is equivalent to 900 jar.

Having said that, in recent years we have seen the development of a type of an exceptionally high capacitance capacitor termed a '**supercapacitor**' (also termed '**ultracapacitor**'), whose capacitance values are, indeed, measured in farads rather than microfarads or picofarads. In fact, some of the biggest commercial supercapacitor arrays (supercapacitors connected together) are measured in *thousands* of farads!

So what factors actually determine the capacitance of a capacitor?

### Factors affecting capacitance

The capacitance of a capacitor is, to some extent, dependent upon the physical size of that capacitor, being *directly proportional* to the **area of overlap** (symbol; *A*) of its plates, and *inversely proportional* to the **distance** (symbol; *d*) between the plates – as illustrated in Figure 22.17.

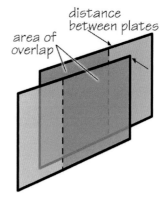

**Figure 22.17**

$$C \propto \frac{A}{d}$$

In other words, the capacitance *increases* with a *larger* area of overlap, and *decreases* if the plates are moved further apart. Again, we can change the proportional symbol, in the above expression, to an equals symbol by introducing a constant; in this case, we call the constant the **absolute permittivity** (symbol: ε, pronounced *'epsilon'*), which is a physical property of the dielectric.

The capacitance of a capacitor, then, is expressed as:

$$C = \varepsilon \frac{A}{d} \qquad \text{—(equation 2)}$$

**Important!** It is *not* the total area of either plate that affects the capacitance but, rather, the area by which the two plates *overlap* each other.

Changing the dielectric can have a *significant* affect on altering a capacitor's capacitance.

This is summarised in Table 22.1.

## How a dielectric affects capacitance

The capacitor's dielectric has *three* important functions. It must

- keep the two plates apart
- insulate the plates from each other
- improve the capacitance of the capacitor.

The first point is obvious; the plates mustn't be allowed to come into contact with each other or they will short-circuit.

The second point is that the insulating property of the dielectric must not be allowed to fail when the rated voltage of the capacitor is applied across its plates. We express this property in terms of its **dielectric strength**, which is measured in volts per metre (in practice, kilovolts per millimetre). An ideal dielectric should have a very high dielectric strength.

The third point is less obvious but, as we shall see, it can have a dramatic effect on increasing the capacitance of the capacitor. So, *why does the dielectric affect a capacitor's capacitance?* Well, we know that there are relatively few free electrons in an insulating material, with the overwhelming majority of electrons being strongly tied within valence shells surrounding their atoms' nucleii. When they are not exposed to an electric field, their shells behave quite normally – as illustrated by the three atoms represented in the left-hand diagram in Figure 22.21.

In the right-hand diagram in Figure 22.21, however, these same atoms are subject to the electric field due to the potential difference between the plates of a charged capacitor. This field acts to 'stretch', or to distort, the electron shells so that they become 'elongated' along the direction of the lines of electric flux, with the 'positive centre' of each atom biased towards the negative plate, and the 'negative centre' biased towards

Table 22.1

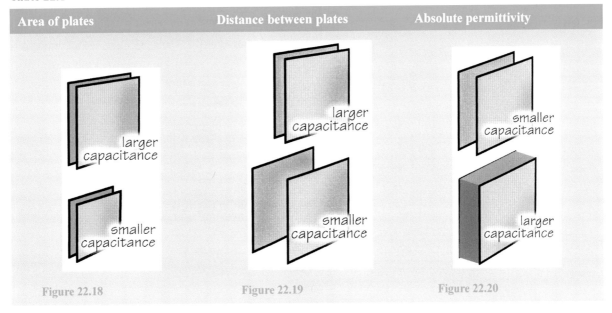

| Area of plates | Distance between plates | Absolute permittivity |

Figure 22.18    Figure 22.19    Figure 22.20

the positive plate. We say that the dielectric's atoms have become 'polarised'.

The amount of polarisation – i.e. the amount by which the electron orbit 'stretches' – depends on the type of dielectric and can vary considerably from one material to another.

In Figure 22.21, only three dielectric atoms are shown for the sake of clarity. In reality, of course, this behaviour applies to *each* of the billions of atoms within the dielectric.

When the capacitor is fully charged, all the polarised atoms within the dielectric will have orientated themselves so that their positive centres are attracted towards the negatively charged plate, while their negative centres are attracted towards the positively charged plate. So the direction of electric field *within each polarised atom itself* effectively acts in the opposite sense (direction) to the *main* electric field between the two plates.

The effect of this is to weaken the main field and, therefore, to *reduce the resultant voltage* across the dielectric *without affecting the charge* on the plates – *the resulting increase in the ratio of charge to voltage, therefore, acts to cause a corresponding increase in the capacitance of the capacitor.*

So, for a capacitor of given physical dimensions, the dielectric ultimately determines its final capacitance, by modifying the ratio of its charge to voltage. This increase in a capacitor's capacitance can be considerable, depending on the dielectric material used.

The 'reference dielectric material' is 'free space' (a term used by scientists to describe a vacuum or, in practice, air), whose permittivity is termed the **'permittivity of free space'** (symbol: $\varepsilon_o$). This is numerically equal to $(8.85 \times 10^{-12})$ F/m.

*All* other dielectric materials are compared with the dielectric of free space – for example, mica has five times the permittivity of free space, so we say that mica has a **'relative permittivity'** (symbol: $\varepsilon_r$) of 5. We can think of 'relative permittivity' as a measure of how much a dielectric material can be polarised (stretched) when subjected to an electric field. So a capacitor using a mica dielectric, for example, will have *five* times the capacitance of a capacitor, of identical dimensions, with an air dielectric.

So the type of dielectric material used can result in dramatic increases in capacitance, as can be seen from Table 22.2.

An older, alternative, name for **relative permittivity** is 'dielectric constant'.

So, **absolute permittivity** ($\varepsilon$) is the product of the **permittivity of free space** ($\varepsilon_o$) and **relative permittivity** ($\varepsilon_r$). So we can now rewrite equation (2) in its final form, as follows:

$$C = \varepsilon_o \varepsilon_r \frac{A}{d}$$
—(equation 3)

where:

$C$ = capacitance, in farads

$\varepsilon_o$ = permittivity of free space (farads per metre)

$\varepsilon_r$ = relative permittivity (no units)

$A$ = overlapping area of plates (square metres)

$d$ = distance between plates (metres)

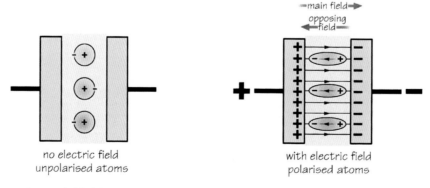

no electric field
unpolarised atoms

with electric field
polarised atoms

electric field within polarised dielectric atoms opposes main electric field
–reducing effective potential difference between plates

Figure 22.21

**Worked example 1** The plates of a capacitor measure 25 mm square, and are placed 1 mm apart. Calculate (a) its capacitance if the dielectric is air, and (b) its capacitance if the dielectric is mica. Assume that the relative permittivity of air is 1, and the relative permittivity of mica is 5.

**Solution** We first need to determine the cross-sectional area of the capacitor's plates:

$$A = (25 \times 10^{-3}) \times (25 \times 10^{-3})$$
$$= 25 \times 25 \times 10^{-6} = 625 \times 10^{-6} \, \text{m}^2$$

a  For an **air** dielectric:

$$C = \varepsilon_o \varepsilon_r \frac{A}{d} = \left(1 \times 8.85 \times 10^{-12}\right) \times \frac{625 \times 10^{-6}}{1 \times 10^{-3}}$$
$$= 5.53 \times 10^{-12} = 5.53 \, \text{pF Answer (a)}$$

b  For a **mica** dielectric:

$$C = \varepsilon_o \varepsilon_r \frac{A}{d} = \left(5 \times 8.85 \times 10^{-12}\right) \times \frac{625 \times 10^{-6}}{1 \times 10^{-3}}$$
$$= 27.66 \times 10^{-12} = 27.66 \, \text{pF Answer (b)}$$

## Relative permittivity versus dielectric strength

Table 22.2 compares the average relative permittivities with the average dielectric strengths of a range of dielectrics commonly used in the manufacture of capacitors.

**Table 22.2**

| Dielectric material | Relative permittivity | Dielectric strength (kV/mm) |
|---|---|---|
| Air | 1.0006 | 3 |
| Teflon | 2 | 60 |
| Paper | 2.5 | 20 |
| Rubber | 3 | 28 |
| Transformer oil | 4 | 16 |
| Mica | 5 | 200 |
| Glass | 6 | 120 |
| Porcelain | 6 | 8 |
| Bakelite | 7 | 16 |
| Barium-strontium titanate | 7500 | 3 |

You will notice from the table that there is absolutely *no* correlation whatsoever between a dielectric's **relative permittivity**, and its **dielectric strength**. For example, barium-strontium titanate (termed a 'ferroelectric ceramic') has a *huge* relative permittivity of 7500, but a *very low* dielectric strength of just 3 kV/mm which is no better than that of air.

This can be a problem for manufacturers who need to design capacitors, particularly those for use in high voltage systems, who find themselves unable to use a dielectric with a high relative permittivity if its dielectric strength is too low. So the selection of an appropriate dielectric is an exercise in compromise – i.e. deciding on which is the most important for a particular application; its relative permittivity or its dielectric strength!

**Absolute permittivity vs absolute permeablity**

In the chapter on *magnetic circuits*, we met the term 'absolute permeability' and learnt that it was a measure of the ease with which a medium (air or a ferromagnetic material) allows the formation of magnetic flux.

We also learnt that absolute permeability was the ratio of **magnetic flux density** *(B)* to **magnetic field strength** *(H), that is:*

$$\mu = \frac{B}{H}$$

An electric circuit's 'equivalent' of magnetic field strength, you will recall, is a 'voltage gradient' which, when applied to a dielectric, is called its **dielectric strength** *(D)*.

**Absolute permittivity** (symbol: $\varepsilon$) compares with absolute permeability, as it is the ratio of **electric flux density** (symbol: *D*) to **dielectric strength** (symbol: *E*), that is:

$$\varepsilon = \frac{D}{E}$$

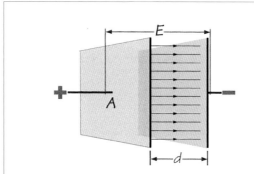

**Figure 22.22**

If you imagine every line of electric flux represented to eminate from a charge of one coulomb, then the unit of measurement for electric flux density will be a **coulomb per square metre** (C/m²). And, of course, dielectric strength will be measured in **volts per metre** (V/m).

For *air* (or, more strictly, for a vacuum), absolute permittivity corresponds to the **permittivity of free space** ($\varepsilon_o$), which is a constant and equal to:

**$8.85 \times 10^{-12}$ F/m** (farads per metre)

To find the absolute permittivity for any other dielectric, we must multiply the absolute permittivity of free space by the relative permittivity of that particular dielectric:

$$\varepsilon = \varepsilon_r \varepsilon_o$$

This equates with the following for magnetic circuits:

$$\mu = \mu_r \mu_o$$

## Voltage rating of capacitors

Unless specifically stated otherwise, a capacitor's **voltage rating** is *always* a **d.c. rating**. So if, for example, the voltage rating of a capacitor is 12 V, then any applied voltage in excess of 12 V might cause its dielectric to break down and fail.

Care, therefore, needs to be taken if the same capacitor is to be used in an *a.c.* circuit, to ensure the circuit's a.c. *peak value* of voltage does not exceed the capacitor's d.c. voltage rating.

Unfortunately, a.c. voltages are *always* expressed as *root-mean-square* (*r.m.s.*) values, *never* as peak values – where, $E_{peak} = 1.414 E_{rms}$.

The nominal '230 V' which supplies your residence is an r.m.s. value, *not* a peak value; its peak value is 325 V!

This means that an a.c. voltage, having an r.m.s. value of 12 V, will actually *peak* to very nearly 17 V, which is significantly more than the dielectric of a capacitor rated at 12 V is designed to withstand.

We will learn about r.m.s. values in a later chapter on *alternating current*, but for now it's good enough to understand that an a.c. peak value is $1.414 E_{rms}$.

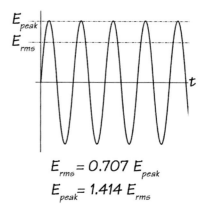

$$E_{rms} = 0.707 \, E_{peak}$$
$$E_{peak} = 1.414 \, E_{rms}$$

**Figure 22.23**

So, installing a 12-V capacitor into a 12-V a.c. circuit will subject that capacitor to nearly 17 V which will likely damage that capacitor!

**So, the safe a.c. (r.m.s.) voltage which can be applied to a capacitor is no more than 0.707 × the capacitor's rated d.c. voltage.**

**Important!** Unless otherwise stated, a capacitor's rated voltage is a d.c. voltage. Care must, therefore, be taken when using a capacitor in an a.c. circuit, because a.c. voltages are always root-mean-square voltages, which peaks to 1.41× the r.m.s. value.

**Worked example 2** Can a capacitor rated at 250-V(d.c.) be used in parallel with a 230-V(a.c.) supply?

**Solution** Remember, the peak value of the a.c. voltage must not exceed the rated d.c. value of the capacitor, or there is a risk that the capacitor's dielectric may be damaged. So, in this case, the supply's peak value must not exceed 250 V.

Since 230 V is the supply's r.m.s voltage,

$$E_{rms} = 0.707\,E_{max}$$

$$\text{so, } E_{max} = \frac{E_{rms}}{0.707} = \frac{230}{0.707} = 325 \text{ V}$$

So, this capacitor *cannot* be connected across the 230-V load without the risk of its dielectric failing.

## Energy stored by a capacitor

We learnt, in the chapter on *potential and potential difference*, that whenever we isolate an electric charge, there must be an equal charge, of opposite polarity, left behind, and that an electric field is set up between them.

The **energy** stored within a capacitor's electric field originates from the work done establishing that electric field.

As the potential difference *(U)* across a capacitor's plates is directly proportional to the charge *(Q)*, if we were to draw a graph of potential difference against charge throughout the charging process, the result would be a straight-line graph, as illustrated in Figure 22.24.

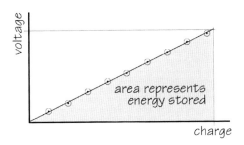

Figure 22.24

The triangular area (shaded grey in Figure 22.24) under the graph line represents the quantity of energy stored in that field. The area of a triangle, you will recall, is half its base times its height; so the equation for the energy stored in the field may be expressed as:

$$W = \frac{1}{2} U Q \qquad \text{—(equation 4)}$$

If we apply simple dimensional analysis to this, it should become apparent *why* this area represents the stored energy:

$$U \times Q = \text{volt} \times \text{coulomb} = \left(\frac{\text{joule}}{\text{coulomb}}\right) \times \text{coulomb} = \text{joule}$$

So, as you can see, the product of voltage and charge is *energy*, expressed in joules.

If we now substitute for $Q$ in equation (4), we can rewrite the equation as follows:

$$W = \frac{1}{2} U (CU)$$

$$\boxed{W = \frac{1}{2} CU^2} \qquad \text{—(equation 5)}$$

where:

$W$ = energy, in joules

$C$ = capacitance, in farads

$U$ = voltage, in volts

**Worked example 3**    A fully charged 250 μF capacitor has a potential difference of 100 V across its plates. What is the amount of charge transferred?

**Solution**

$$Q = CU = \left(250 \times 10^{-6}\right) \times 100 = 2500 \times 10^{-6}$$

$$= 25 \text{ mC (Answer)}$$

**Worked example 4**    How much energy is stored within the electric field of the above capacitor when it is fully charged?

**Solution**

$$W = \frac{1}{2} CU^2 = \frac{1}{2} \times (250 \times 10^{-6}) \times 100^2$$

$$= 1.25 \text{ J (Answer)}$$

## Construction of practical capacitors

Up until now, we've rather assumed that a capacitor consists of two *flat* metal plates, separated by a dielectric – just like its circuit symbol. However, this does *not* reflect the actual construction of practical capacitors.

First of all, as we have already learnt, there are basically *two* types of capacitor; **fixed value capacitors** and **variable value capacitors**.

As the names indicate, fixed-value capacitors are manufactured with a fixed value of capacitance, while variable-value capacitors allow their capacitance to be varied by the user.

As we have also learnt, the plates of a capacitor assume the same polarities as the external potentials to which they are attached, so it doesn't really matter how they are connected to their voltage source. The exception to this rule is a type of fixed-value capacitor, called an '**electrolytic capacitor**' – this capacitor *must* be connected according to the polarity marks printed on its container. Failure to do so will result in the destruction of the electrolytic capacitor.

### Fixed-value capacitors

Capacitors are manufactured in a huge range of physical sizes, ranging from the size of a match head to the size of a distribution transformer.

Capacitors are generally named after the material from which their dielectrics are manufactured. These include *paper capacitors, mica capacitors, ceramic capacitors*, etc.

An exception to this rule is the '**electrolytic capacitor**'. For these capacitors, the dielectric is an *oxide* which is formed on the surface of its positive plate. An electrolytic gel then forms the *negative electrode*. So, the term 'electrolytic' refers to its negative plate, *not* its dielectric.

**Paper capacitors** typically consist of two long strips of metal foil, interlaced with two strips of oil-impregnated paper dielectric, rolled together rather like a 'Swiss roll', and sealed inside an aluminium or plastic protective tubular container (Figure 22.25). These days, plastics have tended to replace paper, but there are still lots of paper capacitors about.

Figure 22.25

Electricians are most likely to come across **oil-impregnated paper capacitors**. This type of capacitor exists in a great many shapes and sizes, from relatively small types used to start some types of single-phase a.c. motor, to much larger types used for **power factor improvement** (see the chapter on *power factor improvement*) in industrial installations.

Capacitors used for these industrial purposes are very much larger (as illustrated in Figure 22.26), partly because they often have to operate at relatively high voltages, so their plates need to be relatively far apart and consist of interleaved rectangular plates, connected in parallel, and immersed in a mineral oil.

Figure 22.26

Instead of a simple pair of plates, some capacitors (such as mica capacitors) often have a number of plates/dielectrics stacked in such a way as to create the equivalent of several individual capacitors in parallel which, of course, *increases* the overall capacitance by that number. This arrangement is illustrated in Figure 22.27.

**Stacked-Plate Capacitor**

Figure 22.27

### Variable-value capacitors

A **variable-value capacitor** is an air-dielectric capacitor designed to allow its capacitance to be varied over a given range. One of the most common applications for a variable-value capacitor, of the type shown in Figure 22.28, is the tuning control of an analogue radio. It consists of a number of fixed fan-shaped plates, interleaved with moving plates, or 'vanes', controlled by a rotating knob. The capacitance is varied according to the area of overlap of the two sets of plates.

Figure 22.28

## Behaviour of capacitors in d.c. circuits

In the circuit shown in Figure 22.29, a capacitor is connected in series with a resistor across a battery of voltage, $E$ volts.

Figure 22.29

With the switch in position *a*, the battery will act to *charge* the capacitor. As the capacitor charges, its voltage, $U_C$, 'gradually' increases until it equals (but opposes) the supply voltage – following the curve shown in Figure 22.30. Had the capacitor continued to charge at its initial rate, then its voltage would have followed the dotted line – reaching its fully charged voltage in $CR$ seconds. This is known as the circuit's **time constant**, expressed in seconds (s):

Figure 22.30

$$\text{time constant} = CR$$

where:

$C$ = capacitance (farads)

$R$ = resistance (ohms)

However, by following the curve, it actually reaches its fully charged voltage in **5 time constants**:

$$\text{time to fully discharge} = 5\,CR$$

The chain line represents the **charging current,** which is maximum *before* the potential difference across the capacitor $(U_C)$ starts to build up. In fact, *the charging current is **directly proportional to the rate of change of** $U_C$*, which means that it is greatest when the voltage's curve is at its steepest.

We will return to the fact that a capacitor's charging current is proportional to the rate of change of voltage when we study a.c. circuits, later in this book.

With the switch moved to position **b** (Figure 22.31), the capacitor will discharge. As it discharges, its voltage, $U_C$, 'gradually' decreases until it reaches zero volts – following the curve shown in Figure 22.32.

Figure 22.31

**Figure 22.32**

Had the capacitor continued to discharge at its *intial* rate then, again, it would fully discharge in *CR* seconds – again, known as the circuit's **time constant**. However, by following the actual curve, the capacitor will fully discharge in **5 time constants**.

> time to fully discharge = 5*CR*

*The **time constant** is unaffected by the circuit's voltage, and depends solely upon the capacitance and resistance of the circuit.*

The chain line, again, represents the **discharge current** which, of course, acts *in the opposite direction to the charging current*.

If we compare the two *voltage* curves, above, with the *current* curves for an inductor, we see that they have *identical* shapes. So, whereas an inductor opposes any change in current, *a capacitor will oppose any change in voltage*.

> So from this we can see that, in the same way that an inductor will oppose any change in *current*, a **capacitor always acts to oppose any change in voltage**.

**Worked example 5**   A 50 μF capacitor is connected in series with a 100-Ω resistor. If this circuit is suddenly connected across a 100-V (d.c.) supply, how long will the capacitor take to fully charge?

**Solution**   The 100-V (d.c.) supply is irrelevant to this question, as the time taken for the capacitor to reach its fully charged state will be five time constants. So,

$$\text{time to fully charge} = 5\,CR$$
$$= 5 \times (50 \times 10^{-6}) \times 100$$
$$= 25000 \times 10^{-6}$$
$$= 25 \text{ ms (Answer)}$$

**Worked example 6**   A circuit, comprising a fully charged 50 μF capacitor in series with a 100-MΩ resistor, is suddenly short-circuited. How long will the capacitor take to fully discharge?

**Solution**

$$\text{time to fully discharge} = 5CR$$
$$= 5 \times (50 \times 10^{-6}) \times (100 \times 10^{6})$$
$$= 25000\text{s (Answer)}$$
$$\text{or} \quad \frac{25000}{60} = 416 \text{min } 40 \text{ s (Answer)}$$

As the above examples show, a capacitor's charging and discharging times can vary considerably, depending on the resistance of the circuit.

## Supercapacitors

Supercapacitors, or 'ultracapacitors' (see Figure 22.33) as they are also known, are a type of electrolytic capacitor which differ from regular electrolytic capacitors in three important ways: (1) the 'effective' area of their plates is much bigger than their 'actual' area, (2) the distance between their plates is very much smaller, and (3) the material between them behaves quite differently from normal dielectrics.

**Figure 22.33**

A supercapacitor's plates are coated with a porous material which makes them behave as though they have a much larger area than their actual, physical, area. This is what we mean when we say the 'effective area' of the plates (due to their irregular surface) is larger than their 'actual area'.

The plates of most normal capacitors are separated by a dielectric which is typically between a few micrometres to around a millimetre or so thick. Supercapacitors, on the other hand, have no dielectric! Instead, both plates are soaked in an electrolyte and separated by an extremely-thin insulator equivalent to the width of one molecule. When the capacitor is charged, opposite electrical charges appear on either side of the separator resulting in what is, effectively, two capacitors in series, but each having an incredibly-narrow plate separation!

This incredibly-small plate separation means that supercapacitors can only be used, individually, with very low values of voltage (typically, depending on type, 2.5 V, 2.7 V, or 5.5 V). So, for larger voltages, numerous supercapacitors must be connected in series. The drawback with this, of course, is that the effective capacitance when capacitors are connected in series is reduced. So, in practice, a combination of series and parallel connections is used to form what is known as an 'array'.

This combination of a plate area which is effectively very much bigger than their physical area, combined with an incredibly-narrow separation, is what gives supercapacitors their huge capacitance value.

So how are supercapacitors used?

Well, they are starting to be used as temporary energy 'reservoirs', in situations where it's necessary to store more energy than can be stored with a regular capacitor, but when there is simply insufficient time to charge a battery.

Applications include regenerative braking in electric and hybrid vehicles in which the energy, which would otherwise be wasted during braking, is briefly stored before being re-used when the vehicle moves off; a sort of 'electronic flywheel', if you like. It is claimed that this can lead to energy savings of up to 50%.

Wind turbines and solar panels are also using supercapacitors to stabilise their energy output when wind gusts or clouds would otherwise cause fluctuations.

## Natural capacitance

It's not just **capacitors** that exhibit **capacitance**. Both **overhead lines** and **underground cables** behave as 'natural capacitors'!

In the case of overhead lines, adjacent conductors (and each conductor and the earth) behave like the plates of a capacitor, and the air acts as the dielectric.

In the case of underground cables (in fact, *any* long length of cable), adjacent conductors (or individual conductors and earth) act like plates, while the conductors' insulation acts as the dielectric.

The 'capacitance' of cables is significantly higher than for overhead lines, due to the extreme closeness of their conductors ('plates'), and **care must be taken to fully discharge long lengths of cable that may have accumulated appreciable charge separation during an insulation test using a Megger** (a high voltage test instrument), in order to avoid a shock hazard!

Electrical cables used for residential wiring have a capacitance of around 100 pF per metre length and the resulting capacitive currents can be responsible for some odd behaviours that occasionally occur in electrical installations, such as CFLs (compact fluorescent lamps) which continue to flicker *after* they have been switched off.

## Capacitors in series and parallel

### Capacitors in series

Figure 22.34

In accordance with Kirchhoff's Voltage Law, the sum of the voltage drops across each capacitor will equal the supply voltage:

$$E = U_1 + U_2 + U_3 \qquad \text{—(equation 6)}$$

But we know that voltage is charge divided by capacitance, so we can substitute for voltage in equation (6):

$$\frac{Q}{C} = \frac{Q}{C_1} + \frac{Q}{C_2} + \frac{Q}{C_3} \qquad \text{—(equation 7)}$$

As we learnt earlier, the charge on capacitors in series is simply the amount of charge accumulated on the negative plate of the outermost capacitor ($C_3$, in the above example), so it is the same *regardless of the number of capacitors*.

So, we can divide equation (7) throughout by Q, giving:

$$\frac{\cancel{Q}}{\cancel{Q}C} = \frac{\cancel{Q}}{\cancel{Q}C_1} + \frac{\cancel{Q}}{\cancel{Q}C_2} + \frac{\cancel{Q}}{\cancel{Q}C_3}$$

$$\boxed{\frac{1}{C} = \frac{1}{C_1} + \frac{1}{C_2} + \frac{1}{C_3}}$$

To summarise for a **series** circuit:

$$U = U_1 + U_2 + U_3$$

$$Q = Q_1 = Q_2 = Q_3$$

$$\frac{1}{C} = \frac{1}{C_1} + \frac{1}{C_2} + \frac{1}{C_3}$$

**Worked example 7**   Three capacitors, of 25 μF, 50 μF and 75 μF, are connected in series across a 100-V D.C. supply. Calculate each of the following:

a   the circuit's total capacitance
b   the total charge
c   the voltage drop across each capacitor
d   the total energy stored.

**Solution**   As always, you should start by sketching the circuit (Figure 22.35), and inserting all the values given in the question.

**Figure 22.35**

a   $\dfrac{1}{C} = \dfrac{1}{C_1} + \dfrac{1}{C_2} + \dfrac{1}{C_3}$

$\dfrac{1}{C} = \dfrac{1}{25} + \dfrac{1}{50} + \dfrac{1}{75}$

$\dfrac{1}{C} = \dfrac{6+3+2}{150} = \dfrac{11}{150}$

$C = \dfrac{150}{11} = 13.64\ \mu\text{F (Answer a.)}$

b   $Q = CE = 13.64 \times 10^{-6} \times 100 = 1.364\ \text{mC}$
(Answer b.)

c   $U_1 = \dfrac{Q}{C_1} = \dfrac{1.364 \times 10^{-3}}{25 \times 10^{-6}} = 54.56\ \text{V (Answer c.1)}$

$U_2 = \dfrac{Q}{C_2} = \dfrac{1.364 \times 10^{-3}}{50 \times 10^{-6}} = 27.28\ \text{V (Answer c.2)}$

$U_3 = \dfrac{Q}{C_3} = \dfrac{1.364 \times 10^{-3}}{75 \times 10^{-6}} = 18.19\ \text{V (Answer c.3)}$

check your answer:

$$(U_1 + U_2 + U_3 = 54.56 + 27.28 + 18.19 \approx 100\ \text{V})$$

d   $W = \dfrac{1}{2}CE^2$

$= \dfrac{1}{2} \times 13.64 \times 10^{-6} \times 100^2 = 68200 \times 10^{-6}$

$= 68.2\ \text{mJ (Answer)}$

## Capacitors in parallel

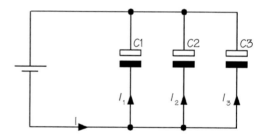

**Figure 22.36**

In accordance with Kirchhoff's Current Law, the sum of the individual branch currents will equal the supply current:

$$I = I_1 + I_2 + I_3 \qquad \text{—(equation 8)}$$

This time, the total charge must be the sum of the negative charge accumulated on the right-hand plate of each of the individual capacitors:

$$Q = Q_1 + Q_2 + Q_3 \qquad \text{—(equation 9)}$$

Since charge is the produce of capacitance and voltage, we can rewrite equation (9) as follows:

$$CE = C_1U + C_2U + C_3U$$

Since the supply voltage is common to each branch of a parallel circuit, we can now divide throughout by voltage:

$$\frac{CE}{E} = \frac{C_1 U}{U} + \frac{C_2 U}{U} + \frac{C_3 U}{U}$$

$$\boxed{C = C_1 + C_2 + C_3}$$

To summarise for a **parallel** circuit:

$$I = I_1 + I_2 + I_3$$

$$Q = Q_1 + Q_2 + Q_3$$

$$C = C_1 + C_2 + C_3$$

**Worked example 8**    Three capacitors, of 25 μF, 50 μF and 75 μF, are connected in parallel across a 100-V D.C. supply. Calculate each of the following:

a    the circuit's total capacitance
b    the charge on each capacitor
c    the voltage drop across each capacitor
d    the energy stored on capacitor $C_1$.

**Solution**    As always, you should start by sketching the circuit (see Figure 22.37), and inserting all the values given in the question.

Figure 22.37

a    $C = C_1 + C_2 + C_3 = 25 + 50 + 75 = 150 \ \mu\text{F}$
(Answer a.)

b    $Q = CU = 150 \times 10^{-6} \times 100 = 15000 \times 10^{-6}$
$= 15 \ \text{mC}$ (Answer b.)

c    $Q_1 = C_1 U = 25 \times 10^{-6} \times 100 = 2.5 \ \text{mC}$
(Answer c.1)

$Q_2 = C_2 U = 50 \times 10^{-6} \times 100 = 5.0 \ \text{mC}$
(Answer c.2)

$Q_3 = C_3 U = 75 \times 10^{-6} \times 100 = 7.5 \ \text{mC}$
(Answer c.3)

check your answer:

$$(Q_1 + Q_2 + Q_3 = 2.5 + 5.0 + 7.5 = 15.0 \ \text{mC})$$

d    $W = \frac{1}{2} C_1 U^2$

$= \frac{1}{2} \times 25 \times 10^{-6} \times 100^2 = 125 \ \text{mJ}$ (Answer)

## Capacitors in series-parallel

We resolve series-parallel capacitive circuits in much the same way we learnt to logically solve resistors in series-parallel. By way of example, let's solve the series-parallel circuit shown in Figure 22.38.

Figure 22.38

We start by finding the equivalent capacitance of the two capacitors, $C_1$ and $C_2$ that are connected in parallel, which we'll label $C_A$:

$$C_A = C_1 + C_2 = 100 + 200 = 300 \ \text{pF}$$

This leaves $C_A$ and $C_3$ in series, from which we can find the equivalent capacitance of the complete circuit, $C$:

$$\frac{1}{C} = \frac{1}{C_A} + \frac{1}{C_3} = \frac{1}{300} + \frac{1}{300} = \frac{2}{300}$$

$$\therefore C = \frac{300}{2} = 150 \ \text{pF} \quad (\text{Answer})$$

Now, let's find the total charge of the circuit:

$$Q = CU = 150 \times 10^{-12} \times 200 = 30 \times 10^{-9} \ \text{C} \quad (\text{Answer})$$

This charge will appear on capacitor $C_3$ and on the parallel part of the circuit. So, we can now work out the voltage drops. First of all, the voltage drop, $U_2$, across $C_3$ can be found from:

$$U_2 = \frac{Q}{C_3} = \frac{30 \times 10^{-9}}{300 \times 10^{-12}} = 100 \text{ V} \quad \text{(Answer)}$$

$$\text{So, } U_1 = E - U_2 = 200 - 100 = 100 \text{ V (Answer)}$$

Finally, we can determine the charge on each of the capacitors $C_1$ and $C_2$:

$$Q_1 = U_1 C_1 = 100 \times 100 \times 10^{-12} = 10 \times 10^{-9} \text{ C (Answer)}$$
$$Q_2 = U_1 C_2 = 200 \times 100 \times 10^{-12} = 20 \times 10^{-9} \text{ C (Answer)}$$

(We can confirm our answer from: $Q_1 + Q_2 = Q$)

## Review your learning

Now that we've completed this chapter, we need to examine the **objectives** listed at its start. Placing '*Can I...*' at the beginning, and a question mark at the end, of each objective turns that objective into a **test item**. If we can answer each of those test items, then we've met the objectives of this chapter.

**Online resources**

The companion website to this book contains further resources relating to this chapter. The website can be accessed via the following link:
www.routledge.com/cw/waygood

# Chapter 23

# Capacitors

On completion of this chapter, you must be able to

1  identify fixed-value, variable-value, polarised and trimmer capacitors by their circuit symbol.
2  identify the factors that determine the physical size of a capacitor.
3  identify types of capacitor listed in this chapter.
4  identify the capacitance of a capacitor, using
   a  colour codes
   b  alpha-numeric codes.

## Introduction

The natural capacitance of any circuit can be changed by using circuit components called **capacitors**.

Because capacitors oppose the change in voltage, we can also use capacitors, together with other components such as resistors, to create useful circuits, such as timing circuits or smoothing circuits.

A **capacitor** is defined as *'a passive two-terminal electrical component that stores electrical energy in an electric field'*.

Capacitors may be **fixed-value** or **variable-value** (Figures 23.1–23.3).

capacitors

**ffixed-value capacitors**    **variable-value capacitors**

Figure 23.1 (IEC symbol)    Figure 23.2 (IEC symbol)

Figure 23.3 (polarised capacitor IEC symbol)

A special type of variable capacitor is the **trimmer capacitor**. This is a variable capacitor that can be pre-set to a particular value of capacitance by the manufacturer when setting up a particular circuit, and is not intended to be adjusted by the end user.

Figure 23.4 (IEC symbol)

## Types of capacitor

### Comparing capacitors in terms of physical size

In the chapter on capacitance, we learnt that the **capacitance** *(C)* of a capacitor depends upon the area *(A)* of overlap of its plates and the distance *(d)* between them, and the absolute permittivity *(ε)* of its dielectric material:

$$C = \varepsilon \frac{A}{d}$$

The **absolute permittivity** is the constant in this equation, and is determined by the nature of a capacitor's dielectric. So, for a *given* dielectric material, a capacitor's capacitance is proportional to the area of overlap of its plates and their distance apart. In other words, *for a given dielectric and voltage rating, the larger the **volume** of a capacitor, the higher its capacitance.*

However, as the absolute permittivity varies enormously, it is impossible to compare the capacitance of capacitors with *different dielectrics* in terms of their physical size.

A capacitor's voltage rating also affects its physical size. For a given dielectric, the higher the voltage rating, the further the plates have to be apart – i.e. the *thicker* the dielectric. So, for a given capacitance and dielectric, the higher the rated voltage, the greater the physical size of the capacitor.

So, from the preceding, we need to understand that *it is impossible to compare the capacitance of different types of capacitor simply by comparing their physical sizes.*

## Fixed-value capacitor

Most, but not all, capacitors are named according to the **dielectric** used to separate their plates. For example, **paper capacitors** use oil- or wax-impregnated paper, **mica capacitors** use sheets of mica, etc.

**Electrolytic capacitors**, however, are named according to (1) *the metal which forms their positive plate* (e.g. aluminium, tantalum, etc.) and (2) the fact that *an electrolyte acts as their negative 'plate'*. Hence, 'aluminium electrolytic' or 'tantalum dielectric' capacitors.

The most commonly used types in power-level (i.e. mains voltage and above) applications are **paper** capacitors, while in electronics circuits the most common are **ceramic**, **plastic-film** and **electrolytic** types.

### Paper capacitors

Figure 23.5     Figure 23.6

**Paper capacitors** vary enormously in physical size, as shown in Figures 23.5 and 23.6, and are used for a wide range of applications at low, medium and high voltage – including induction motor starting and power-factor correction. They use oil- or wax-impregnated paper

as a dielectric and are available for a wide range of voltage levels. Power-factor improvement capacitors (to be discussed later in this book) are rated in **reactive volt amperes (var)**, *never* farads. A 'reactive volt ampere' is a unit of measurement used in alternating-current circuits for measuring reactive power (i.e. the 'power' of a purely-inductive circuit). We will discuss this unit, further, in the chapters on alternating current.

### Mica capacitors

Mica capacitors use the mineral mica as a dielectric. They are manufactured by plating the surface of the mica with silver, and layering them in order to achieve the desired capacitance. To protect the assembled capacitor, it is usually encapsulated in epoxy resin.

Mica capacitors typically range within the picofarad – nanofarad range, and their voltage rating is usually between 100 V and 1000 V.

### Ceramic capacitors

Figure 23.7

**Ceramic capacitors** are used in radio-frequency and audio applications. They vary in shape, including the disc type shown in Figure 23.7. They vary in capacitance from around 1 nF to around 100 μF, depending on their physical size.

Physically large ceramic capacitors can be made to withstand voltages in the range of 2 kV – 100 kV.

### Plastic-film capacitors

Figure 23.8

**Plastic-film capacitors** (Figure 23.8) are general-purpose capacitors which use thin dielectric films made from materials such as polycarbonate, polyester and polystyrene. They vary in capacitance from less than 1 nF to around 30 μF, and have rated voltages between 50 V and around 2000 V, making them ideal for power applications such as power-factor improvement in fluorescent lamps.

### Electrolytic capacitors

Figure 23.9

An **electrolytic capacitor** (Figure 23.9) is a polarised capacitor which uses an electrolyte (a conducting fluid), usually in a paste or gel form, in order to achieve a larger capacitance than most other capacitor types.

Aluminium electrolytic capacitors are manufactured from two layers of aluminium foil, with a spacer soaked in an electrolytic gel, such as manganese dioxide, rolled and inserted into a cylindrical enclosure. The positive plate is formed by depositing a thin layer of oxide on one of the aluminium foils. This oxide then becomes the dielectric, while the electrolyte acts as the negative plate. The second aluminium foil simply acts to connect the electrolyte 'plate' to the external lead.

Electrolytic capacitors are polarised, and it is very important to connect them into any circuit the correct way around, observing the polarity markings on the capacitor – failure to do so will certainly damage the capacitor, and may even cause it to explode! They should also be operated well below their rated voltage. The capacitance of these capacitors is typically between 1 μF – 50 mF.

A **tantalum capacitor** (Figure 23.10) is a type of electrolytic capacitor which uses the metal tantalum, instead of aluminium, and is physically much smaller than an equivalent aluminium electrolytic capacitor described above.

Figure 23.10

Its positive plate is made from tantalum on which a thin film of oxide is allowed to form and which then acts as a dielectric. An electrolyte gel that covers the oxide then acts as the negative plate. The combination of the very thin and high permittivity dielectric results in a very high capacitance (up to around 70 mF) for its small volume.

In the case of the capacitor illustrated in Figure 23.10, the lead nearest the shaped side connects to the positive plate.

A **supercapacitor** is a type of electrolytic capacitor which uses a 'double-layer' (see previous chapter) technique to achieve amazingly high values of capacitance, well into the thousands of farads.

### Variable-value capacitors

Variable-value capacitors use air as a dielectric. There are *two* ways of making a variable-value capacitor: (1) by *varying the area of overlap of the capacitor's plates*, or (2) by *varying the distance between the plates*. Of these, the most common method is by varying the area of overlap – as illustrated in Figure 23.11.

control knob

moving plates

fixed plates

Figure 23.11

A very common example of a variable resistor is an analogue radio's tuning capacitor. It comprises two sets of interleaved plates (called 'vanes'), separated by air (the dielectric); by rotating one set of plates, the area of overlap and, therefore, the capacitance, varies.

Trimmer capacitors are miniature variable-value capacitors. They need to be preset to a particular value, using a screwdriver. Some types use rotating 'vanes', as described above, while others allow their plates to be 'squeezed' together.

## Ratings

### Voltage rating

Unless specifically stated otherwise, a **capacitor's voltage rating is** *always* **a d.c. rating**. Care, therefore, needs to be taken if the capacitor is to be used in an *a.c.* circuit, to ensure the circuit's a.c. *peak-value* of voltage does not exceed the capacitor's d.c. voltage rating. This topic was covered in the preceding chapter, *Capacitance*, so there is little point in repeating it here.

## Identifying the capacitance of capacitors

Remember, the SI unit of capacitance, the farad (symbol: F) is an *enormous* unit; until the advent of supercapacitors, we would *never* have seen any capacitor expressed in farads! However, some of the biggest commercial supercapacitor arrays now have capacitances rated up to an incredible *several thousand farads!*

Most capacitors, however, are measured in **microfarads** (symbol: **µF**), **picofarads** (symbol: **pF**) or **nanofarads** (symbol: **nF**). There are *two* systems for identifying the capacitance of a capacitor:

### Colour code

The **colour code** is rarely used, these days. However, we are quite likely to come across colour-coded capacitors in older circuits or in parts boxes! Unfortunately, there are a number of different ways in which the colour code is applied to capacitors although, in all cases, the colours themselves represent the same values as for resistors. So, whichever mnemonic we learnt for resistor colour codes will also apply to capacitors.

However, the capacitance indicated by the colour code is *always* in **picofarads** (symbol: **pF**).

### *Stripes method – 1 (in picofarads)*

Figure 23.12

With this method, the colours are read from the top downwards (Figure 23.12). We need to be aware that there *are usually no gaps between the colours*, so when two or more bands are the same colour they appear as a single, thick band!

### *Stripes method – 2 (in picofarads)*

Figure 23.13

In the case of the disc-shaped ceramic capacitors, illustrated in Figure 23.13, coloured stripes are applied around the circumference of the capacitor. There are two systems: 5-colour and 3-colour. Both systems are read in a clockwise sequence – but note the position of the 'multiplier' stripe in the 3-band system, which can be confusing!

### *Dot methods (in picofarads)*

Figure 23.14

With this method (applied, in this case, to a mica capacitor), the capacitor must be orientated so that the direction-chevrons or arrows point to the right, and the colours are then read starting with the top left dot and working *clockwise* around the capacitor (Figure 23.14).

Table 23.1 shows the standard colour codes for capacitors.

### Alpha-numeric codes

Physically larger, modern capacitors may have their capacitance value printed on to their body or enclosure.

> **Note!** High-voltage and power-factor improvement capacitors are *not* rated in terms of their capacitance, but by their **reactive power** (measured in **reactive volt amperes**), together with their **operating voltage**.

Table 23.1 Standard colour codes for capacitors (picofarads)

| colour | first colour | second band | third band | (fourth band) |
| --- | --- | --- | --- | --- |
| | first sig. figure | second sig.figure | tolerance | voltage |
| Black | 0 | 0 | | – |
| Brown | 1 | 1 | 1% | 100 |
| Red | 2 | 2 | 2% | 200 |
| Orange | 3 | 3 | – | 300 |
| Yellow | 4 | 4 | – | 400 |
| Green | 5 | 5 | 5% | 500 |
| Blue | 6 | 6 | – | 600 |
| Violet | 7 | 7 | – | 700 |
| Grey | 8 | 8 | – | 800 |
| White | 9 | 9 | – | 900 |
| Gold | – | – | – | 1000 |
| Silver | – | – | 10% | 2000 |
| no colour | – | – | – | |

Electrolytic capacitors often have their capacitance and voltage rating (as well as polarity) printed on their aluminium container, as illustrated in Figure 23.15.

Figure 23.15

In this example, the capacitor's capacitance is 1000 µF, its rated voltage is 35 V and the negative terminal is indicated as being to the right.

We might experience some confusion over the way in which the units of measurement are shown. For example, the two oil-impregnated paper capacitors, shown in Figure 23.16, labelled '**50MFD**' and '**500mmFd**', are actually **50 µF** and **500 pF** respectively.

Figure 23.16

On older capacitors, '**microfarads**' may be seen printed as *any* of the following ways:

**µF µFd mF mFd mFD MFd MFD**

Similarly, '**picofarads**' may be seen printed in *any* of the following ways:

**pF µµF µµFd mmFd mmFd mmFD MMFd MMFD**

## Modern alpha-numeric code

For modern capacitors, the following alpha-numeric code is sometimes used on the capacitors themselves and, more usually, as labels on circuit diagrams.

As the name suggests, the alpha-numeric code uses a combination of **letters** and **numerals** to identify resistance values. The letters used are:

- **µ** – which represents **microfarads** ($\times 10^{-6}$)
- **n** – which represents **nanofarads** ($\times 10^{-9}$)
- **p** – which represents **picofarads** ($\times 10^{-12}$)

The letter also represents the *position of the decimal point*, as shown in Figure 23.17.

Figure 23.17

For the examples shown in Figure 23.17:

| μ50 = 0.50 μF | 5μ0 = 5.0 μF | 50μ = 50 μF |
|---|---|---|
| n50 = 0.50 nF | 5n0 = 5.0 nF | 50n = 50 nF |
| p50 = 0.50 pF | 5p0 = 5.0 pF | 50p = 50 pF |

In this system, *tolerances* are indicated as follows:

- **C** – which represents ±0.25%
- **D** – which represents ±0.50%
- **F** – which represents ±1.00%
- **G** – which represents ±2.00%
- **J** – which represents ±5.00%
- **K** – which represents ±10.00%

Any digits used *after* the tolerance code *usually* indicate the capacitor's **voltage rating** in volts.

An example of the alpha-numeric code, described above, is shown printed on the ceramic capacitor illustrated in Figure 23.18.

Figure 23.18

In this example, the alpha-numeric code indicates:

**560p** = **560 pF** capacitance
**K** = **10%** tolerance
**50** = **50-V** working voltage

## Review your learning

Now that we've completed this chapter, we need to examine the **objectives** listed at its start. Placing *'Can I ...'* at the beginning, and a question mark at the end, of each objective turns that objective into a **test item**. If we can answer each of those test items, then we've met the objectives of this chapter.

# Chapter 24

# Geometry and trigonometry revision for a.c.

On completion of this unit, you must be able to

1  recognise a right-angled triangle.
2  identify the adjacent, opposite and hypotenuse sides, relative to an internal angle of a right-angled triangle.
3  apply Pythagoras's Theorem to a right-angled triangle.
4  define the sine, cosine and tangent of the internal angles of a right-angled triangle, in terms of the ratios of its sides.
5  use a scientific calculator to determine the sine, cosine and tangent of any angle.
6  use a scientific calculator to determine an angle, given the value of its sine, cosine or tangent.
7  solve right-angled triangles using the sine, cosine and tangent ratios.

## Introduction

As we shall learn in the following chapters, the key to understanding alternating-current theory is to be able to construct and manipulate **phasor diagrams**.

'**Phasors**' and '**phasor diagrams**' are, to electrical engineering, what '**vectors**' and '**vector diagrams**' are to mechanical engineering. In fact, the terms 'phasors' and 'phasor diagrams' weren't introduced until around the 1960s; prior to that they were known as 'electrical vectors' and 'electrical vector diagrams'.

There are important differences between phasors and vectors, but we will not be discussing those differences in this chapter.

But, in the same way that we can use vectors to determine, for example, the resultant of two or more *forces* or *velocities*, we can use phasors to determine the resultant of two or more *a.c. voltages* or *currents*. The same rules apply to both vectorial addition and the addition of phasors.

We will learn more about phasor diagrams, and their vital importance in understanding the behaviour of alternating current, in the following chapter. However, before that, we need to remind ourselves of the rules of basic **geometry** and **trigonometry** as they relate to triangles – in particular, how these rules relate to **right-angled triangles**. We will be making a great deal of these rules when we learn how to use phasors in the following chapters.

## Terminology

To understand and apply phasor diagrams, we *must* be familiar with triangles. In particular, we *must* be able to 'solve' **right-angled triangles**. By 'solve', we mean that we have to be able to apply simple geometry and trigonometry in order to determine the internal angles of triangles, as well as the lengths of their sides.

A '**right-angled triangle**' (known as a '**right triangle**' in American English) is a triangle in which one of its internal angles is a right angle (90°).

The longest side of a right-angled triangle is opposite its right angle, and is termed its '**hypotenuse**'.

The other two sides are termed the '**adjacent**' and the '**opposite**', and are named in relation to either one of the two remaining internal angles – as illustrated in Figure 24.1.

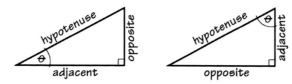

**Figure 24.1**

## Pythagoras's Theorem

Probably the best-known theorem in geometry, '**Pythagoras's Theorem**' applies to right-angled triangles. According to legend, the Greek mathematician and philosopher Pythagoras was so excited at discovering his theorem that he sacrificed an ox to the Gods! We may not find the theorem quite as exciting as Pythagoras seems to have, but we *must* know how to apply it – particularly if we wish to understand alternating current!

> **Pythagoras's Theorem** states that *'for a right-angled triangle, the square of the hypotenuse is equal to the sum of the squares of the other two sides'*.

This theorem can be expressed as:

$$\left(\text{hypotenuse}\right)^2 = \left(\text{adjacent}\right)^2 + \left(\text{opposite}\right)^2$$

The square of the hypotenuse is equal
to the sum of the squares of the other two sides.

**Figure 24.2**

The truth of this theorem can be demonstrated using a '**three-four-five**' triangle, that is a triangle whose sides measure 3 units, 4 units and 5 units in length, respectively – as illustrated in Figure 24.2. A 'three-four-five' ratio triangle *always* forms a right-angled triangle.

**Worked example 1**    A right-angled triangle has an adjacent measuring 2 m and an opposite measuring 8 m. What is the length of its hypotenuse?

**Solution**

$$\left(\text{hypotenuse}\right)^2 = \left(\text{adjacent}\right)^2 + \left(\text{opposite}\right)^2$$
$$\text{hypotenuse} = \sqrt{2^2 + 8^2} = \sqrt{4 + 64} = \sqrt{68}$$
$$= 8.25\,\text{m (Answer)}$$

**Worked example 2**    A right-angled triangle has a hypotenuse measuring 6 m and an opposite measuring 4 m. What is the length of its adjacent?

**Solution**

$$\left(\text{hypotenuse}\right)^2 = \left(\text{adjacent}\right)^2 + \left(\text{opposite}\right)^2$$
$$\left(\text{adjacent}\right)^2 = \left(\text{hypotenuse}\right)^2 - \left(\text{opposite}\right)^2$$
$$\text{adjacent} = \sqrt{6^2 - 4^2} = \sqrt{36 - 16} = \sqrt{20}$$
$$= 4.47\,\text{m (Answer)}$$

**Worked example 3**    An object is subjected to two forces, acting at right angles to each other. What is the value of the resultant force if the two forces are 25 N and 15 N respectively?

**Solution**

**Figure 24.3**

If the two forces are acting at right angles to each other, then they are equivalent to the adjacent and opposite sides of a 'triangle of forces', with the hypotenuse representing the resultant force, so:

$$\text{hypotenuse}^2 = \text{adjacent}^2 + \text{opposite}^2$$
$$\text{hypotenuse} = \sqrt{\text{adjacent}^2 + \text{opposite}^2}$$
$$= \sqrt{25^2 + 15^2} = \sqrt{625 + 225} = \sqrt{850};$$
$$= 29\,\text{N (Answer)}$$

## Trigonometry

**Trigonometry** is the branch of mathematics which is concerned about the relationship between the *sides* and *angles* of **triangles**. In this chapter, we are only concerned with right-angled triangles because, as we shall see, when we deal with phasors (in single-phase a.c. circuits, at least) we treat them as right-angled triangles.

The sum of a triangle's internal angles is 180°.

### Trigonometric ratios

In order to solve right-angled triangles, we need to understand those **trigonometric ratios**, called **sine (sin)**, **cosine (cos)** and **tangent (tan)**. It would be nice to know where these terms came from, but their origins (well, for sine and cosine, anyway) are obscure and have been handed down to us via several ancient languages such as ancient Greek and Latin. So, it is far easier to simply accept them rather than to worry about where they came from or what they originally meant.

The sine, cosine and tangent of any angle, $\theta$, are defined in terms of the ratios of two of the triangle's sides, as follows:

$$\sin\angle\theta = \frac{\text{opposite}}{\text{hypotenuse}} \qquad \cos\angle\theta = \frac{\text{adjacent}}{\text{hypotenuse}}$$

$$\tan\angle\theta = \frac{\text{opposite}}{\text{adjacent}}$$

To help us remember these ratios, we can use the mnemonic, '**SOH-CAH-TOA**', as shown in Figure 24.4.

Figure 24.4

We can determine the sine, cosine and tangent of any angle using a scientific calculator. The first step is to ensure that the calculator is set to measure **degrees** (rather than radians or grads), then hit the '**sin**', '**cos**' or '**tan**' key, followed by the angle (some calculators may require you to enter the angle first).

Although angles are traditionally measured in **degrees**, **minutes** and **seconds**, we would normally enter angles in *decimal* form – e.g. 50.5°, rather than as 50°30'. Refer to the calculator's user manual for details.

For sine and cosine, it's useful (but *not* essential) to try to remember the following:

| angle | sine | cosine |
|-------|-------|--------|
| 0° | 0.000 | 1.000 |
| 30° | 0.500 | 0.866 |
| 60° | 0.866 | 0.500 |
| 90° | 1.000 | 0.000 |

As we will learn in the next chapter, the most commonly used of these, when it comes to working with phasor diagrams, is the **cosine ratio**.

We can also use our scientific calculators to determine the *angle* if we know the value of its sine, cosine or tangent. Again, we need to ensure that the calculator is set to measure degrees, then hit the '**sin⁻¹**', '**cos⁻¹**' or '**tan⁻¹**' key, followed by the value (some calculators may require you to enter the value first).

Note that the expressions '**sin⁻¹**', '**cos⁻¹**' or '**tan⁻¹**' simply mean *'the angle whose sine, cosine or tangent is …'*

Calculators normally output their angles in *decimal* form, rather than in degrees, minutes and seconds. There is no problem with this, as we normally work with angles in their decimal form so, generally speaking, there is no need to convert the decimal form into minutes and seconds. Refer to the calculator's user manual for details.

**Worked example 4**  A right-angled triangle's adjacent, opposite and hypotenuse sides are 150 mm, 200 mm and 250 mm, respectively. Calculate the size of its internal angles.

**Solution**

**Figure 24.5**

This can be solved, using the sine, cosine *or* tangent ratios.

$$\sin \angle \theta = \frac{\text{opposite}}{\text{hypotenuse}} = \frac{200}{250} = 0.8$$

$$\angle \theta = \sin^{-1} 0.8 = 53.13° \text{ Answer}$$

(**Remember**, the expression sin⁻¹ means *'the angle whose sine is ...'*)

$$\cos \angle \theta = \frac{\text{adjacent}}{\text{hypotenuse}} = \frac{150}{250} = 0.6$$

$$\angle \theta = \cos^{-1} 0.6 = 53.13° \text{ Answer}$$

(**Note**, the expression cos⁻¹ means *'the angle whose cosine is ...'*)

$$\tan \angle \theta = \frac{\text{opposite}}{\text{adjacent}} = \frac{200}{150} = 1.33$$

$$\angle \theta = \tan^{-1} 1.33 = 53.13° \text{ Answer}$$

(**Remember**, the expression tan⁻¹ means *'the angle whose tangent is ...'*)

Since the sum on the internal angles of a triangle is 180°, the remaining angle must be equal to:

$$\angle \phi = 180 - \left(90° + 53.13°\right) = 36.87° \left(\text{Answer}\right)$$

**Worked example 5**  A small boat is sailing at 9 m/s in a particular direction, and a current of 4 m/s is acting at right angles to the boat. By how many degrees off its intended course will the boat be sailing due to the cross current?

**Solution**

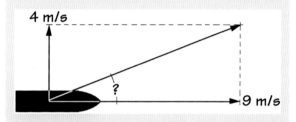

**Figure 24.6**

As shown in Figure 24.6, in this case, the forward velocity of the boat is represented by the adjacent side of a triangle, and the current is represented by the opposite side. So, relative to its intended course, the direction of the boat will be:

$$\tan \theta = \frac{\text{opposite}}{\text{adjacent}} = \frac{4}{9} = 0.44$$

$$\angle \theta = \tan^{-1} 0.44 = 23.7° \text{ Answer}$$

## Review your learning

Now that we've completed this chapter, we need to examine the **objectives** listed at its start. Placing *'Can I ...'* at the beginning, and a question mark at the end, of each objective turns that objective into a **test item**. If we can answer each of those test items, then we've met the objectives of this chapter.

# Chapter 25

# Introduction to alternating current

## Objectives

On completion of this chapter, you should be able to:

1 identify the five main components of the electricity supply industry.
2 describe the function of each of these five main components.
3 explain how the voltage induced into a conductor varies according to the angle at which the conductor moves through a magnetic field.
4 apply Fleming's Right-Hand Rule to determine the direction of the voltage induced into a conductor moving through a magnetic field.
5 explain why a conductor or loop, rotating within a magnetic field, generates a sinusoidal voltage.
6 explain each of the following terms specifying, where applicable, their SI units of measurement:
   a amplitude
   b instantaneous value
   c period
   d wavelength
   e cycle
   f frequency.
7 given the peak value of an a.c. voltage or current, calculate its r.m.s. (or 'effective') value, and *vice versa*.
8 describe the functions of a phasor and a phasor diagram.

## Introduction

Electricity generation, transmission and distribution systems are almost* exclusively **alternating current** (a.c.) systems. The primary reason for this is because, as we learnt in an earlier chapter, high voltages are *essential* for the transmission and distribution of electrical energy, and a.c. voltages can be easily and efficiently changed using transformers.

> *High-voltage, **direct current** transmission systems do exist and are used for very long transmission lines (to reduce losses), to interconnect independent grid systems (to avoid frequency synchronisation difficulties), and for undersea cables (to avoid large capacitive currents).

**High voltages** are *essential* for the transmission and distribution of electrical energy if we are to avoid (a) enormous voltage drops along the lines, (b) conductors with unrealistically high cross-sectional areas and weights, and (c) unacceptably high line losses.

There are *five* components to any national **electricity supply industry**. These are

a **generation**: thermal, hydroelectric, and renewables, often located at remote energy sources (coal fields, mountain reservoirs, offshore wind farms, etc.).
b **transmission**: grid system, providing bulk energy transfer from generating stations to load centres,

An Introduction to Electrical Science, Waygood, ISBN 9780815391821, 2019. © Taylor & Francis

and interconnecting all generating stations and load centres.

c  **distribution**: transfer of energy from load centres to nearby consumers.

d  **system operator**: ensuring generation output meets demand.

e  **supply**: marketing and selling energy to consumers.

The actual organisation of an electricity supply industry varies somewhat from country to country. In some countries, the system is government owned; this was the case, in the United Kingdom, from 1948 until 1990. In other countries, the industry is either wholly, or partially, in the hands of private enterprise.

In this chapter, we will briefly examine how the electricity-supply industry is organised in the United Kingdom.

Prior to its nationalisation, in 1948, there were numerous private electricity supply companies operating in the United Kingdom. In the 1920s, for example, there were nearly 500 companies, supplying electricity at various voltages and frequencies. Furthermore, there was no such thing as a 'national standard' for voltage or frequency.

## England, Wales and Scotland electricity-supply system

From 1948 until 1990, the electricity-supply industry of England and Wales was under public ownership, during which time system voltages and their operating frequency were standardised. From 1957, the **Central Electricity Generating Board (CEGB)** was responsible for the *generation and transmission* of electrical energy in England and Wales, and twelve **Area Boards** were responsible for its *distribution and marketing* to the consumer. In Scotland, electricity generation was provided by the **South of Scotland Electricity Board** and the **North of Scotland Hydro-Electric Board**.

From 1990, the industry was restructured and privatised, with **generation**, **transmission**, **distribution**, and **supply** becoming separate, but inter-related, groups of companies.

The term, '**transmission**' refers to the **national grid** system, which interconnects power stations and load centres, comprising 400-kV and 275-kV overhead lines and underground cables and substations.

The term, '**distribution**' refers to the system of 132-kV, 66-kV, 33-kV, 11-kV high-voltage overhead lines, underground cables, and substations, together with the medium/low voltage system which provides energy from load centres to local industrial, commercial, and residential consumers.

Today, the industry comprises the **Regulator**, the **Generating Companies**, the **Network Companies**, the **System Operator**, and the **Suppliers**.

## Regulator

**Ofgem (Office of Gas and Electricity Markets)** is the UK Government's independent National Regulatory Authority for the electricity and gas industries – its role is '*to promote choice and value for all electricity and gas customers*'.

## Generating companies

The **generating companies** own and operate the country's generating stations. These are:

- *EDF* (Électricité de France).
- *RWE* (Rheinisch-Westfälisches Elektrizitätswerk AG).
- *E·ON* which, as well as being a generating company, is also a **supplier**.

## Network companies

The **network companies** (Table 25.1) *maintain* and *operate* the electricity transmission and distribution networks. These comprise **transmission network owners (TNOs)** and **distribution network owners (DNOs)**:

- **Transmission network owners (TNOs).** The 400-kV and 275-kV *transmission network* (i.e. the 'national grid') within England and Wales is *owned* and *maintained* by *National Grid*. For Scotland, the TNOs are *Scottish Power* and *Scottish Hydro*.
- **Distribution network owners (DNOs)** own, maintain, and operate the 132-kV, 33-kV, 11-kV, and low-voltage *distribution network* of overhead lines and underground cables that distribute electricity to homes and businesses. They do *not* sell electricity to consumers; that is done by the **suppliers**. Currently, there are *seven* DNO companies providing energy to specific regions within the United Kingdom.

## Distribution network owners

| No: | Area |
| --- | --- |
| 1 | Scottish and Southern Electricity Networks |
| 2 | SP Energy Networks |
| 3 | Northern Ireland Electricity Networks |
| 4 | Electricity North West |
| 5 | Northern Powergrid |
| 6 | Western Power Distribution |
| 7 | UK Power Networks |

## System operator

The **system operator** is *National Grid*, which is responsible for controlling the operation of the generation and transmission system throughout Britain – i.e. balancing the energy supply with the energy demand placed on the network, from one second to the next, 24-hours a day, 365 days a year.

## Suppliers

The **suppliers** purchase electrical energy, wholesale, from the generating companies, pay network and government policy costs and charges, and sell it on to the consumers. Suppliers can purchase energy from *any* supplier in the United Kingdom as they are not restricted to any particular region. Furthermore, consumers are free to 'switch' from one supplier to another in order to obtain the best tariff (pricing system).

It should be emphasised that the suppliers do *not* provide or maintain the physical means (i.e. the network of substations, overhead lines, cables, etc.) by which energy reaches the consumer. This is done by the distribution network owners (DNOs). The suppliers simply market and sell the electrical energy provided by the DNOs.

The largest, and oldest, of these companies, commonly referred to as '*The Big Six*', were created from the area Electricity Boards when the industry was privatised in 1990. They are: *British Gas, EDF Energy, NPower, Scottish Power, E·ON UK*, and *SSE*. However, since

1997, numerous, smaller, independent companies have entered the market, and will probably continue to do so, with the intention of providing greater competition.

## Northern Ireland and Channel Islands electricity-supply system

As well as generating its own electricity, Northern Ireland has interconnectors linking with the Republic of Ireland and with Scotland.

In **Northern Ireland**, the regulator is the *Utility Regulator*, the network companies are *Northern Ireland Electricity Networks*, and *Mutual Energy Ltd,*

which operates the Moyle high-voltage d.c. link between Ireland and Scotland. The system operator is *SONI*.

In the **Channel Islands**, the sole provider of electrical energy for Guernsey is *Guernsey Electricity* and, for Jersey, the *Jersey Electricity Company*.

The UK's generation/transmission/distribution system is illustrated in Figure 25.2.

As we shall learn in a later chapter, **three-phase** systems are used throughout the transmission and distribution system, both high voltage and low voltage, because for a given load, less volume of copper is required for a three-phase system than an equivalent single-phase system, making three-phase more economical compared with single phase.

Figure 25.2

## Generation of a sine wave

From earlier chapters, you will recall that whenever there is relative movement between a conductor and a magnetic field, a potential difference (voltage) is *induced* into that conductor.

Faraday's Law tells us that the *magnitude* of that induced voltage is *directly proportional to the rate at which the lines of magnetic flux are cut by the conductor*. For any given velocity, the maximum rate will occur when the conductor cuts the flux perpendicularly (at right angles); on the other hand, if the conductor runs *parallel* with the flux, then no flux is cut, and no voltage will be induced into the conductor.

If the conductor cuts the flux at an angle, $\theta$ (pronounced *'theta'*) at a velocity represented by vector *v*, then we must find the *perpendicular* component of that velocity vector, *v′* – as represented by the broken line in Figure 25.3.

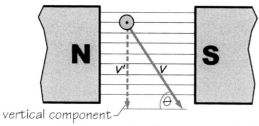

vertical component
conductor cutting flux at an angle θ

Figure 25.3

As we learnt in the chapter on *electromagnetic induction*, the general equation for the voltage induced into a conductor that is cutting lines of magnetic flux at an angle, $\theta$, is given by:

$$E = Blv \sin\theta$$

where:

$E$ = potential difference (V)

$B$ = flux density (T)

$l$ = length of conductor (m)

$v$ = velocity of conductor (m/s)

$\theta$ = angle cutting flux (°)

The direction of the induced voltage can be deduced by applying **Fleming's Right-Hand Rule**. For example, in Figure 25.4, when the conductor moves vertically downwards through the magnetic field, Fleming's Right-Hand Rule tells us that the positive end of the conductor is at the nearest end, and (conventional) current will enter the external circuit from that end.

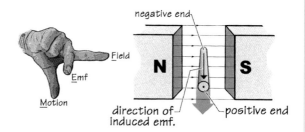

Figure 25.4

## Alternators: basic constructional features

A **simple a.c. generator**, or **alternator**, consists of a single loop of wire (called an **armature**), pivoted so that it can rotate within a magnetic field – as illustrated in Figure 25.5.

Figure 25.5

When it is rotated by an external force, provided by a **prime mover** (steam turbine, a diesel engine, a water turbine, a wind turbine, etc.), a potential difference is induced into each side of the loop. If we apply Fleming's Right-Hand Rule to each side of the loop, we will see that the potential difference induced into the left-hand side will act *towards* us, while the potential difference induced into the right-hand side will act *away* from us – as they are in series with each other, the two act to reinforce each other.

So if the voltage induced into one side is *e* volts, then the voltage induced into the complete loop must be **2e** volts.

In a more practical generator, of course, the armature will be a coil, *not* a single loop. So, for an armature winding with $N$ turns, this voltage will be increased $N$ times. However, for the sake of simplicity, we'll continue for now to illustrate it as a single loop.

When this single loop generator is connected to an external circuit, the resulting (conventional) load current will flow through the armature in the counterclockwise direction shown in Figure 25.5.

Connecting the rotating loop of our simple generator to its external load is achieved using a pair of **slip rings** and spring-loaded **carbon brushes**, as illustrated in Figure 25.6.

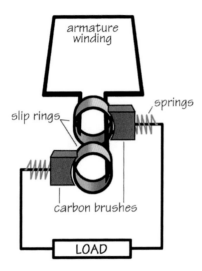

**Figure 25.6**

One end of the armature winding is connected to one slip ring, while the other end is connected to the second slip ring. As the generator's loop rotates past the mid-way point (i.e. when running parallel with the flux), the voltages induced into each side of the loop reverse direction. So, the output voltage from a generator equipped with slip rings also changes direction, and produces (in theory at least) a *sinusoidal alternating voltage.*

## Action and reaction

In the earlier chapter on *electromagnetism*, we learnt that a current-carrying conductor placed between the poles of a magnet will *experience a force acting at right angles to the field's magnetic flux.* In the same chapter, we also learnt that this force is *proportional to the current* – so, the greater the current, the greater the force.

If we were to apply Fleming's Left-Hand Rule to our simple **generator** when it is supplying a load current, we would find that the resulting force due to this 'motor action' always acts *against the direction in which the conductor is moving!*

In other words, whenever the generator is supplying current to the load, the armature loop behaves *both as a generator and as a motor simultaneously!* The greater

the load current, the greater will be the 'motor effect' which acts to oppose the movement of the armature loop. This 'action' and 'reaction' effect plays a very important part in the operation of a generator, requiring it to *react to changes in load* – and explains why, as the load increases (i.e. draws more current), the prime mover 'tends' to slow down and when the load decreases, the speed of the prime mover 'tends' to increase.

In the last sentence, we've deliberately enclosed the word 'tends' in quotation marks for a good reason. This is because the speed of a practical generator's prime mover is actually controlled by a **governor**. So if, for example, the prime mover was a steam turbine, then the governor would control the steam inlet valve, increasing or decreasing the steam flow through the turbine. Then, if the generator does 'tend' to slow down or 'tend' to speed up, for the reasons described in the previous paragraph, the governor will raise or lower the torque on the prime mover to compensate. So, it's the **torque**, rather than the speed, that actually changes with variations in load.

Indeed, it's important that the speed of the generator doesn't change because there's a legal requirement to maintain the output frequency within ±1% of its nominal (stated) value.

For a more-detailed explanation of how a.c. generators react to changes in their load, see the companion book: ***Electrical Science for Technicians.***

This action/reaction behaviour explains why, when the electrical load demands more energy *from* the generator, more energy must be supplied *to* the generator by the prime mover to *drive* the generator. This is why, for example, leaving unnecessary electrical accessory-loads operating in a car (lights on in daylight, the radio when it is not being listened to, etc.) will actually increase its fuel consumption.

For the types of a.c. generator that supply very large currents and generated voltages (typical power station alternators supply currents of the order 1000 A at 11–25 kV), the use of slip rings is quite impractical (they would be unable to handle such large currents, and the heavy arcing involved would require regular shutdowns for maintenance).

To avoid this problem, the *armature windings are installed on the stator* (the stationary part of the machine) while the *field windings are installed on the rotor* (the rotating part of the machine). In other words, the construction of real alternators is *exactly* **opposite** *to that of our simple generator!*

Because only the field windings rotate, the necessary slip rings need only supply the relatively low current (typical power station alternators utilise field currents of the order 350 A at 400 V d.c.) to the windings that produce the magnetic field, while the fixed armature windings can supply the heavy current directly to the load. As a result, the arcing at the slip rings is insignificant, and the machine's maintenance cycle can be significantly increased and its downtime reduced.

Furthermore, practical alternators have **multiple pairs** of armature windings, not just *two* like our simple generator and, so, are called 'multiple-pole' machines. These machines have significantly more efficient magnetic circuits, but the number of poles affects the frequency of the output voltage – in the case of a **two-pole machine**, for example, *one revolution of the rotor will generate one complete cycle*; whereas in a **four-pole machine** *one revolution will generate two complete cycles*, and so on. Machines with a large number of poles, such as those driven by water turbines, can therefore run more slowly to provide the required output frequency.

The relationship between a machine's **speed**, **number of poles** and its output **frequency** is given by:

$$f = pn$$

where:

$f$ = frequency (hertz)

$p$ = number of *pairs* of poles

$n$ = speed (revolutions per second)

**Worked example 1**   At what speed must a two-pole alternator run in order to generate a voltage at 50 Hz?

**Solution**   A two-pole machine has just one *pair* of poles, so:

$$n = \frac{f}{p} = \frac{50}{1} = 50 \text{ rev/s}$$

As it is more usual to express a machine's speed in *revolutions per minute*, so if the machine runs at 50 rev/s, then in one minute (60 s), it will need to run at:

$$n = 50 \times 60 = 3000 \text{ rev/min}  \quad \text{(Answer)}$$

> Interestingly, as two poles are obviously the *minimum* number of poles it is possible to have, *the highest speed at which a 50-Hz alternator can be allowed to operate is at 3000 rev/min.*

**Worked example 2**   The standard nominal mains frequency in Canada is 60 Hz. At what speed must a two-pole alternator run in order to generate a voltage at 60 Hz?

**Solution**   A two-pole machine has just one *pair* of poles, so:

$$n = \frac{f}{p} = \frac{60}{1} = 60 \text{ rev/s}$$

As it is more usual to express a machine's speed in *revolutions per minute*, so if the machine runs at 60 rev/s, then in one minute (60 s), it will need to run at:

$$n = 60 \times 60 = 3600 \text{ rev/min}  \quad \text{(Answer)}$$

**Worked example 3**   At what speed must a six-pole alternator run in order to generate a voltage at 50 Hz?

**Solution**   A six-pole machine has three *pairs* of poles, so:

$$n = \frac{f}{p} = \frac{50}{3} = 16.67 \text{ rev/s}$$

As it is more usual to express a machine's speed in *revolutions per minute*,

$$n = 16.67 \times 60 = 1000 \text{ rev/min}  \quad \text{(Answer)}$$

In the case of power station alternators, in order to maintain the strict frequency requirements laid down, in the United Kingdom, by the **Electricity, Safety, Quality and Continuity Regulations (2002)**, which specify that consumers are to be supplied at a frequency at **50 Hz ± 1%** (i.e. between 49.5 Hz and 50.5 Hz) over a 24-h period, the rotational speed of these alternators is strictly controlled. The machines' terminal voltage, therefore, *cannot be adjusted by changing the speed of the machine* but, instead, by adjusting the flux density of the field, by controlling the field current.

## Electrical v mechanical degrees

In the previous section, we learnt that when the armature of a two-pole machine completes one revolution (360°), it generates one complete cycle (360°) of a.c. voltage but when, for example, a four-pole machine completes one revolution (360°), it generates *two* complete cycles (720°) of a.c. voltage. It's important to understand, therefore, that one complete cycle of a.c. voltage doesn't necessarily correspond to one complete revolution of a generator's armature. To distinguish between the two, whenever we talk about 'degrees', we often use the term '**electrical degrees**' to describe *the measurement of a sine wave*, and '**mechanical degrees**' to describe *the corresponding physical rotation of an armature*.

## Generation of a sinusoidal voltage

Let's now return to our simple model of an a.c. generator, to remind ourselves of how an a.c. voltage can be generated in a rotating armature loop. In Table 25.2, a conductor follows the counterclockwise circular path, shown, cutting the magnetic flux set up from the north towards the south magnetic poles. We'll assume that, when the conductor cuts the flux at right angles, a voltage of 1 V is induced into the conductor.

So, as the conductor (or, in practice, a **loop**) moves through its circular path between the magnetic poles, the value of voltage induced into the conductor continually varies from zero, to some maximum value in one direction, then back to zero and, finally, to the same maximum value but in the opposite direction.

In practice, the armature conductors of an alternator are part of its stator and remain stationary, while the magnetic field rotates, being produced by a field winding wound around the rotor – i.e. exactly *opposite* to that illustrated in Table 25.2. However, this is very difficult to illustrate clearly,

and the actual waveform is exactly the same as shown in the sequence of illustrations!

Let's assume, for a moment, that the *maximum* voltage induced into the conductor is just 1 V. Figure 25.20 shows what we will find if we draw the waveform to scale, and check the values of instantaneous voltages at, say, 30° intervals from 0° to 360°.

The angle, expressed in electrical degrees, of any of these instantaneous values of voltage, is termed the **displacement angle** *(θ)*. The values of each of these instantaneous voltage correspond to 1 V times the sine of the corresponding displacement angle, i.e:

*instantaneous voltage = maximum voltage × **sine** of the displacement angle*

instantaneous voltage = maximum voltage × sine of the displacement angle

$$e = E_{max} \sin \theta$$

For this reason, the generated waveform is called a **sine wave**.

So, the **instantaneous voltage** (symbol: *e*) at any point along the sine wave, then, is given by the equation:

$$\boxed{e = E_{max} \sin \theta}$$

where:

$$e = \text{instananeous voltage}$$
$$E_{max} = \text{peak voltage}$$
$$\theta = \text{displacement angle}$$

Table 25.2

In this position, the conductor is moving parallel to the magnetic flux, so no voltage is induced into the conductor.

$$e = 1 \sin \theta = 1 \sin 0°$$
$$= 0 \text{ V}$$

Figure 25.7

The conductor has moved along its circular path by 30°, and is cutting the flux at 30°, so a voltage is starting to be induced into the conductor.

$$e = 1\sin\theta = 1\sin 30°$$
$$= 0.5\ \text{V}$$

Figure 25.8

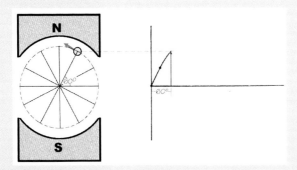

The conductor has now moved along its circular path at 60°, and is now cutting the flux by 60°, and an even greater voltage is induced into the conductor.

$$e = 1\sin\theta = 1\sin 60°$$
$$= 0.866\ \text{V}$$

Figure 25.9

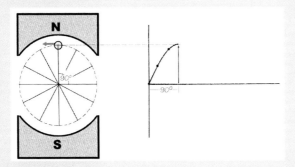

The conductor has now moved along its circular path by 90°, and is now cutting the flux at right angles, so the maximum voltage is induced into the conductor.

$$e = 1\sin\theta = 1\sin 90°$$
$$= 1.0\ \text{V}$$

Figure 25.10

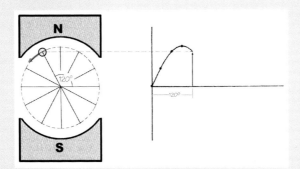

The conductor has now moved along its circular path by 120°, and is now cutting the flux at 60°, so the induced voltage is starting to fall.

$$e = 1\sin\theta = 1\sin 120°$$
$$= 0.866\ \text{V}$$

Figure 25.11

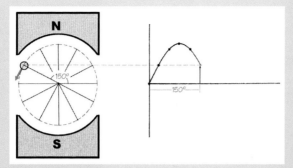

Figure 25.12

The conductor has now moved along its circular path by 150°, and is now cutting the flux at 30°, so the induced voltage has fallen further.

$$e = 1\sin\theta = 1\sin 150°$$
$$= 0.5 \text{ V}$$

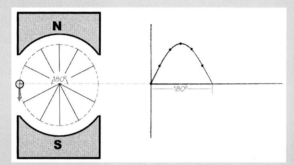

Figure 25.13

The conductor has now moved along its circular path by 180° and, once again, the conductor is moving parallel to the flux, so no voltage is induced into it.

$$e = 1\sin\theta = 1\sin 180°$$
$$= 0 \text{ V}$$

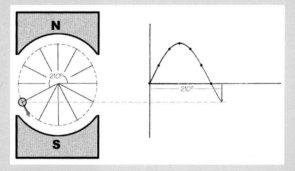

Figure 25.14

The conductor has now moved along its circular path by 210°, and the conductor is cutting the flux at 30° – but this time, in the opposite direction. So the induced voltage is now acting in the opposite sense.

$$e = 1\sin\theta = 1\sin 210°$$
$$= -0.5 \text{ V}$$

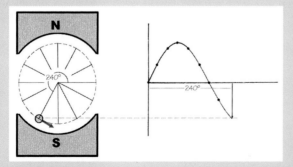

Figure 25.15

The conductor has now moved along its circular path by 240°, and the conductor is cutting the flux at 60°, so the induced voltage is now increasing in the opposite sense.

$$e = 1\sin\theta = 1\sin 240°$$
$$= -0.866 \text{ V}$$

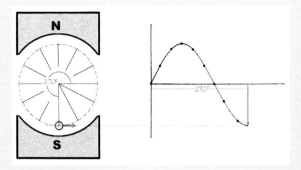

Figure 25.16

The conductor has now moved along its circular path by 270°, and is now cutting the flux at right-angles (but in the opposite direction compared with the 90° position), so the maximum voltage is induced into the conductor.

$$e = 1\sin\theta = 270\sin 30°$$
$$= -1.00 \text{ V}$$

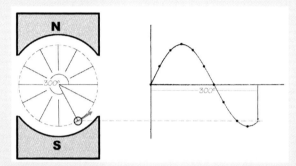

Figure 25.17

The conductor has now moved along its circular path by 300°, and is now cutting the flux at 60°, so the induced voltage is starting to fall again.

$$e = 1\sin\theta = 1\sin 300°$$
$$= -0.866 \text{ V}$$

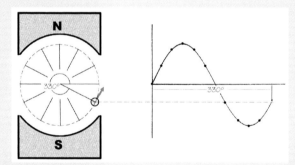

Figure 25.18

The conductor has now moved along its circular path by 330°, and is now cutting the flux at 30°, so the induced voltage is starting to fall further.

$$e = 1\sin\theta = 1\sin 330°$$
$$= -0.5 \text{ V}$$

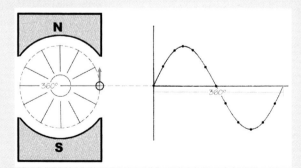

Figure 25.19

The conductor has now moved through 360° and is, once again, moving parallel to the flux and no voltage is induced into the conductor.

$$e = 1\sin\theta = 1\sin 360°$$
$$= 0.00 \text{ V}$$

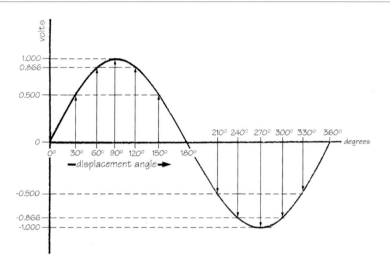

**Figure 25.20**

Since current is directly proportional to voltage, so we can also state:

$$i = I_{max} \sin \theta$$

where:

$i$ = instananeous current

$I_{max}$ = peak current

$\theta$ = displacement angle

**Worked example 4** Calculate the instantaneous voltages at displacement angles of 15°, 40°, 70°, 100° and 240°, if the peak value of voltage is 10 V.

**Solution**

**at 15°:** $e = E_{max} \sin \theta = 10 \times \sin 15 = 10 \times 0.259$
$$= 2.59 \text{ V } (Answer)$$

**at 40°:** $e = E_{max} \sin \theta = 10 \times \sin 40 = 10 \times 0.643$
$$= 6.43 \text{ V } (Answer)$$

**at 70°:** $e = E_{max} \sin \theta = 10 \times \sin 70 = 10 \times 0.94$
$$= 9.40 \text{ V } (Answer)$$

**at 100°:** $e = E_{max} \sin \theta = 10 \times \sin 100 = 10 \times 0.985$
$$= 9.85 \text{ V } (Answer)$$

**at 240°:** $e = E_{max} \sin \theta = 10 \times \sin 230 = 10 \times (-0.766)$
$$= -7.66 \text{ V } (Answer)$$

## Terminology

The **terminology** shown in Table 25.3 is used to describe any sine wave (voltage or current).

## Measuring sinusoidal values

Because the value of a sinusoidal voltage or current is continually changing in both magnitude and direction, how do we assign any meaningful value to them? Well, we *could* simply use the *peak value* – but as the waveform only reaches this value twice during any complete cycle, it is hardly representative of the entire waveform's variation.

What about the *average* value? Well, if we work out the waveform's average value over a complete cycle, it will work out at *zero* – because the average value over the *positive* half-cycle will be cancelled out by its average value over the *negative* half-cycle!

In fact, average values are sometimes assigned to sine waves but only over a half cycle, so are usually applied to rectified a.c. values.

If neither the **peak value**, nor the **average value**, represents a meaningful way of measuring a sinusoidal voltage or current over a complete cycle, *how do we proceed?*

Table 25.3

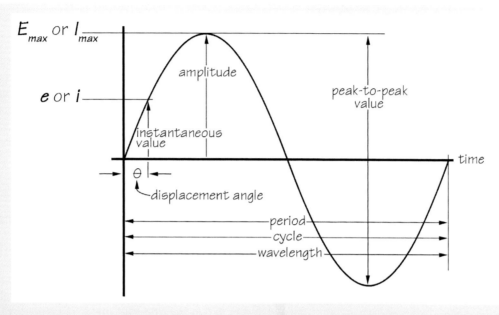

Figure 25.21

| | |
|---|---|
| **Amplitude:** | The peak value of an a.c. waveform, in either the positive *or* in the negative sense. Symbols: $E_{max}$ or $I_{max}$. |
| **Peak-to-peak value:** | Twice the amplitude of an a.c. waveform. Symbols: $E_{p-p}$ or $I_{p-p}$. |
| **Instantaneous value:** | The value of voltage or current, at any specified displacement angle, during a complete cycle. Symbols: $e$ or $i$. |
| **Displacement angle:** | The displacement of any instantaneous value of voltage or current, expressed in electrical degrees and measured from the origin of the sine wave. Symbol: $\theta$ (the Greek letter, *'theta'*). |
| **Period, or Periodic time:** | The time taken to complete one cycle, measured in seconds. Symbol: $T$. |
| **Wavelength:** | The distance between two displacements of the same phase, measured in metres. Symbol: $\lambda$ (the Greek letter, *'lambda'*). |
| **Cycle:** | One complete set of changes in the values of a recurring variable quantity. |
| **Frequency:** | The number of complete cycles per unit time, measured in hertz. Symbol: $f$, where: |

$$f = \frac{1}{T}$$

The problem is a little like trying to compare the speed of a reciprocating saw to that of a circular saw! The reciprocating saw blade is continuously oscillating up and down, whereas the circular saw is continuously rotating – so it's rather difficult to compare the two in terms of their 'speeds', as they are measured in completely different ways. Instead, we could compare them in terms of *the rate at which they each cut timber*. In other words, we can say that if the reciprocating saw cuts timber at exactly the same rate as the circular saw then its reciprocating speed must be *'equivalent'* to (rather than the *'same'* as!) the rotational speed of the circular saw.

We do a similar thing when we measure a.c.; that is, *we compare the* **work** *an a.c. current does with the* **work** *that a d.c. current does*. We know that if a voltage is applied to a resistive circuit, the resulting current will cause the temperature of the resistor to rise, and this will happen *regardless* of whether the current is direct current or alternating current, so we make use of this property.

Let's examine the simple experiment shown in Figure 25.22.

**Figure 25.22**

With the switch in position *a*, the incandescent lamp is connected to an a.c. supply, and the brightness (or, more accurately, 'luminous intensity') of the lamp is measured using an appropriate optical instrument. Next, the switch is moved to position *b*, and the variable resistor is adjusted until *exactly the same brightness is achieved*. The ammeter now indicates the value of d.c. current that has produced *exactly the same brightness* (in other words, produces *exactly the same heating effect*) as that produced by the a.c. current. So if the ammeter indicates a direct current of, say, 0.5 A when identical brightness is achieved, then the effective value of a.c. is also considered to be 0.5 A.

> So, we can say that *'the **effective value** of an alternating current is measured in terms of the direct current that produces exactly the same heating effect in the same resistance'*.

If we were now to compare the a.c. current's effective value with its peak value (using an oscilloscope), we would find (for a sine wave) that the effective value is equal to 0.707× its peak value.

So, the relationship between the effective value of this a.c. current and its peak value (as illustrated in Figure 25.23) would be:

$$I_{effective} = 0.707 \times I_{max}$$

$I_{r.m.s.}$ does exactly the same work as $I_{d.c.}$
so, $I_{r.m.s.}$ is exactly equivalent to $I_{d.c.}$

**Figure 25.23**

Since voltage and current are directly proportional to each other:

$$E_{effective} = 0.707 \times E_{max}$$

The **effective value** of an a.c. current or voltage is more commonly known as its '**r.m.s. value**'. The abbreviation '**r.m.s.**' stands for '**root-mean-square**', a term derived from the *mathematical proof* for obtaining this value of 0.707. We don't need to go into this proof in any great detail in this chapter, but you should be aware of *why* the term **root-mean-square** is used, and *where* it comes from.

In the above explanation of effective value, we said it is based on the *work done* by a current. We know that the equation for work is: $W = I^2Rt$. So, if we start by dividing the a.c. current waveform up into lots and lots of *instantaneous values of current*: $i_1$, $i_2$, $i_3$, $i_4$, etc. (the more the better!) and apply this equation to each value of instantaneous current (Figure 25.24), we'll end up with the following equation:

$$I^2_{effective}Rt = i_1^2 Rt + i_2^2 Rt + i_3^2 Rt + i_4^2 Rt + etc.$$

(for the whole waveform)

By dividing throughout by $Rt$, we can eliminate both $R$ and $t$ (both being constants) from the equation:

$$I^2_{effective}\frac{\cancel{Rt}}{\cancel{Rt}} = i_1^2 \frac{\cancel{Rt}}{\cancel{Rt}} + i_2^2 \frac{\cancel{Rt}}{\cancel{Rt}} + i_3^2 \frac{\cancel{Rt}}{\cancel{Rt}} + i_4^2 \frac{\cancel{Rt}}{\cancel{Rt}} + etc.$$

(for the whole waveform)

$$I^2_{effective} = i_1^2 + i_2^2 + i_3^2 + i_4^2 + etc. \text{ (for the whole waveform)}$$

Next, we find the **mean** (average) for all the individual **squared** instantaneous currents:

$$I^2_{effective} = \frac{i_1^2 + i_2^2 + i_3^2 + i_4^2 + etc. \text{ (for the whole waveform)}}{n}$$

(where *n* represents the number of individual instantaneous currents)

Finally, to eliminate the squares, we can find the **square root** of both sides of the equation:

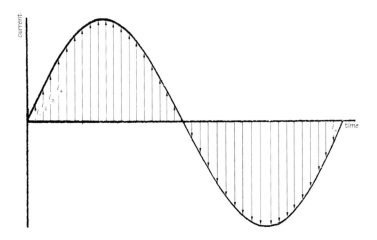

Figure 25.24

$$I_{effective} = \sqrt{\frac{i_1^2 + i_2^2 + i_3^2 + i_4^2 + etc.\,(\text{for the whole waveform})}{n}}$$

So, as you can now see, the **effective value of current** is equal to the **square root** of the **mean** (i.e. 'average') of the **squares** of each of the individual instantaneous currents. Hence, the term: '**root-mean-square**'. If you were to insert actual values into the above equation then, for a *sine wave* it would *always* work out to be **0.707 $I_{max}$** (it would be different for other shaped waveforms).

In practice, this calculation is performed using calculus, rather than by the technique described above – but this is beyond the scope of this text.

To summarise:

> The **effective**, or **r.m.s. value** of a.c. current is given by:
>
> $$I_{rms} = 0.707\,I_{max}$$

And, since voltage and current are proportional to each other:

> The **effective**, or **r.m.s. value** of a.c. voltage is given by:
>
> $$E_{rms} = 0.707\,E_{max}$$

It's important to understand that **a.c. currents and voltages are** *always* **expressed in effective, or r.m.s., values** *unless otherwise specified*. For example,

voltmeters and ammeters are calibrated to indicate r.m.s. values. So, we do *not* normally need to add the subscript 'rms' to the symbols for voltage or current – in other words, we usually simply write '$E$' or '$I$', rather than '$E_{rms}$' or '$I_{rms}$'.

> **Important!**
> Unless otherwise specified, a.c. values are *always* quoted in r.m.s. values, and *all* a.c. ammeters and voltmeters are calibrated to output their readings in r.m.s. values. Because of this, you will rarely see the subscript 'r.m.s.' used, unless clarity is necessary.

**Worked example 5**   What is the peak value of a 230-V a.c. residential supply?

**Solution**   (For clarity, we'll retain the subscript 'r.m.s.' in this example.)

Since *all* a.c. values are normally expressed as r.m.s. values, then 230 V is an r.m.s. value, so:

since   $E_{rms} = 0.707\,E_{max}$

then   $E_{max} = \dfrac{E_{rms}}{0.707} = \dfrac{230}{0.707} \approx 325$ V (Answer)

**Worked example 6**   Using an oscilloscope, we measure the peak value of a sine-wave voltage across a resistor as 250 mV. What is its r.m.s. value?

**Solution** (Again, for clarity, we'll retain the 'r.m.s.' subscript in this example.)

$$E_{rms} = 0.707\, E_{max}$$
$$= 0.707 \times (250 \times 10^{-3}) \approx 177 \times 10^{-3}\ \text{V}$$
$$\text{or}\quad 177\ \text{mV (Answer)}$$

## Representing sinusoidal waveforms with phasors

You are probably already familiar with the concept of **vectors**. Quantities such as **force** or **velocity** can be represented by means of a **vector** – where the *length* of the vector represents the *magnitude* of the force or velocity, and the *direction* in which the vector points represents the *direction* in which the force or velocity acts. Two or more forces may be added or subtracted, by adding or subtracting their corresponding vectors – this can either be done graphically, to scale, or by applying the rules of simple geometry and trigonometry.

'**Phasors**' are, to electrical engineering, what **vectors** are to mechanical engineering. However, while the *length* of a phasor represents the *magnitude* (normally, expressed in terms of r.m.s. values) of an alternating voltage or alternating current, its '*direction*' (or, more accurately, its 'angle') doesn't represent direction but, rather, the *time displacement* of that voltage or current – expressed in terms of *displacement angle measured in a counterclockwise direction*. For this reason, phasors have been described as '**rotating vectors**'.

Although, in Figure 25.25, the length of the phasor is equal to the peak value of voltage, *in practice, they usually represent the root-mean-square of that voltage.*

A 'phasor' is a line, whose length represents the magnitude of an a.c. voltage or current, and whose angle, measured counterclockwise from the real positive axis, represents its displacement angle.

Figure 25.25

The phasor's arrow head serves two functions: (1) it indicates the 'rotating' end of the phasor, and (2) it helps distinguish phasors from each other, when different phasors lie at the same angle.

Phasors that represent the *same* quantities (voltages *or* currents) may be added or subtracted in exactly the same way as vectors are added or subtracted, and by using exactly the same techniques – graphically or or mathematically – as illustrated in Figure 25.26.

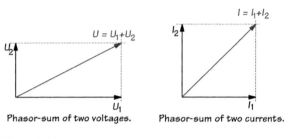

Phasor-sum of two voltages.    Phasor-sum of two currents.

Figure 25.26

Phasors that represent *different* quantities (voltages *and* currents) obviously *cannot* be added or subtracted, but the angle between them represents the phase relationship between those quantities, as illustrated in Figure 25.27:

Phase angle between current and voltage

Figure 25.27

Fortunately, as you will see, *phasors are much easier to use than they are to explain!* There are, though, numerous animations available on *YouTube*, for example, which will help us understand phasors.

A **phasor** is an arrowed line whose *length* represents the *magnitude* of an alternating voltage or current and whose *angle*, measured counterclockwise from the real positive axis, represents the *displacement angle* of that voltage or current.

A phasor's arrow head represents the 'rotating' end of that phasor and *not* its direction.

The rules of **vector addition** and **subtraction** apply to phasors representing like quantities (i.e. two or more voltage phasors, or two or more current phasors).

As you will discover in the following chapters, it's very much easier to *use* phasors than it is to describe what they are!

PHASORS
- the <u>length</u> of a phasor represents a waveform's amplitude.
- the <u>angle</u> of a phasor represents the time-angle between a waveform and a second (reference) waveform. Angles are measured counterclockwise.

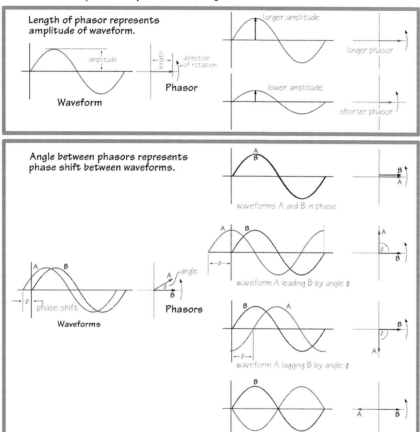

**Figure 25.28**

## All you need to know to understand alternating current

The chapters that follow explain how phasors can be used to understand (and solve problems on) the behaviour of alternating-current circuits. If you

- are already able to add or subtract vectors
- understand how to apply Pythagoras's Theorem
- can apply the trigonometric ratios, sine, cosine and tangent, to simple right-angled triangles,

. . . then you are already well on your way to being able to solve most problems relating to behaviour of series, parallel and series-parallel single-phase a.c. circuits, as well as three-phase a.c. circuits!

And you will be able to do all of this *without having to remember more than just **two** of the many equations associated with a.c. theory!*

Let's expand on the promise made in the last sentence of the previous paragraph.

*To understand the behaviour of a.c. circuits, both single-phase and three-phase, you will need to:*

- recall, and be able to apply, Pythagoras's Theorem.
- recall, and be able to apply to right-angled triangles, each of the following trigonometric ratios:

$$\sin \phi = \frac{\text{opposite}}{\text{hypotenuse}}; \quad \cos \phi = \frac{\text{adjacent}}{\text{hypotenuse}};$$

$$\tan \phi = \frac{\text{opposite}}{\text{adjacent}}$$

- remember, and be able to apply, the following two equations:

$$X_L = 2\pi f L \quad \text{and} \quad X_C = \frac{1}{2\pi f C}$$

- remember, and be able to apply, the mnemonic **CIVIL** – but more of this, later.

Really! That's *all* there is to it! Well . . . not quite! Of course, there really *is* more to a.c. theory than just that – but armed with only the knowledge described above, you will soon be well on your way to gaining a tremendous understanding of the behaviour of single-phase and three-phase a.c. circuits, and you will quickly be able to solve a great many of the problems that you will be faced with both in the classroom and on the job.

**Tip!** As you progress through the remaining chapters of this book, you'll come across an increasing number of equations. *Don't bother to try to remember them!* You don't actually *need* to remember them, because *you are going to learn how to derive them*. Once you know how to derive them, you'll *never* need to remember any individual equations. However, after a while, you'll probably find yourself remembering these equations without even realising it!

## Review your learning

Now that we've completed this chapter, we need to examine the **objectives** listed at its start. Placing '*Can I...*' at the beginning, and a question mark at the end, of each objective turns that objective into a **test item**. If we can answer each of those test items, then we've met the objectives of this chapter.

**Online resources**

The companion website to this book contains further resources relating to this chapter. The website can be accessed via the following link:
www.routledge.com/cw/waygood

# Chapter 26

# Series alternating-current circuits

An Introduction to Electrical Science, Waygood, ISBN 9780815391821, 2019. © Taylor & Francis

## Objectives

On completion of this chapter, you should be able to:

1 sketch a waveform for each of the following, showing the phase relationship between the current and supply voltage:
   a purely resistive circuit
   b purely inductive circuit
   c purely capacitive circuit
   d series *R-L* circuit
   e series *R-C* circuit.
2 state the phase relationship between the current and supply voltage for
   a a purely resistive circuit
   b a purely inductive circuit
   c a purely capacitive circuit.
3 sketch the phasor diagram representing a
   a purely resistive circuit
   b purely inductive circuit
   c purely capacitive circuit
   d series *R-L* circuit
   e series *R-C* circuit
   f series *R-L-C* circuit.
4 develop an impedance diagram for a
   a series *R-L* circuit
   b series *R-C* circuit
   c series *R-L-C* circuit.
5 from impedance diagrams, derive equations for resistance, inductive reactance, capacitive reactance and impedance, in terms of voltages and currents.
6 state the equation for inductive reactance, in terms of inductance and frequency.
7 state the equation for capacitive reactance, in terms of capacitance and frequency.

8 explain what is meant by the term 'series resonance'.
9 list the effects of series resonance.
10 solve problems on series a.c. circuits, including series-resonant circuits.

## Introduction

**Important!** The key to understanding and solving a.c. circuits, whether they are series circuits, parallel circuits or even three-phase circuits, is your ability to sketch a phasor diagram which represents that circuit and, then, use this phasor diagram to generate all the equations you need, simply by treating it as a simple exercise in geometry or trigonometry.

All 'real' alternating-current circuits exhibit combinations of **resistance** (symbol: $R$), **inductance** (symbol: $L$) and **capacitance** (symbol: $C$). The amount of each of these quantities appearing in any particular circuit is determined by the *configuration* and *design characteristics* of that particular circuit.

For example, *all* conductors (by virtue of their length, cross-sectional area and resistivity) exhibit *natural* amounts of resistance. Overhead power lines, due to the configuration of their individual conductors, exhibit relatively high *natural* values of inductance, as well as some capacitance and resistance. Underground cables, because of the closeness of their individual conductors, exhibit relatively high *natural* values of capacitance as well as some inductance and resistance.

'Real' a.c. circuits, then, are relatively complicated because they contain a *combination* of resistance, inductance and capacitance. So, in order to understand the behaviour of an a.c. circuit, it is necessary to start by considering how an 'ideal' circuit would behave. In this context, an 'ideal' circuit is one which is '**purely resistive**', '**purely inductive**' or '**purely capacitive**'.

'Ideal' circuits *only exist theoretically*. But if we are able to understand how these relatively simple, ideal circuits would behave if they *did* exist, then we will be able to move on to combine these behaviours in order to understand how *real* and *more complicated* circuits behave.

> **Important!** In the circuit diagrams that follow, it is important to understand that the symbols represent quantities *not* components. That is, **resistance** *not* resistors; **inductance** *not* inductors; **capacitance** *not* capacitors.

Throughout the rest of this chapter, the voltages and currents we will be referring to are 'phasor' quantities. In order to remind ourselves of this, the symbols for voltage and current will be shown with small 'bars' above them (i.e. $\bar{E}, \bar{U}_R, \bar{U}_L, \bar{U}_C$ and $\bar{I}$). This is one way of indicating that these are phasor quantities. It's unnecessary to do this when labelling phasor diagrams as the quantities involved are obviously phasors.

Resistance, inductive reactance, capacitive reactance and impedance are *not* phasor quantities and, so, will not have bars placed above their symbols.

## Purely resistive circuit

So let's start with the simplest of all 'ideal' circuit: the **purely resistive circuit**.

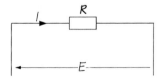

**Figure 26.1**

In a purely resistive circuit, the current $(\bar{I})$ and the potential difference $(\bar{E})$ across that resistance are said to be **in phase** with each other – i.e. the peak and zero

points of their two separate waveforms correspond exactly throughout each complete cycle. This is exactly how we would instinctively expect a circuit to behave but, as we shall learn, this is *not* the case with other types of circuits.

current and voltage in phase

**Figure 26.2**

The waveforms shown in Figure 26.2 can be represented by means of what is known as a '**phasor diagram**', with the current and voltage phasors each lying alongside each other. In this case, it's usual to draw them both along the horizontal positive axis (i.e. horizontally, pointing to the right). In Figure 26.3, and in those that follow, the small curved arrow to the right is used simply to remind ourselves that, by common consent, phasors 'rotate' in a counterclockwise direction.

> Whenever we measure the angles between phasor quantities, such as currents and voltages, counterclockwise is *always* considered to be the positive direction, and clockwise is *always* considered to be the negative direction.

**Figure 26.3**

The *length* of the voltage phasor $(\bar{E})$ represents the *r.m.s. value of the supply voltage*, and the *length* of the current phasor $(\bar{I})$ represents the *r.m.s. value of the current*. There is absolutely *no* relationship between the scales of the two phasors, because they each represent different quantities (for example, the voltage phasor could be drawn to a scale of 10 volts per millimetre, while the current phasor is drawn to a scale of 2 amperes per millimetre). What *is* important, however, is the **angle** (or, in this case, the *lack* of any angle) between the two phasors, which indicates that the two quantities are 'in phase' with each other.

As is the case for *any* circuit, the ratio of voltage to current represents the *opposition* to current. In a **purely resistive circuit**, this 'opposition' is, of course, the **resistance** (symbol: *R*) of the circuit, measured in ohms:

$$R = \frac{\bar{E}}{\bar{I}}$$

However, as we have already learnt, this equation tells us what the resistance happens to be for that particular ratio of voltage to current – the resistance itself being determined by the load's physical factors (length, cross-sectional area and resistivity).

## Purely inductive circuit

Figure 26.4

You will recall, from the section on *inductance* that, whenever the current through an inductor varies, a voltage is induced into that inductor through the process of *self-induction*. By Lenz's Law, the direction of this induced voltage *(u)* always acts to *oppose any change in current*. And, in accordance with Faraday's Law, this induced voltage is *directly proportional to the rate of change of current*, as expressed below:

$$u \propto -\frac{\Delta i}{\Delta t}$$

The Greek letter 'delta' (Δ) simply means 'change in', so the expression $\Delta i / \Delta t$ means 'rate of change of current'.

The greatest rate of change in current occurs when the current waveform is at its steepest. As can be seen in Figure 26.5, this occurs whenever the current waveform passes through the zero axis. So, at point **A**, for example, the current is *increasing* at its greatest rate of change, so the maximum self-induced voltage (point **B**) will occur at the same time but, in accordance with Lenz's Law, must act in the negative sense (i.e. opposing the increase in current). This induced voltage waveform is shown as a broken line in Figure 26.5.

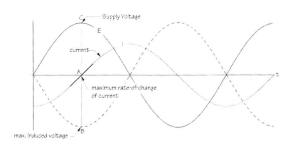

Figure 26.5

By Kirchhoff's Voltage Law, the induced voltage must be equal but *opposite* to the supply voltage (point **C**). The supply-voltage waveform is shown as a solid line. So, the current is clearly one-quarter of a wavelength behind the supply voltage, *(Ē)*. We say, therefore, that '*the current **lags** the supply voltage by 90°*'.

In a **purely inductive** circuit, then, the current *(Ī)* is said to **lag** the supply voltage *(Ē)* by 90° (and it is equally true to say that the *voltage leads the current by 90°*). If we now redraw the waveform, ignoring the self-induced voltage, it will look like Figure 26.6.

current lags voltage by 90°

Figure 26.6

Again, the waveforms can be represented by means of a **phasor diagram**, with the current and voltage phasors *lying at right angles to each other*. By convention, phasors are considered to 'rotate' counterclockwise, so the voltage phasor *(Ē)* is drawn 90° counterclockwise from the current phasor *(Ī)*, as shown below. It would be equally correct to place the voltage phasor horizontally, with the current phasor drawn 90° clockwise but, for consistency throughout this chapter, we'll draw the current phasor horizonally.

Figure 26.7

As before, the *length* of the voltage phasor represents the *r.m.s. value of voltage*, and the *length* of the current phasor represents the *r.m.s. value of the current*. Again, there is absolutely *no* relationship between the lengths of the two phasors, as one represents voltage and the other current. What *is* important, however, is that *the voltage phasor is drawn 90° counterclockwise relative to the current phasor.*

Once again, the ratio of voltage to current represents the *opposition* to current. In a **purely inductive circuit**, of course, *there is no resistance*, so we call this 'opposition' to current **inductive reactance** (symbol: $X_L$) measured in ohms, with the term, 'reactance', meaning 'reacting against' the passage of current.

$$X_L = \frac{\bar{E}}{\bar{I}}$$

Once again, it's important to understand that the ratio of voltage to current tells us what the inductive reactance happens to be for that particular ratio of voltage to current. The **inductive reactance** itself is determined by the inductance of the load, and the frequency of the supply, as shown in the following equation:

$$X_L = 2\pi f L$$

where:

$X_L$ = inductive reactance (ohms)

$f$ = supply frequency (hertz)

$L$ = inductance (henrys)

Unfortunately, in order to understand how the above equation is derived, it is necessary to have some knowledge of calculus, which is beyond the scope of this text. *So it is necessary to commit this equation to memory.*

Figure 26.8

You will recall, from the section on *capacitance* that, for a capacitor, the current *(i)* is *directly proportional to the rate of change of voltage*, as expressed below:

$$i \propto \frac{\Delta e}{\Delta t}$$

Where $\Delta e / \Delta t$ simply means 'rate of change of voltage'.

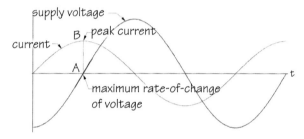

Figure 26.9

In Figure 26.9, the greatest rate of change of voltage occurs when the supply voltage waveform passes through the zero axis and is at its *steepest* – for example, at point **A**. This is the point at which the maximum current occurs (point **B**). So, the current is clearly one-quarter of a waveform *ahead* of the supply voltage. We say that *'the current **leads** the supply voltage by 90°'.*

In a **purely capacitive** circuit, then, the current $(\bar{I})$ is said to **lead** the voltage drop $(\bar{E})$ by 90° (and it is equally true to say the *voltage lags the current by 90°*).

current leads voltage by 90°

Figure 26.10

Again, the waveforms in Figure 26.10 can be represented by means of a **phasor diagram**, with the current phasor and voltage phasor each lying at right angles to each other. The voltage phasor $(\overline{E})$ is drawn 90° clockwise relative to the current phasor $(\overline{I})$, as illustrated in Figure 26.11. Again, it would be equally correct to draw the voltage phasor horizontally, with the current phasor 90° counterclockwise.

Figure 26.11

As with all phasor diagrams, the *length* of the voltage phasor represents the *r.m.s. value of voltage*, and the *length* of the current phasor represents the *r.m.s. value of the current*, but there is absolutely *no* relationship between the scales of the two phasors. What *is* important, however, is that *the current phasor is drawn 90° counterclockwise from the voltage phasor.*

As always, the ratio of voltage to current determines the opposition to current. In a **purely capacitive circuit**, there is no resistance, so we call the 'opposition' to current the **capacitive reactance** (symbol: $X_C$) of the circuit, measured in ohms:

$$X_C = \frac{\overline{E}}{\overline{I}}$$

The ratio of voltage to current tells us what the capacitive reactance happens to be for that particular ratio of voltage to current. The **capacitive reactance** itself being inversely proportional to the capacitance of the load, and the frequency of the supply, as specified in the following equation:

$$X_C = \frac{1}{2\pi f C}$$

where:

$X_C$ = capacitive reactance (ohms)

$f$ = supply frequency (hertz)

$C$ = capacitance (farads)

Once again, in order to understand how the above equation is derived, it is necessary to have some knowledge of calculus, which is beyond the scope of this text. So, again, *it is necessary to commit this equation to memory.*

Don't worry, though, as this equation, together with that for inductive reactance, are the *only two equations you will be asked to commit to memory* in these chapters on alternating current. From now on, **all other equations can be derived from phasor diagrams!**

### Does alternating current flow through a capacitor?

Not really, although it certainly *appears* to be doing so from the behaviour of the current in the external circuit. However, if you refer back to the chapter on *capacitors and capacitance*, you will recall that there can be no conduction current within a capacitor's dielectric due to the lack of any free electrons. However, the changing voltage applied across the plates causes a **displacement current** to take place.

The term 'displacement current' describes the distortion and polarisation of the electron orbits around the fixed nuclei of the dielectric's atoms. As the applied voltage increases, the orbits 'stretch' more and more, causing an increase in the displacement current; as the voltage decreases and reverses direction, so too does the direction of the displacement current. You will also recall that the direction of the field set up by these polarised atoms is such that it always opposes and, therefore, reduces the electric field set up within the dielectric by the applied voltage – 'reacting' against any change in the external voltage.

To summarise, for a capacitive circuit, a 'conduction' current (electron flow) takes place *around the external circuit*, whereas a 'displacement' current (distortion of electron orbits) takes place *within the dielectric*.

## CIVIL

It is **absolutely essential** to remember the phase relationships between currents and voltages for purely resistive, purely inductive and purely capacitive circits. *Unless you do so, you will **not** be able to construct phasor diagrams!*

**And if you cannot construct phasor diagrams, you will never understand the behaviour of alternating current!**

To help you, you should learn the mnemonic 'CIVIL', in which 'C' stands for 'capacitive circuit', and 'L' stands for 'inductive circuit' (see Figure 26.12).

An alternative mnemonic you might wish to consider is '**ELI the ICEman**', which (considering their winters!) is popular with American and Canadian students, where '*ELI*' indicates that, in an inductive (**L**) circuit, voltage (**E**) is before (leads) current (**I**); and where '*ICE*' indicates that, in a capacitive (**C**) circuit, current (**I**) is before (leads) voltage (**E**).

capacitive (C) circuits:     inductive (L) circuits:
I *before* ('leads') Voltage     I *after* ('lags') Voltage

**Figure 26.12**

### 'Real' circuits

Now that we have learned how these three 'ideal' (theoretical) circuits behave, let's now turn our attention to 'real' circuits.

'Real' circuits exhibit *combinations* of resistance, inductance and capacitance.

Since *all* circuits exhibit resistance, we will look at **series Resistive-Inductive (series *R-L*)** circuits, then at **series Resistive-Capacitive (series *R-C*)** circuits and, finally, at **series Resistive-Inductive-Capacitive (series *R-L-C*)** circuits.

Again, it's worth reminding ourselves that the circuit symbols used throughout this chapter represent the *quantities* resist*ance*, induct*ance* and capacit*ance* – *not* resistors, inductors and capacitors.

### Series R-L circuits

**Figure 26.13**

The circuit diagram shown in Figure 26.13 represents a **series resistive-inductive** circuit. It does *not* necessarily represent a resistor in series with an inductor, but a load (such as a coil) which exhibits both resistance and inductance. We have, if you like, 'separated out' the resistance and the inductance from the coil or other inductive component, so that they can be considered separately.

We now know that, in a purely resistive circuit, *the current and voltage are in phase* with each other and, in a purely inductive circuit, *the current lags the voltage by 90°*. So, what happens in a series *R-L* circuit? Well, clearly, our instincts tell us that *the current is likely to lag the voltage by some angle between 0° and 90°* – this angle being called the circuit's **phase angle** (symbol: φ, pronounced '*phi*').

The general definition of **phase angle** is '*the angle by which the **current** leads or lags the supply voltage*'. Note, we *always* measure phase angles in terms of what the load *current* is doing, relative to the supply voltage, *never the other way around*.

So, for an *R-L* circuit, because the current *always* lags the supply voltage, the phase angle is *always* described as **lagging**. Whenever a phase angle is quoted, *it is usual to specify whether it's leading or lagging*.

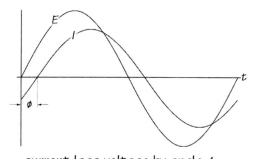

current lags voltage by angle φ

**Figure 26.14**

Of course, whenever a current $(\bar{I})$ flows through a **series *R-L* circuit**, a voltage drop, $\bar{U}_R$, will appear across the resistive component of the circuit, and a voltage drop, $\bar{U}_L$, will appear across the inductive component of the circuit – as shown in the schematic diagram in Figure 26.13.

> In the following step-by-step construction of a phasor diagram, the step being described is illustrated in **blue**, while the previous steps are shown in black.

## Drawing the phasor diagram

### *Step 1*

In a series *R-L* circuit, the **current** is common to both the resistive and the inductive components and, so, current is *always* chosen as the **reference phasor**. The reference phasor is *always drawn first, and always along the horizontal positive axis*. It's also drawn fairly long in order to distinguish it from the other phasors. In the following diagrams, we will further distinguish it by using an outline, rather than solid, arrow head (although this is not absolutely necessary) (Figure 26.15).

Figure 26.15

### *Step 2*

Since the voltage drop, $\bar{U}_R$, across the resistive component is *in phase with the current,* it is also drawn along the horizontal positive axis. Some textbooks show the phasor $\bar{U}_R$ *superimposed* over the reference phasor; others show it drawn *very close and parallel* with the reference phasor – the method preferred in this text (Figure 26.16).

Figure 26.16

### *Step 3*

The voltage drop, $\bar{U}_L$, across the inductive component *leads* the current by 90° (remember **CIVIL**), so is drawn 90° counterclockwise ('leading') from the reference phasor. Both $\bar{U}_R$ and $\bar{U}_L$, of course, represent voltage drops and, so, are both drawn to the same scale as each other (Figure 26.17).

Figure 26.17

### *Step 4*

We know, from Kirchhoff's Voltage Law, that in a series circuit, the supply voltage is the sum of the individual voltage drops. However, because, in this case, the two voltage drops, $\bar{U}_R$ and $\bar{U}_L$, lie at right angles to each other, we have to add them *vectorially* (Figure 26.18).

Figure 26.18

From the completed phasor diagram (Figure 26.18), we can see that the supply voltage, $\bar{E}$, is the **phasor sum** (or **vector sum**) of $\bar{U}_R$ and $\bar{U}_L$, which can be determined using Pythagoras's Theorem:

$$\bar{E} = \sqrt{\bar{U}_R^2 + \bar{U}_L^2}$$

It's completely unnecessary to commit this equation to memory, because it has been derived from the phasor diagram, using Pythagoras's Theorem. If you can draw the phasor diagram, and know Pythagoras's Theorem, then you don't need to bother to remember this equation!

**Worked example 1**   The voltage drop across the resistive component of a series *R-L* circuit is 30 V, and the voltage drop across the inductive component is 40 V. What is the value of the supply voltage?

**Solution**   *Always* start by sketching the circuit diagram (Figure 26.19), and inserting all the values given to you in the question.

Figure 26.19

Next, sketch the phasor diagram, following the steps described above. You *don't* have to draw the phasor diagram to scale (Figure 26.20).

**Figure 26.20**

Now, you can apply Pythagoras's Theorem to solve the problem:

$$\bar{E} = \sqrt{\bar{U}_R^2 + \bar{U}_L^2}$$
$$= \sqrt{30^2 + 40^2}$$
$$= \sqrt{2500} = 50 \text{ V (Answer)}$$

## Impedance diagram

The current flowing through a series *R-L* circuit will be opposed by *both* resistance (*R*) *and* inductive reactance ($X_L$). The combination of these two 'oppositions' is called the **impedance** (symbol: **Z**) of the circuit, also measured in ohms. 'Impedance' is yet another word, meaning to 'oppose' or 'impede' the passage of current.

But we *cannot* simply add the resistance and inductive reactance – so how can we find the impedance? The answer is by means of an **impedance diagram.** As before, in the following series of diagrams, the step being described is shown in **blue**; those previously described are shown in black.

## Drawing the impedance diagram

### Step 1

We start by drawing the circuit's phasor diagram, following the steps already explained (Figure 26.21).

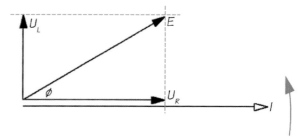

**Figure 26.21**

### Step 2

Next, we *divide each of the voltage phasors by the reference phasor (I)* (Figure 26.22).

The resulting diagram is known as an '**impedance diagram**' (sometimes called an '*impedance triangle*'), and is useful because it generates the following important equations:

$$R = \frac{\bar{U}_R}{\bar{I}} \qquad X_L = \frac{\bar{U}_L}{\bar{I}} \qquad Z = \frac{\bar{E}}{\bar{I}}$$

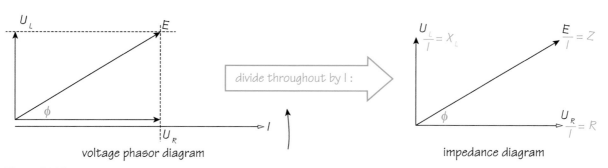

voltage phasor diagram          impedance diagram

**Figure 26.22**

Again, you don't have to commit these equations to memory, because they are derived when you convert a voltage phasor diagram into an impedance diagram!

Also from the impedance diagram, you can also see that the impedance is also the *vector sum of resistance and inductive reactance*, which can be calculated by simply applying Pythagoras's Theorem (Figure 26.23).

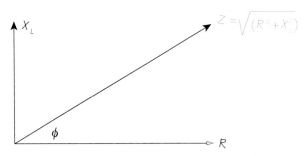

applying Pythagoras's Theorem

Figure 26.23

$$Z = \sqrt{(R^2 + X_L^2)}$$

If required, you can also write similar equations for resistance and inductive reactance, by applying Pythagoras's Theorem. That is:

$$R = \sqrt{(Z^2 - X_L^2)} \qquad X_L = \sqrt{(Z^2 - R^2)}$$

Once again, you don't need to commit any of these equations to memory, *providing* you can draw a phasor diagram, convert it to an impedance diagram, and apply Pythagoras's Theorem!

We can also find the circuit's **phase angle**, using basic trigonometry, utilising either the *sine*, *cosine* or *tangent* ratios. In practice, for a reason we'll see later in this text, the best choice is always to use the *cosine*:

$$\cos\phi = \frac{\text{adjacent}}{\text{hypotenuse}} = \frac{R}{Z}$$

$$\angle\phi = \cos^{-1}\frac{R}{Z}$$

**Important!** Dividing a voltage phasor diagram by current produces an impedance diagram which generates each of the equations shown above. *So you don't have to learn any of these equations* – they can all be generated provided you learn how to draw the phasor and impedance diagrams!

**Worked example 2** An inductor, of resistance 5 Ω and inductance 0.02 H, is connected across a 230-V, 50 Hz a.c. supply. Calculate each of the following:

a inductive reactance
b impedance
c current
d voltage drop across the resistive component of the circuit
e voltage drop across the inductive component of the circuit
f phase angle of the circuit.

**Solution** The first set in solving any a.c. circuit problem is to sketch the circuit diagram, and label it with all values supplied in the problem (Figure 26.24).

Figure 26.24

The next step is to draw the voltage phasor diagram, following the steps described earlier (Figure 26.25).

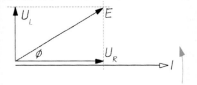

Figure 26.25

As the problem relates to inductive reactance, impedance, etc., the next step is to convert the voltage phasor diagram into an impedance diagram by dividing throughout by the reference quantity – i.e. by the current. This generates all the equations that we need to solve the problem (Figure 26.26).

a To find the inductive reactance (XL) of the circuit, we start by looking at the equations generated when we constructed the impedance diagram. There's only one equation for inductive reactance, $X_L = \dfrac{\bar{U}_L}{I}$. Unfortunately, we don't know the value of $\bar{U}_L$, so we cannot use this formula. What about applying Pythagoras's Theorem $\left( X_L = \sqrt{Z^2 - R^2} \right)$? Could we use this equation to find XL? Unfortunately, no, because we don't know the value of Z. If we can't use any of the equations generated by the impedance diagram, then we must fall back on the basic equation for inductive reactance, as follows:

$$X_L = 2\pi f L = 2\pi \times 50 \times 0.02 = 6.28\ \Omega\ \text{(Answer a.)}$$

b  To find the impedance, we *can* use an equation generated by the impedance diagram:

$$Z = \sqrt{R^2 + X_L^2}$$
$$= \sqrt{5^2 + 6.28^2}$$
$$= \sqrt{25 + 39.44}$$
$$= \sqrt{64.44} = 8.03\ \Omega\ \text{(Answer b.)}$$

c  To find the current, we use the following equation that was generated by the impedance diagram:

$$\bar{I} = \dfrac{\bar{E}}{Z} = \dfrac{230}{8.03} = 28.64\ \text{A (Answer c.)}$$

d  Again, using the equation generated by the impedance diagram:

$$\bar{U}_R = \bar{I}R = 28.64 \times 5 = 143.20\ \text{V (Answer d.)}$$

e  Again, using the equation generated by the impedance diagram:

$$\bar{U}_L = \bar{I}X_L = 28.64 \times 6.28 = 179.86\ \text{V (Answer e.)}$$

f  Using the cosine function:

$$\angle\phi = \cos^{-1}\dfrac{R}{Z} = \cos^{-1}\dfrac{5}{8.03} = \cos^{-1} 0.6227$$
$$= 51.48°\ \text{lagging (Answer f.)}$$

(*'Lagging'*, because the supply current lags the supply voltage in an inductive circuit.)

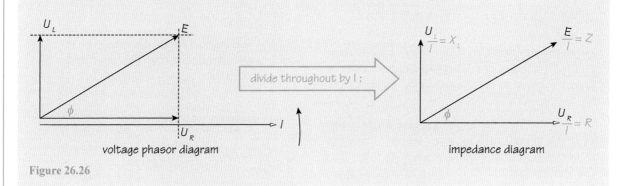

Figure 26.26

## Series R-C circuits

Figure 26.27

Again, it's important to realise that the circuit shown in Figure 26.27 does not necessarily represent a resistor

and a capacitor; rather, it simply represents resistance in series with capacitance. For example, it could represent the resistance and capacitance of a very long underground cable.

We know that in a purely resistive circuit, the current and voltage are in phase with each other; and, in a purely capacitive circuit, the current leads the voltage by 90°. So, what happens in a *series R-C* circuit? Well, clearly this time, we instinctively know that the *current must lead the voltage by some angle between 0° and 90°* – this angle is called the circuit's **phase angle** (symbol: $\phi$, pronounced 'phi').

Remember, the general definition of **phase angle** is *the angle by which the current leads or lags the supply voltage*. So for *resistive-capacitive circuits*, because the current *always* leads the supply voltage, the phase angle is *always* described as **leading** (see Figure 26.28).

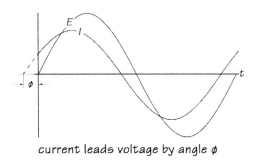

current leads voltage by angle φ

Figure 26.28

When a current *(I)* flows through a **series R-C circuit**, a voltage drop $\bar{U}_R$ will appear across the resistive component of the circuit, and a voltage drop $\bar{U}_C$ will appear across the capacitive component of the circuit.

## Drawing the phasor diagram

### Step 1

In a series circuit, the **current** is common to each component and, so, current is *always* chosen as the **reference phasor**. The reference phasor is *always drawn along the horizontal positive axis*, and it's also normally drawn fairly long in order to distinguish it from the others (Figure 26.29).

Figure 26.29

### Step 2

The voltage drop, $\bar{U}_R$, across the resistive component is *in phase with the current* and, so, is also drawn along the horizontal positive axis (Figure 26.30).

Figure 26.30

### Step 3

The voltage drop, $\bar{U}_C$, across the capacitive component *lags the current by 90°* (remember **CIVIL**), so is drawn 90° clockwise from the reference phasor (Figure 26.31).

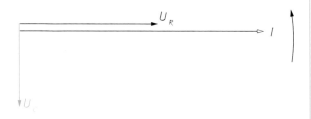

Figure 26.31

### Step 4

We know, from Kirchhoff's Voltage Law, that in a series circuit, the total voltage drop is the sum of the individual voltage drops. Because, in this case, the two voltage drops, $\bar{U}_R$ and $\bar{U}_C$, lie at right angles to each other, we have to add them *vectorially*.

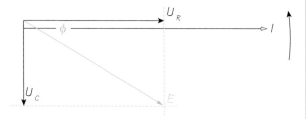

Figure 26.32

From the completed phasor diagram shown in Figure 26.32, we can see that the supply voltage, $\bar{E}$, is the **phasor sum** (or **vector sum**) of $\bar{U}_R$ and $\bar{U}_C$, which can be found using Pythagoras's Theorem:

$$\bar{E} = \sqrt{\bar{U}_R^2 + \bar{U}_C^2}$$

**Worked example 3**    The voltage drop across the resistive component of a series R-C circuit is 40 V, and the voltage drop across the capacitive component is 30 V. What is the value of the total voltage drop?

**Solution**    *Always* start by sketching the circuit diagram (Figure 26.33), and inserting all the values given to you in the question.

Figure 26.33

Next, sketch the phasor diagram, following the steps described above. You *don't* have to draw the phasor diagram to scale (Figure 26.34).

Figure 26.34

Now, you can apply Kirchhoff's Voltage Law, and use Pythagoras's Theorem to solve the problem:

$$\bar{E} = \sqrt{\bar{U}_R^2 + \bar{U}_C^2}$$
$$= \sqrt{40^2 + 30^2}$$
$$= \sqrt{2500} = 50 \text{ V (Answer)}$$

## Impedance diagram

The current flowing through a series *R-C* circuit will be opposed by both its resistance (*R*) *and* by its capacitive reactance ($X_C$). The combination of these two 'oppositions' is called the **impedance** (symbol: **Z**) of the circuit, and is measured in ohms. Again, we *cannot* simply add the resistance and capacitive reactance – so how can we find the impedance? Again, the answer is by means of an **impedance diagram**.

## Drawing the impedance diagram

### Step 1

We start by drawing the circuit's phasor diagram, following the steps already explained (Figure 26.35).

Figure 26.35

### Step 2

Next, we *divide each of the voltage phasors by the reference phasor* ($\bar{I}$).

The resulting diagram (Figure 26.36) is an **impedance diagram** (or *'impedance triangle'*), and is useful because it generates the following equations:

$$R = \frac{\bar{U}_R}{\bar{I}} \qquad X_C = \frac{\bar{U}_C}{\bar{I}} \qquad Z = \frac{\bar{E}}{\bar{I}}$$

Also from the impedance diagram, you can see that the impedance is the *vector sum of resistance and capacitive reactance*, which can be calculated by applying Pythagoras's Theorem (Figure 26.37).

$$Z = \sqrt{R^2 + X_C^2}$$

You can, if necessary, write similar equations for the resistance and capacitive reactance, by applying Pythagoras's Theorem. That is:

$$R = \sqrt{(Z^2 - X_C^2)} \qquad X_C = \sqrt{(Z^2 - R^2)}$$

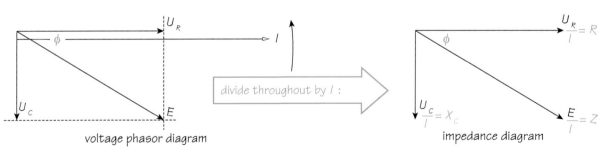

voltage phasor diagram          divide throughout by *I* :          impedance diagram

Figure 26.36

applying Pythagoras's Theorem

Figure 26.37

We can also find the circuit's **phase angle**, using basic trigonometry, utilising either the *sine*, *cosine* or *tangent* ratios – as before, the best choice is to use the *cosine*:

$$\cos\phi = \frac{\text{adjacent}}{\text{hypotenuse}} = \frac{R}{Z}$$

$$\angle\phi = \cos^{-1}\frac{R}{Z}$$

**Important!** Dividing a voltage phasor diagram by current produces an impedance diagram which generates each of the equations shown above. *So you don't have to learn any of these equations – they can all be generated provided you learn how to draw the phasor and impedance diagrams!*

Worked example 4   A capacitor, of resistance 40 Ω and capacitance 50 μF is connected across a 110-V, 50 Hz, a.c. supply. Calculate each of the following:

a   capacitive reactance
b   impedance
c   current
d   voltage drop across the resistive component of the circuit
e   voltage drop across the capacitive component of the circuit
f   phase angle of the circuit.

Solution   The first step in solving any a.c. circuit problem is to sketch the circuit diagram, and label it with all values supplied in the problem (Figure 26.38).

Figure 26.38

The next step is to draw the voltage phasor diagram, following the steps described earlier (Figure 26.39).

Figure 26.39

As the problem relates to inductive reactance, impedance, etc., the next step is to convert the voltage phasor diagram into an impedance diagram, by dividing throughout by the reference quantity – i.e. current. This generates the equations that we need to solve the problem (Figure 26.40).

a   To find the **capacitive reactance** ($X_C$) of the circuit, we start by looking at the equations generated when we constructed the impedance diagram. There's only one equation for capacitive reactance, $X_C = \dfrac{\bar{U}_C}{I}$. Unfortunately, we don't know the value of $\bar{U}_C$ so we can't use this formula. What about applying Pythagoras's Theorem ($X_C = \sqrt{Z^2 - R^2}$)? Could we use this to find $X_C$? Unfortunately, no, because we don't know the value of Z. Clearly, then, we must fall back on the basic equation for capacitive reactance, as follows:

$$X_C = \frac{1}{2\pi f C} = \frac{1}{2\pi \times 50 \times (50 \times 10^{-6})}$$

$$= 63.67\ \Omega\ (\text{Answer a.})$$

b   To find the impedance, we can now use the equation generated by the impedance diagram:

$$Z = \sqrt{R^2 + X_L^2}$$

$$= \sqrt{40^2 + 63.67^2}$$

$$= \sqrt{1600 + 4054}$$

$$= \sqrt{5654} = 75.2\ \Omega\ (\text{Answer b.})$$

c   To find the current, we use the following equation generated by the impedance diagram:

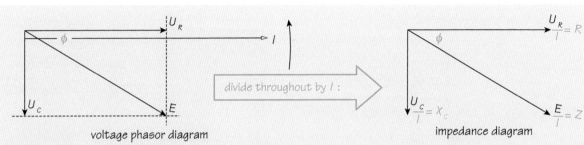

voltage phasor diagram          impedance diagram

Figure 26.40

$$\bar{I} = \frac{\bar{E}}{Z} = \frac{110}{75.2} = 1.46 \text{ A (Answer c.)}$$

d   Again, using the equation generated by the impedance diagram:

$$\bar{U}_R = \bar{I}R = 1.46 \times 40 = 58.53 \text{ V (Answer d.)}$$

e   Again, using the equation generated by the impedance diagram:

$$\bar{U}_C = \bar{I}X_C = 1.46 \times 63.67 = 93.14 \text{ V (Answer e.)}$$

f   Using the cosine function:

$$\angle\phi = \cos^{-1}\frac{R}{Z} = \cos^{-1}\frac{40}{75.2} = \cos^{-1} 0.5319$$
$$= 57.87° \text{ leading (Answer f.)}$$

('*Leading*', because the supply current leads the supply voltage in a capacitive circuit.)

## Series R-L-C circuits

Figure 26.41

We have learnt that:

- in a series *R-L* circuit, *the current lags the supply voltage by some angle* and
- in a series *R-C* circuit, *the current leads the supply voltage by some angle*.

So, what happens in a *series R-L-C* circuit? Well, clearly the *current could either lag* **or** *lead the voltage* – depending on the values of the inductive reactance and the capacitive reactance! Whatever value this angle happens to be, it will be the circuit's **phase angle** (symbol: $\phi$, pronounced 'phi').

When a current $(\bar{I})$ flows through a series *R-L-C* circuit, a voltage drop $\bar{U}_R$ will appear across the

resistive component of the circuit, a voltage drop $\bar{U}_L$ will appear across the inductive component, and a voltage drop $\bar{U}_C$ will appear across the capacitive component of the circuit.

### Drawing the phasor diagram

#### Step 1

In a series circuit, the **current** is common to each component and, so, current is again chosen as the **reference phasor**. The reference phasor is *always drawn along the horizontal positive axis*, and it's also normally drawn fairly long in order to distinguish it from the others (Figure 26.42).

Figure 26.42

#### Step 2

The voltage drop, $\bar{U}_R$, across the resistive component is *in phase with the current* and, so, is also drawn along the horizontal positive axis (Figure 26.43).

Figure 26.43

## Step 3

The voltage drop, $\bar{U}_L$, across the inductive component *leads the current by 90°* (remember C**IV**IL), so is drawn 90° counterclockwise from the reference phasor (Figure 26.44).

Figure 26.44

## Step 4

The voltage drop, $\bar{U}_C$, across the capacitive component *lags the current by 90°* (remember CI**VL**), so is drawn 90° clockwise from the reference phasor (Figure 26.45).

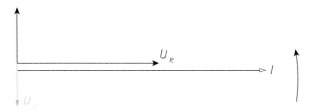

Figure 26.45

## Step 5

From Kirchhoff's Voltage Law, the total voltage drop in a series circuit is the sum of the individual voltage drops, and we have to add them *vectorially*.

It's a little more difficult to add *three* phasors. As $\bar{U}_L$ and $\bar{U}_C$ lie in *opposite* directions, the simplest thing to do is to start by subtracting them and, *then*, add the difference to phasor $\bar{U}_R$. (Figure 26.46).

The snag is, of course, that we might not know whether $\bar{U}_L$ is bigger than $\bar{U}_C$, or vice versa! Fortunately, *it doesn't matter!* The purpose of the phasor

diagram is simply to *generate equations*, not to accurately represent the actual conditions in the circuit to scale. And the phasor diagram will *always* generate the correct equations whether $\bar{U}_L$ is actually bigger than $\bar{U}_C$, or vice versa!

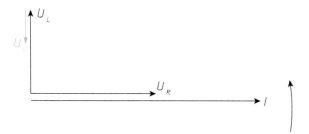

Figure 26.46

So, the simplest solution is to get into the habit of **always** *drawing $\bar{U}_L$ longer than $\bar{U}_C$* – or the other way around, if you prefer! But, for a reason that will be revealed later, whatever you do, *never* **ever** *draw them the same length!!*

Figure 26.47 shows what the finished phasor diagram will look like.

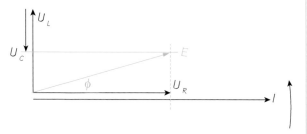

Figure 26.47

From the completed phasor diagram, we can see that $\bar{E}$ is the **phasor sum** (or **vector sum**) of $\bar{U}_R$, $\bar{U}_L$ and $\bar{U}_C$, which can be found using Pythagoras's Theorem:

$$\bar{E} = \sqrt{\bar{U}_R^2 + (\bar{U}_L - \bar{U}_C)^2}$$

If, when we draw the phasor diagram, we make $\bar{U}_L$ bigger than $\bar{U}_C$ when, really, it's the other way around, *it doesn't really matter*. When we subtract the two, we'll end up with a negative quantity inside the brackets which, when squared, will result in a positive value.

**Worked example 5** In a series *R-L-C* circuit, the voltage drop across the resistive component is 4 V, the voltage drop across the inductive component is 10 V, and the voltage drop across the capacitive component is 7 V. What is the value of the total voltage drop?

**Solution** *Always* start by sketching the circuit diagram, and inserting all the values given to you in the question (Figure 26.48).

Figure 26.48

Next, sketch the phasor diagram, following the steps described above. You *don't* have to draw the phasor diagram to scale (Figure 26.49).

Figure 26.49

Now, you can apply Kirchhoff's Voltage Law, and use Pythagoras's Theorem to solve the problem:

$$\bar{E} = \sqrt{\bar{U}_R^2 + (\bar{U}_L - \bar{U}_C)^2}$$
$$= \sqrt{4^2 + (10-7)^2}$$
$$= \sqrt{4^2 + 3^2} = \sqrt{25} = 5 \text{ V (Answer)}$$

## Impedance diagram

The current flowing through a series *R-L-C* circuit will be opposed by resistance (*R*) *and* by inductive reactance ($X_L$), *and* by capacitive reactance ($X_C$). The combination of these three 'oppositions' is called the **impedance** (symbol: *Z*) of the circuit, measured in ohms. But, again, we *cannot* simply add the resistance, inductive reactance and capacitive reactance – so how can we find the impedance? Again, the answer is by means of an **impedance diagram**.

## Drawing the impedance diagram

### Step 1

We start by drawing the circuit's phasor diagram, following the steps already explained (Figure 26.50).

Figure 26.50

### Step 2

Next, we *divide each of the voltage phasors by the reference phasor* ($\bar{I}$) (Figure 26.51).

The resulting diagram is an **impedance diagram** (or '*impedance triangle*'), and is useful because it generates the following equations:

$$R = \frac{\bar{U}_R}{\bar{I}} \qquad X_L = \frac{\bar{U}_L}{\bar{I}} \qquad X_C = \frac{\bar{U}_C}{\bar{I}} \qquad Z = \frac{\bar{E}}{\bar{I}}$$

Also from the impedence diagram, you can see that the impedance is also the *vector sum of resistance,*

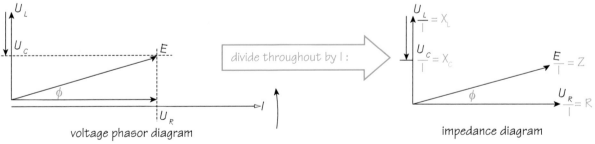

voltage phasor diagram          impedance diagram

Figure 26.51

*inductive reactance* and *capacitive reactance*, which can be calculated by applying Pythagoras's Theorem (Figure 26.52).

$$Z = \sqrt{R^2 + (X_L - X_C)^2}$$

applying Pythagoras's Theorem

Figure 26.52

Once again, we can also find the circuit's **phase angle**, using basic trigonometry:

$$\cos\phi = \frac{\text{adjacent}}{\text{hypotenuse}} = \frac{R}{Z}$$

$$\angle\phi = \cos^{-1}\frac{R}{Z}$$

**Important!** Dividing a voltage phasor diagram by current produces an impedance diagram which generates each of the equations shown above. *So you don't have to learn any of these equations – they can all be generated* **provided you learn how to draw the phasor and impedance diagrams!**

Worked example 6    A circuit of resistance of 1.5 Ω, inductance of 16 mH, and capacitance 500 µF, is connected across a 230-V, 50 Hz a.c. supply. Calculate each of the following:

a    inductive reactance
b    capacitive reactance
c    impedance
d    current
e    voltage drop across the resistive component of the circuit
f    voltage drop across the inductive component of the circuit
g    voltage drop across the capacitive component of the circuit
h    phase angle of the circuit.

Solution    The first step in solving *any* a.c. circuit problem is to sketch the circuit diagram, and label it with all values supplied in the problem (Figure 26.53).

Figure 26.53

The next step is to draw the voltage phasor diagram, following the steps described earlier (Figure 26.54).

Figure 26.54

As the problem relates to inductive reactance, capacitive reactance, impedance, etc., the next step is to convert the voltage phasor diagram into an *impedance diagram* (Figure 26.55), by dividing throughout by the reference quantity – i.e. current. This generates the equations that we need to solve the problem.

a    To find the **inductive reactance ($X_L$)** of the circuit, we start by looking at the equations generated when we constructed the impedance diagram. There's only one equation for inductive reactance, $X_L = \bar{U}_L/\bar{I}$. Unfortunately, we don't know the value of $\bar{U}_L$ so we can't use this formula. What about applying Pythagoras's Theorem? Could we use this to find $X_L$? Unfortunately, no, because we don't know the value of $Z$. Clearly, then, we must fall back on the basic equation for inductive reactance, as follows:

$$X_L = 2\pi f L = 2\pi \times 50 \times 16 \times 10^{-3}$$
$$= 5.03\ \Omega\ \text{(Answer a.)}$$

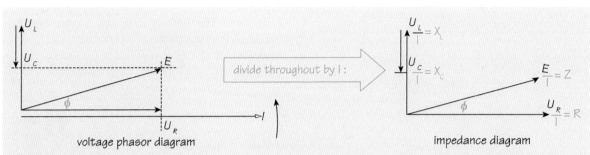

Figure 26.55

b To find the **capacitive reactance** $(X_C)$ of the circuit, we start by looking at the equations generated when we constructed the impedance diagram. There's only one equation for capacitive reactance, $X_C = \bar{U}_C / \bar{I}$. Unfortunately, we don't know the value of $\bar{U}_C$ so we can't use this formula. What about applying Pythagoras's Theorem? Could we use this to find $X_C$? Unfortunately, no, because we don't know the value of Z. Clearly, then, we must fall back on the basic equation for capacitive reactance, as follows:

$$X_C = \frac{1}{2\pi f C} = \frac{1}{2\pi \times 50 \times (500 \times 10^{-6})}$$
$$= 6.37 \ \Omega \ \text{(Answer b.)}$$

c To find the impedance, we can now use the equation generated by the impedance diagram:

$$Z = \sqrt{R^2 + (X_L - X_C)^2}$$
$$= \sqrt{1.5^2 + (5.03 - 6.37)^2}$$
$$= \sqrt{1.5^2 + (-1.34)^2}$$
$$= \sqrt{2.25 + 1.8}$$
$$= \sqrt{4.05} = 2.01 \ \Omega \ \text{(Answer c.)}$$

**Note!** Despite $X_C$ being larger than $X_L$, the above equation still delivers the correct answer, as the negative sign inside the bracket becomes positive when the bracket is squared!

d To find the current, we use the following equation generated by the impedance diagram:

$$\bar{I} = \frac{\bar{E}}{Z} = \frac{230}{2.01} = 114.43 \ \text{A (Answer d.)}$$

e Again, using the equation generated by the impedance diagram:

$$\bar{U}_R = \bar{I}R = 114.43 \times 1.5 = 171.65 \ \text{V (Answer e.)}$$

f Again, using the equation generated by the impedance diagram:

$$\bar{U}_L = \bar{I}X_L = 114.43 \times 5.03 = 575.38 \ \text{V (Answer f.)}$$

g Again, using the equation generated by the impedance diagram:

$$\bar{U}_C = \bar{I}X_C = 114.43 \times 6.37 = 728.92 \ \text{V (Answer g.)}$$

h Using the cosine function:

$$\angle\phi = \cos^{-1}\frac{R}{Z} = \cos^{-1}\frac{1.5}{2.01} = \cos^{-1} 0.7462$$
$$= 41.74° \ \text{leading (Answer h.)}$$

('Leading' because, despite how we have actually drawn the phasor diagram, $\bar{U}_C$ is, in this case, *larger* than $\bar{U}_L$, so the supply current must *lead* the supply voltage.)

## Series resonance

From:

$$X_L = 2\pi f L \quad \text{and} \quad X_C = \frac{1}{2\pi f C}$$

... we know that the inductive reactance of a circuit is directly proportional to the supply frequency, whereas the capacitive reactance is inversely proportional to the supply frequency. So, if we gradually increase the supply frequency, the inductive reactance will gradually *increase*, while the capacitive reactance will *decrease*.

Eventually, a point will be reached for any *R-L-C* circuit, when the values of the inductive reactance and capacitive reactance will equal one another – as illustrated in Figure 26.56.

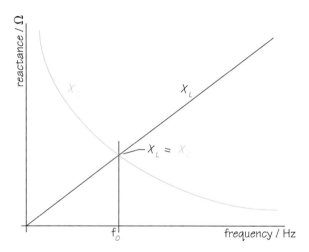

**Figure 26.56**

The frequency at which this occurs is called the '**resonant frequency**' $(f_o)$ of the circuit, and the circuit is said to be at '**series resonance**'.

'**Series resonance**', then, is a unique condition which occurs whenever a series *R-L-C* circuit's inductive reactance is *exactly* equal to its capacitive reactance. That is, resonance occurs when:

$$\boxed{X_L = X_C}$$

At resonance, rather strange things happen to a circuit!

*Any R-L-C* circuit can be made to resonate at its unique resonant frequency $(f_o)$, and determining what that frequency is for any circuit is straightforward. Starting with the basic condition for resonance:

$$X_L = X_C$$

and expanding:

$$2\pi f_o L = \frac{1}{2\pi f_o C}$$

$$f_o^2 = \frac{1}{(2\pi)^2 LC}$$

$$\boxed{f_o = \frac{1}{2\pi\sqrt{LC}}}$$

**Worked example 7**    What is the resonant frequency for a circuit having an inductance of 16 mH and a capacitance of 100 µF?

**Solution**

$$f_o = \frac{1}{2\pi\sqrt{LC}} = \frac{1}{2\pi\sqrt{(16\times10^{-3})\times(100\times10^{-6})}}$$

$$= \frac{1}{2\pi\sqrt{1600\times10^{-9}}} = \frac{1}{2\pi\times(1.26\times10^{-3})}$$

$$= 126\text{ Hz (Answer)}$$

## Impedance diagram for series resonance

If we were to draw an **impedance diagram** for a series *R-L-C* circuit at resonance, it would look like Figure 26.57.

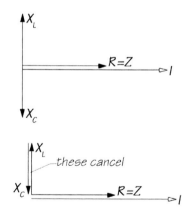

**Figure 26.57**

As you can see, the inductive reactance and the capacitive reactance are identical, but *act in opposite directions*. So, their vector sum must be zero.

This leaves **resistance** as the circuit's *only opposition to the flow of current*, from which we can say:

At **resonance**, a circuit's impedance is equal to its resistance.

So if, at resonance, the vector sum of a circuit's inductive reactance and capacitive reactance is zero, leaving only its resistance to oppose current, then it follows that the circuit's current will reach its maximum value at resonance.

A circuit's current reaches its maximum value when resonance occurs.

If we were to conduct a simple experiment, by adjusting the supply frequency of a series *R-L-C* circuit, until resonance occurs, while measuring its current, the resulting graph would look something like Figure 26.58.

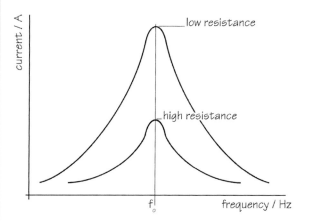

**Figure 26.58**

As the frequency approaches the resonant frequency, the reactance (i.e. the combined inductive and capacitive reactance) falls towards zero ohms, and the corresponding current increases. The value which the current reaches, at resonance, is limited by the resistance of the circuit. If the resistance is low, then the resulting current will be high; if the resistance is high, then the resulting current will be low. As the applied frequency passes beyond the resonant frequency, the reactance starts to increase again, and the current starts to fall again.

Now, let's look at the **voltage phasor diagram** for an *R-L-C* circuit at resonance, shown as Figure 26.59.

**Figure 26.59**

As the voltage drops, $\bar{U}_L$ and $\bar{U}_C$ are equal and opposite, then the phasor sum of $\bar{U}_L$, $\bar{U}_C$ and $\bar{U}_R$ will simply be $\bar{U}_R$. In other words, $\bar{U}_R = \bar{E}$ or, to put it another way: the *entire supply voltage will appear across the resistive component of the circuit.*

You will recall that when we learnt how to draw general voltage phasor diagrams for *R-L-C* circuits, we were warned *always* to make the $\bar{U}_L$ phasor larger than the $\bar{U}_C$ phasor *(or vice versa)*. Well, now you know why! Making the two phasors the same lengths results in resonance, which is a *special* condition, not the normal condition!

It's very important to understand that we are *not* saying that the voltages $\bar{U}_L$ and $\bar{U}_C$ don't exist. They most certainly *do* exist! What we *are* saying is that, if we tried to measure the voltage drop across *both* the inductive *and* capacitive components, then a voltmeter would read zero because the two voltage drops are in *antiphase* – that is, they are equal in magnitude, but 180° out of phase with each other, as illustrated in Figure 26.60.

**Figure 26.60**

Table 26.1

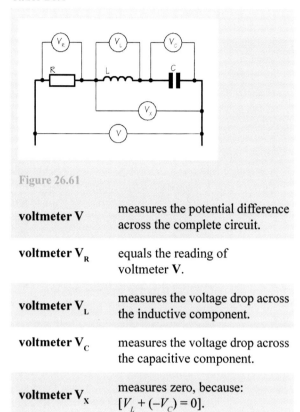

Figure 26.61

| **voltmeter V** | measures the potential difference across the complete circuit. |
| --- | --- |
| **voltmeter $V_R$** | equals the reading of voltmeter **V**. |
| **voltmeter $V_L$** | measures the voltage drop across the inductive component. |
| **voltmeter $V_C$** | measures the voltage drop across the capacitive component. |
| **voltmeter $V_X$** | measures zero, because: $[V_L + (-V_C) = 0]$. |

So, at any instant, the sum of the instantaneous voltages, $u_L + (-u_C)$, must be zero. Or, in general:

$$\bar{U}_L + (-\bar{U}_C) = 0$$

This can be summarised as shown in Table 26.1.

The voltage phasor diagram also tells us another important thing about resonance. At resonance, *the circuit's phase angle is zero.*

At resonance the circuit's phase angle is zero.

The final 'strange' thing which occurs, or rather, which *could* occur, at resonance is revealed in the following worked example.

**Worked example 8**    A circuit has a resistance of 10 $\Omega$. At resonance, the values of the circuit's inductive reactance and capacitive reactance are each 100 $\Omega$. If the supply voltage is 200 V, calculate (a) the current, (b) the voltage drops across the circuit's resistive component and (c) across its inductive and capacitive components.

**Solution**   At resonance, the *only* opposition to current is the resistance of the circuit. So, the resulting current must be:

$$\bar{I} = \frac{\bar{E}}{R} = \frac{200}{10} = 20 \text{ A (Answer a.)}$$

At resonance, the voltage drop across the resistive component will be equal to the supply voltage, i.e. 200 V (Answer b.)

At resonance, the voltage drop across the inductive component is as follows:

$$\bar{U}_L = \bar{I} \, X_L = 20 \times 100 = 2000 \text{ V (Answer c.)}$$

At resonance, the voltage drop across the capacitive component is as follows:

$$\bar{U}_C = \bar{I} \, X_C = 20 \times 100 = 2000 \text{ V (Answer c.)}$$

Yes! The above answers *are* correct! Even though the supply voltage to the circuit is just 200 V, the voltage drop across each of the reactive components is indeed 2000 V!

If the circuit's resistance was just 1 $\Omega$, instead of 10 $\Omega$, then the current would rise to 200 A, and the voltage drops across the reactive components would become 20 000V!

Don't forget, these very large voltages are what will appear across the inductive component and the capacitive component, if they are measured *separately*. If we measure the voltage drop across *both* components, then the result would be zero, because the two voltage drops are in antiphase with each other!

So the final thing we can say about resonance is that, *if the resistance of the circuit is low compared with the inductive and capacitive reactances*, then very large voltages can appear across those components – *voltages that are many times the value of the supply voltage!*

At resonance, if the resistance of the circuit is low compared to the inductive and capacitive reactances, then the individual values of voltage drops, $\bar{U}_L$ and $\bar{U}_C$ can be *many* times larger than the supply voltage.

There are *two* ways of looking at this strange phenomenon. The first is from the electronics engineer's point of view; the second is from the transmission/distribution engineer's point of view.

Electronics engineers have to deal with signals that are frequently in the microvolt range. Resonance can boost these tiny voltages hundreds of times, providing what is, essentially, 'free amplification' of those signals.

Electricity transmission/distribution engineers, on the other hand, are already dealing with very high, and very dangerous, voltages. Unintentional resonance could boost these voltages far higher, with disastrous results to system insulation. For example, resonance could occur when a highly capacitive underground cable feeds a highly inductive transformer. If the resulting values of capacitive reactance and inductive reactance happen to be such that resonance, or even near-resonance, conditions result, then the resulting voltages could well exceed the dielectric strengths of the cable/transformer winding insulation, leading to failure.

## Summary

To summarise what we have learnt about **series resonance**, we can say:

- series resonance occurs when: $X_L = X_C$.
- a series circuit will resonate at a frequency, called its 'resonant frequency' $(f_o)$, which is determined from:

$$f_o = \frac{1}{2\pi\sqrt{LC}}$$

- at resonance, a circuit's impedance will equal its resistance: $Z = R$.
- a circuit's current will reach its maximum value at resonance, and it will be in phase with the supply voltage.
- at resonance, the voltage drop across a circuit's resistive component will equal its supply voltage.
- at resonance, the sum of the voltage drops across a circuit's inductive and capacitive components will be zero.
- at resonance, the individual voltage drops appearing across the inductive or capacitive components can be very much higher than the supply voltage.

## Review your learning

Now that we've completed this chapter, we need to examine the **objectives** listed at its start. Placing '*Can I...*' at the beginning, and a question mark at the end, of each objective turns that objective into a **test item**. If we can answer each of those test items, then we've met the objectives of this chapter.

---

**Online resources**

The companion website to this book contains further resources relating to this chapter. The website can be accessed via the following link:

www.routledge.com/cw/waygood

# Chapter 27

# Power in alternating-current circuits

On completion of this chapter, you should be able to:

1 define the terms energy, work, heat and power, and specify their SI units of measurement.
2 explain the behaviour of power in
   a a purely resistive a.c. circuit
   b a purely inductive a.c. circuit
   c a purely capacitive a.c. circuit
   d an *R-L* a.c. circuit
   e an *R-C* a.c. circuit.
3 state the relationship between true (or 'active') power, reactive power and apparent power.
4 state the units of measurement for
   a true (or 'active') power
   b reactive power
   c apparent power.
5 change the voltage phasor diagram into a power diagram, and derive equations for true, reactive and apparent power for *R-L*, *R-C* and *R-L-C* circuits.
6 define the term 'power factor'.
7 solve problems on power in a.c. circuits.

## Introduction

In this chapter, we are going to examine the behaviour of **energy** and **power** in alternating-current circuits.

Before we do so, though, we need to remind ourselves of how energy can be manipulated, so you may want to revise the earlier chapter on *energy, work, heat and power*.

As we have learnt, energy can be 'manipulated' in either of *two* ways: it can be *changed from one form into another*, or it can be *transferred from one body to another*.

When energy is *changed from one form into another*, we say that '**work**' is being done. For example, when an electric motor converts electrical energy into kinetic energy, the motor is doing work.

Energy is *transferred between objects* whenever those objects are at different temperatures. We call this process '**heat transfer**'. You will recall that the term 'heat' describes *'the transfer of energy from a warmer body to a cooler body'*.

Both work *and* heat are expressed in joules.

> **Work** is defined as *'the conversion of energy from one form into another'*.

> **Heat** is defined as *'the transfer of energy from a body at a higher temperature to one at a lower temperature'*.

Finally, we should also remind ourselves that '**power**' is defined as *'the rate of doing work'* or 'the rate of heat transfer', expressed in watts.

> **Power** is defined as *'the rate of doing work'*, or *'the rate of heat transfer'*.

## Behaviour of energy in a.c. circuits

### Behaviour of energy in purely resistive circuits

Whenever an electric current overcomes the resistance of a conductor, it does **work** on that conductor ($W = I^2Rt$), causing its internal energy (the vibration of its atoms) to increase. An increase in internal energy is always accompanied by an increase in temperature and, if the temperature of the conductor exceeds that of its surroundings, then energy will be transferred *away* from the conductor into its surroundings through **heat transfer**.

As you may recall, from the chapter on *resistance*, the *consequence of resistance is heat.*

Heat transfer away from a conductor is *completely irreversible*. That is, you cannot transfer that energy back into the conductor and send it as electrical energy back to the supply. Once it's gone, it's gone for good!

The *rate* of heat transfer away from a conductor to its surroundings is termed 'power', and is expressed in watts. For reasons that will shortly become clear, in a purely resistive circuit, it is traditional to refer to this as either '**true power**', '**real power**' or '**active power**' (symbol: $P$). Throughout this rest of this book, we will stick with the term 'true power'.

### Behaviour of energy in a purely inductive circuit

You will recall that a purely inductive circuit is an 'ideal' circuit, in which there is no resistance.

If a circuit has no resistance, then no heating can take place. So there can be no expenditure of energy, through heat transfer, away from the circuit.

However, due to the circuit's inductance, the presence of a current creates a *magnetic field*. Because the a.c. current is continuously changing in both magnitude and direction, the resulting magnetic field also varies in magnitude and direction and induces a voltage into the circuit which always acts to oppose the change in current.

During the first quarter-cycle, then, the *increasing* current is *opposed* by this induced voltage. Energy is required to overcome this opposition, and this energy is drawn from the supply and then stored within the magnetic field. During the second quarter-cycle, as the current *falls*, the direction of the induced voltage reverses, and acts to oppose the reduction of the current – that is, it tries to *sustain* the current. In other

words, the energy that was previously *stored* within the magnetic field, is now being *returned* to the supply.

This process repeats itself during subsequent quarter-cycles, with energy being alternately *stored* in the magnetic field and, then, *released* back to the supply. So, although there *is* energy conversion taking place *within* the circuit, there is no net loss of energy away *from* the circuit.

As always, the *rate* at which this energy conversion is taking place, is power. But to distinguish it from 'true power', we call it '**reactive power**' (symbol: $Q$). It's also traditional to measure reactive power in **reactive volt amperes** (symbol: **var**), rather than in watts.

Some textbooks describe reactive power as the rate at which energy is continually *'sloshing back and forth, between the supply and the load'.*

Some textbooks refer to reactive power as *'wattless power'* or *'imaginary power'*. The term *'imaginary power'*, though, *doesn't* mean it exists only in the mind! In this sense, the word 'imaginary' is used by mathematicians to mean 'quadrature' (i.e. 'at right angles'), in other words it is created by a load current that lags (or, as we shall see, leads) the supply voltage by 90°.

### Behaviour of energy in a purely capacitive circuit

A purely capacitive circuit also has no resistance so, again, no heating effect can take place to cause any transfer of energy away from the circuit.

The behaviour of energy in a purely capacitive circuit is almost identical to its behaviour within a purely inductive circuit, except that energy from the circuit is being alternately stored in, and returned from, an *electric* field rather than a magnetic field.

Once again, although there *is* energy conversion taking place within the circuit, there is no net loss of energy away *from* the circuit.

As with a purely inductive circuit, the rate at which this energy conversion is taking place is also called 'reactive power', which is also measured in reactive volt amperes to distinguish it from true power.

### Behaviour of energy in R-L, R-C and R-L-C circuits

In **R-L**, **R-C** and **R-L-C** circuits, the behaviour of energy is a *combination* of the behaviour of energy in purely resistive and purely reactive circuits. That is *some* energy is being irreversibly lost from the circuit through heat transfer or by the work done by the load, while *some* energy is being alternately stored, and returned to the supply from, the magnetic or electric fields (or *both* magnetic *and* electric fields, in the case of an *R-L-C* circuit).

So resistive-reactive circuits have *both* 'true power' *and* 'reactive power' occuring at the same time. As we shall see, shortly, we *cannot* simply add these two quantities together, but the *combination* of the two we call '**apparent power**' (symbol: *S*) and, to distinguish apparent power from true power and reactive power, it is traditional to measure it in **volt amperes***** (symbol: **V·A**).

*****It's worth pointing out that the units, **reactive volt amperes (var)** and **volt amperes (V·A)**, are *traditional*, in order to easily distinguish reactive power, apparent power and true power from each other. SI doesn't recognise either, and uses the watt to measure all 'forms' of power.

### Further explanation of true power

For a *purely resistive* circuit, we learnt that **true power** is the rate at which energy is irreversibly lost from the circuit through heat transfer to its surroundings. However, in some circuits, it can be a little more complicated that this.

Motors, for example, do work by converting electrical energy into kinetic energy, in order to drive their mechanical loads. As the output power of a motor is expressed in **watts**, the rate at which a motor does work driving its load must represent its '**true power**'. Since motors are obviously not 'purely resistive' loads that dissipate energy through heat transfer, how do we account for this?

Because of its windings, a motor is an *R-L* circuit and, as we have learnt ('**CIVIL**'), the

motor's supply current will lag the supply voltage by some angle, φ.

As can be seen in Figure 27.1, the motor's (lagging) supply current can actually be resolved into two 'components', called an *'in-phase'* (or *'active'*) component, and a *'quadrature'* (meaning 'at 90°') component.

The in-phase component represents the machine's **mechanical-load current** (i.e. the component of current responsible for the work done by the motor in driving its **load**), whereas the quadrature component represents its **magnetising current**.

These two components of the motor's current don't *actually* exist as two individual currents, but the circuit *behaves as if they do*. In the equivalent circuit, and corresponding waveforms, for a motor, shown in Figures 27.2 and 27.3, we show the winding and its equivalent resistance as two separate branches of the same circuit although, in reality, they are one and the same.

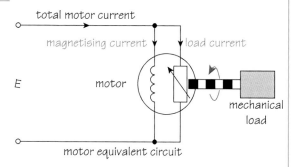

**Figure 27.2**

The quadrature (**magnetising current**) component of current (which accounts for the reactive power providing the motor's magnetic field) is constant, but the in-phase component (*mechanical-load* **current**) *varies with load* – if

**Figure 27.1**

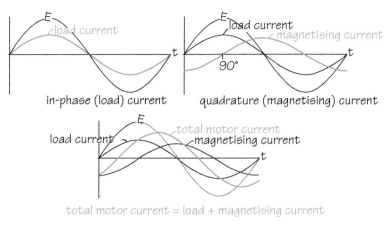

in-phase (load) current   quadrature (magnetising) current

total motor current = load + magnetising current

**Figure 27.3**

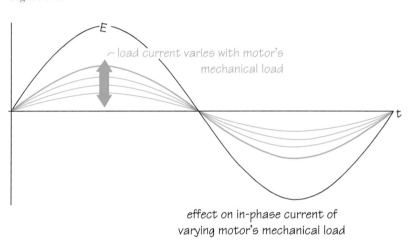

effect on in-phase current of
varying motor's mechanical load

**Figure 27.4**

the motor's mechanical load *increases*, then the in-phase component of the current will also *increase* in order to supply more 'true power' to drive that load. So, in a sense, the varying mechanical load on the motor behaves like a varying *resistive* load.

This 'equivalent resistance' of the motor is *not* the same as its *actual* resistance, and it *varies* (which is why we have shown it as a variable resistor in the schematic diagram, Figure 27.2) as the mechanical load varies – accounting for why the in-phase component of the current varies with load.

## Summary

To summarise our explanations for **true power, reactive power** and **apparent power**, we can say that

- the *in-phase* or *active* component of a lagging or leading load current is responsible for **true power**, which represents the rate at which energy is expended by the supply – i.e. the rate at which energy is either permanently lost through heat transfer, or the rate at which work is done by a motor, driving its mechanical load (e.g. when a motor converts electrical energy into kinetic energy), or a combination of the two.
- the *reactive* or *quadrature* component of the lagging or leading load current is responsible for **reactive power**, which is the rate at which energy is alternately stored in, and returned from, a magnetic or electric field.
- the load current is responsible for **apparent power**, which is the combination (but *not* the sum) of true power and reactive power.

As we shall learn, the apparent power is the vector sum of true power and reactive power.

## A.C. power waveforms

Another approach to understanding the behaviour of power in a.c. circuits is to examine the **power waveform**, which is derived from the corresponding voltage and current waveforms.

Throughout the following, it is conventional to describe the rate at which energy is delivered *from the supply to the load* as being '**positive**' power, while the rate at which energy is *returned from the load to the supply* as being '**negative**' power. In this context, the terms 'positive' and 'negative' describe *flow directions*, and *not* polarities.

### Purely resistive circuits

For a purely resistive circuit, the load current and supply voltage are *in phase* with each other. In Figure 27.5, the power waveform is constructed by multiplying those values of instantaneous voltage and current, that occur at the same instant in time, over a complete cycle. That is: $p = e\,i$.

To clarify this process, we'll consider just two instants, occurring at points **A** and **B**.

At point **A**, then, the resulting point on the power waveform is the product of the instantaneous voltage, $e$, and the instantaneous current, $i$. Both of these are positive, so the corresponding point on the power

curve is, therefore, also positive. The power waveform is thus plotted by repeating this process for numerous instantaneous values of current and voltage throughout one complete cycle of voltage/current.

At point **B**, the point on the power waveform is, again, the product of the instantaneous voltage and current at that particular point along the axis. However, this time, both the instantaneous voltage *and* the instantaneous current are *negative*. The resulting point on the power curve is, therefore, positive, because the product of two negatives is a positive – that is:

$$-e\,(-i) = +p$$

Throughout the second half-cycle, of course, we will continue to multiply *negative* instantaneous voltages by *negative* instantaneous currents, which result in *positive* instantaneous power.

The resulting power waveform is illustrated in Figure 27.6.

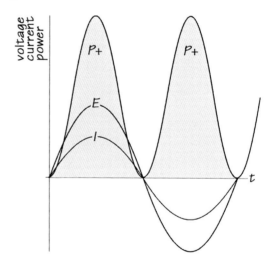

Figure 27.6

The resulting power waveform is called a 'sine-squared' (sine$^2$) waveform, meaning that it is a sine wave that varies entirely *above* the horizontal axis (it is also double the frequency of the voltage frequency), so that it is entirely *positive* – indicating, by convention, that the energy flow is *from* the supply *towards* the load. The *amount* of power is represented by the grey areas enclosed between the power waveform and the horizontal axis.

So, for a purely resistive load, energy is being *expended* by the supply and is being entirely *consumed* by the load or *lost* to the surroundings, through heat transfer, and the rate at which this is happening is termed the **true power** of the circuit, expressed in **watts**.

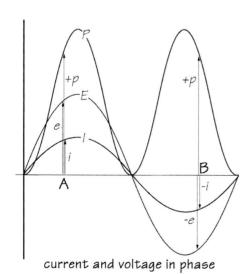

current and voltage in phase

Figure 27.5

Some students are confused as to why, when a.c. current reverses direction every half-cycle, the direction of power doesn't *also* reverse every half-cycle. Hopefully, the 'sense arrows' (described in the chapter on *series, parallel and series-parallel circuits*) will provide the answer.

During the first half-cycle (top illustration in Figure 27.7), the instantaneous potential differences and currents are both acting in the same *directions* as the assumed sense arrows and, so, their product represents 'positive power': $e \times i = +p$.

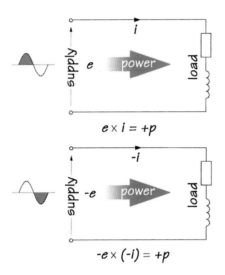

**Figure 27.7**

During the second half-cycle (bottom illustration), the instantaneous potential differences and currents are both acting in the *opposite* directions to the assumed sense arrows and, so, are each considered to be negative in direction. Their product of two negatives, of course, is a positive – so, once again, we have 'positive power': $-e \times (-i) = +p$.

## Purely inductive circuits

For a purely inductive circuit, the load current lags the supply voltage by 90°. In Figure 27.8, the power waveform is, again, constructed by multiplying those values of instantaneous voltage and current, that occur at the same instant of time, throughout a complete cycle. Again, to clarify this process, we'll consider just two instants, at points **A** and **B**.

At point **A**, the point on the power waveform is the product of the instantaneous voltage, $e$, and the instantaneous current, $-i$, because at this point along the axis, the instantaneous voltage is positive, whereas the corresponding instantaneous current is negative. The corresponding point on the power curve, therefore, is *negative*:

$$e(-i) = -p$$

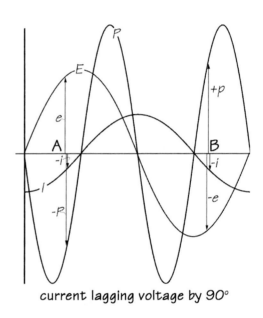

current lagging voltage by 90°

**Figure 27.8**

At point **B**, the point on the power waveform is, again, the product of the instantaneous voltage and current at that particular point along the axis. However, this time, both the instantaneous voltage *and* the instantaneous current are *negative*. The resulting point on the power curve is, therefore, positive, because the product of two negatives is a positive – that is:

$$-e(-i) = +p$$

This process is repeated for numerous values of instantaneous voltages and currents throughout a complete cycle of voltage/current, and the resulting power waveform, which is sinusoidal with twice the frequency of the voltage, is illustrated in Figure 27.9. You will notice, the grey areas enclosed by the power waveform *above* the horizontal axis ('positive' power) are exactly equal to the grey areas enclosed by the power waveform *below* that axis

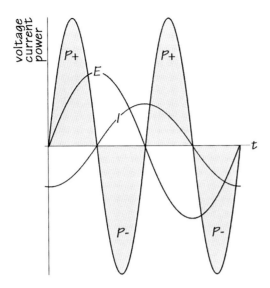

Figure 27.9

('negative' power). So, the rate at which energy is *delivered to* the circuit (the 'positive' power) during each quarter-cycle, and stored in a magnetic field, is *balanced exactly* by the rate at which energy is *returned to* the supply from the collapsing magnetic field (the 'negative' power) during the following quarter-cycle.

So, as already explained, the energy in this circuit is being alternately *stored in*, and *returned from*, a magnetic field. No energy is being consumed by the load or lost to the surroundings through heat transfer, so the 'true power' is zero. On the other hand, the rate at which this energy is being transferred back and forth represents the '**reactive power**' of the circuit, expressed in **reactive volt amperes**.

### Purely capacitive circuits

If we were to draw the power waveform for a purely capacitive circuit, then we would end up with a similar situation to that of a purely inductive circuit. That is, the amount of 'positive' power would be exactly balanced by the amount of 'negative' power. So, once again, the rate at which energy is being *delivered* every quarter-cycle to the load and stored in an electric field is balanced exactly by the rate at which energy is being *returned* to the supply when that field collapses during the following quarter-cycle.

So, as for a purely capacitive circuit, with no energy being consumed by the load or lost through heat transfer, the 'true power' is zero. But the rate at which

energy is being transferred back and forth represents the '**reactive power**' of the circuit.

## Power in resistive-reactive circuits

In the chapter on *a.c. series circuits*, we learnt that 'real' circuits are a *combination* of resistance, inductance and/or capacitance. So, in a 'real circuit' – i.e. resistive-reactive circuit – then:

- some energy is being *permanently lost* due to the resistive component of the circuit, through heat transfer and/or the work done by loads such as motors, while
- some energy is being alternately stored, and returned to the load from the magnetic and/or electric fields associated with the circuit's *reactive (inductive and/or capacitive) components*.

In 'real' circuits, therefore, there exists both **true power** (the rate at which energy is *permanently* lost) *and* **reactive power** (the rate at which energy is continually stored in, and returned from, magnetic or electric fields).

In the power waveform diagram for an *R-L* circuit, Figure 27.10, the current lags the supply voltage by some phase angle, φ. You will notice that the amount of 'positive' power *above* the horizontal axis, is larger than the amount of 'negative' power, *below* that axis. This means that the rate of energy transfer *to* the load is *greater* than the rate of energy transfer (temporarily stored in the magnetic field) from the load back to the supply.

To summarise, we can say that **true power** is the rate at which energy is lost through heat transfer or

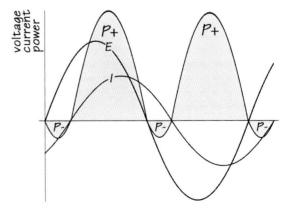

Figure 27.10

by the work done by a load such as a motor, whereas **reactive power** is the rate at which energy must be supplied in order to *sustain the magnetic and/or electric fields*.

> We should not assume that reactive power is unimportant; in fact, reactive power is *essential* to the operation of an electrical transmission and distribution system, in order to maintain magnetic/electric fields, and to maintain the system voltages required to 'push' the power demanded by loads along the power lines.

## Apparent power, true power and reactive power

Now let's move on to examine the mathematical relationship between **apparent power**, **true power** and **reactive power**.

As we have already learnt, apparent power is the *combination* of a circuit's true power and reactive power. However, the relationship between apparent power, true power and reactive power is *not* a simple algebraic relationship but, rather, a *vectorial* relationship, as we will see shortly.

### Units of measurement

Whether we are discussing **apparent power**, **true power** *or* **reactive power**, we should bear in mind that 'power' is *always* the rate at which energy is being transferred – *regardless* of whether that work is reversible or irreversible, useful or useless! So, there is absolutely no technical reason, therefore, why each of these quantities shouldn't be measured using the *same* unit of measurement – the **watt** (symbol: **W**). In fact,

**Table 27.1**

| Quantity | Symbol | Unit of measurement | Symbol |
|---|---|---|---|
| apparent power | $S$ | volt ampere | V·A |
| true power | $P$ | watt | W |
| reactive power | $Q$ | reactive volt ampere | var |

that is precisely what SI does! SI doesn't acknowledge the following units.

*Traditionally*, however, *in order to clearly differentiate between each of these quantities*, different units of measurement have been allocated to them, and it is unlikely we will ever see them being replaced by the watt! In fact, using these distinctive units of measurement is very useful, as there can then be absolutely no doubt as to which 'form' of power is then being referred to. This is summarised in Table 27.1.

## Power in a series R-L circuit

You will recall that to convert a phasor diagram into an *impedance diagram*, we divided throughout by the reference phasor, $\bar{I}$. Doing this generated various useful equations for *resistance*, *inductive reactance* and *impedance*, as well as for the *phase angle* (Figure 27.11).

**Figure 27.11**

Well, by *multiplying* the phasor diagram by the reference phasor, $\bar{I}$, we can produce a power **diagram** and, at the same time, generate various equations that will allow us to determine the circuit's **true power**, **reactive power**, **apparent power** and **power factor** (more on this, later).

**Step 1**: We start by drawing the voltage phasor diagram for the circuit, just as we did in the chapter on *a.c. series circuits* (Figure 27.12).

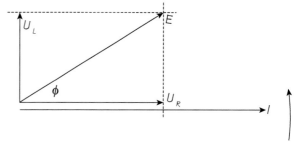

**Figure 27.12**

**Step 2**: *Multiply* throughout by the reference phasor, $\bar{I}$ (Figure 27.13).

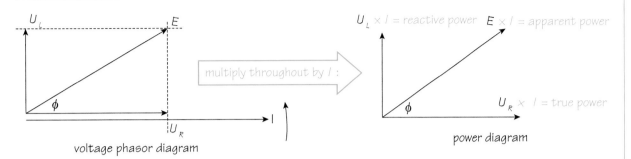

voltage phasor diagram

**Figure 27.13**

power diagram

So, by simply *multiplying the voltage phasor diagram by $\bar{I}$*, we have created a **power diagram**, from which the following equations have been derived:

$$\text{apparent power} = \bar{I}\,\bar{E}$$ $$\text{true power} = \bar{I}\,\bar{U}_R$$
$$\text{reactive power} = \bar{I}\,\bar{U}_L$$

If we know *any two* out of these three, then *we can find the third*, by applying *Pythagoras's Theorem*. For example, to find apparent power in terms of true and reactive power (Figure 27.14).

applying Pythagoras's Theorem

**Figure 27.14**

$$(\text{apparent power})^2 = (\text{true power})^2$$
$$+ (\text{reactive power})^2$$

$$S = \sqrt{P^2 + Q^2}$$

To demonstrate further how useful phasor diagrams are in generating equations for power, let's look at another example of using this technique.

We know that power in a resistive circuit can be obtained from $P = \bar{I}^2 R$, so if we redraw the impedance diagram, and multiply throughout by $\bar{I}^2$, we'll end up with another version of the power diagram, as shown in Figure 27.15.

Thus generating the following additional equations for power!

$$\text{apparent power} = \bar{I}^2 Z$$ $$\text{true power} = \bar{I}^2 R$$
$$\text{reactive power} = \bar{I}^2 X_L$$

If a load is 'static' such as, say, an **inductor** having a resistance $R$ and inductive reactance, $X_L$, then the above equation for **true power** may be manipulated to determine the *actual* **resistance** of that circuit, if the true power is known. That is,

$$\text{since true power} = \bar{I}^2 R \text{ then } R = \frac{\text{true power}}{\bar{I}^2}$$

However, if the load is 'dynamic', such as a **motor**, then its true power will vary as its mechanical load varies. So if you use the same equation to determine the

impedance diagram

power diagram

**Figure 27.15**

'resistance' of the motor circuit, what you will obtain is the '**equivalent resistance**' of the machine, *not* its true resistance. As we learnt, earlier in this chapter, the equivalent resistance of a motor will vary as the mechanical load varies.

**Note!** Hopefully, you are now beginning to realise how important it is to be able to construct a voltage phasor diagram and to be able to change this into an impedance diagram and into a power diagram. These diagrams will generate all the equations that you will ever need to solve a.c. problems, without the need to commit any of these equations to memory!

## Power factor

In a d.c. circuit, *regardless of the type of load*, we can determine the power simply by multiplying together the readings of a voltmeter and an ammeter.

In a resisitive-reactive a.c. circuit, however, the product of the supply voltage and the load current gives us the **apparent power** of the load, *not* its true power.

To determine its **true power**, in practice, we must use a *wattmeter*, which is designed specifically to measure true power, by monitoring the supply voltage together with the *in-phase (resistive) component* of the load current (Figure 27.16).

**Figure 27.16**

The *ratio of a load's true power to its apparent power* is *very* important, and is called the '**power factor**' of the load:

$$\text{power factor} = \frac{\text{true power}}{\text{apparent power}}$$

If we re-examine the power diagram, derived from the voltage phasor diagram, it should be obvious that the ratio of true power to apparent power (adjacent over hypotenuse) is the cosine of the phase angle:

$$\text{power factor} = \frac{\text{true power}}{\text{apparent power}} = \frac{\text{adjacent}}{\text{hypotenuse}} = \cos\phi$$

So an alternative definition for power factor is that it is '*the cosine of the load's phase angle*'.

Power factor is usually expressed as a *per-unit* value (e.g. 0.85), although it may still (usually in older textbooks) occasionally be seen expressed as a *percentage* value (e.g. 85%).

A circuit's power factor *must* also be specified as '**leading**' or '**lagging**' (e.g '0.85 lagging'). These terms refer to *where the* **load current** *phasor lies in relation to the supply voltage phasor* (never the other way around!). Resistive-inductive circuits, therefore, always have 'lagging' power factors, while resistive-capacitive circuits always have 'leading' power factors.

The reason that power factor is so important is that it may be thought of as an indication of the 'efficiency' of the rate of energy conversion, and it might prove helpful to think of it in terms of the mechanical analogy, shown in Figure 27.17.

**Figure 27.17**

The two diagrams in Figure 27.17 represent a plan view of a barge being pulled along a canal by a horse, walking along the canal's towpath.

The force in the tow-rope can be resolved into two (imaginary) forces: one *in the direction of the barge*, and the other *towards the canal's bank*.

Of these two resolved forces, it is the one acting in the direction of the barge's motion that contributes the *useful* force, as it's acting to pull the barge forward. On the other hand, the resolved force acting towards the bank (quadrature force) contributes nothing to the forward motion to the barge. Yet neither force can exist without the other!

In the second illustration, the canal is wider, and the barge is further from the bank, so the tow-rope assumes a somewhat greater angle than before.

If we, again, resolve the force in the tow-rope (assuming that the same forward resolved force is required to keep the barge moving at the same velocity), we notice that the resolved force towards the canal bank is *greater* than it was before.

In this new situation, *in order to provide the same amount of forward force*, a greater force *has to be made available in the direction of the bank*.

This situation may be likened to the power supplied to an resistive-reactive circuit. The **true power** is equivalent to *the forward force acting on the barge*; the **reactive power** is equivalent to the '*wasted*' *force acting towards the canal bank* (which, despite being *apparently* 'wasted', is still *essential* to the behaviour of the barge). The greater the phase angle (equivalent to the angle between the forward force on the barge and the force in the tow-rope), the more 'apparently wasted' (but, nonetheless essential!) reactive power that must be provided!

By expressing the angle between the tow-rope and the forward-resolved force in terms of its cosine, the 'efficiency' of the arrangement is indicated. The maximum efficiency (×1 or 100%) occurs when the tow-rope acts in the *same direction as the necessary forward force* (cos 0° = 1); and the minimum efficiency (×0 or 0%) occurs when the tow-rope is at 90° (cos 90° = 0) to the forward direction of the barge – i.e. no forward motion whatsoever!

In terms of **power factor**, 'efficiency' isn't really the appropriate term to describe what it represents. It would be more accurate to say that

> **Power factor** is '*the percentage of apparent power that represents the rate of doing real work*'.

In other words,

- when the power factor is unity (1), the apparent power is equal to the true power.
- when the power factor is zero (0), the apparent power is equal to the reactive power.

For most circuits, the power factor lies somewhere in between these two extremes.

## Power in a series R-C circuit

Figure 27.18

As with the *R-L* circuit, described above, to convert a phasor diagram into a *power diagram*, we simply multiply the phasor diagram by the reference phasor, $\overline{I}$, and generate equations that will allow us to determine the circuit's true power, reactive power, apparent power and power factor (Figure 27.18).

**Step 1:** Draw the voltage phasor diagram for the circuit, as we did in the chapter on *a.c. series circuits* (Figure 27.19).

Figure 27.19

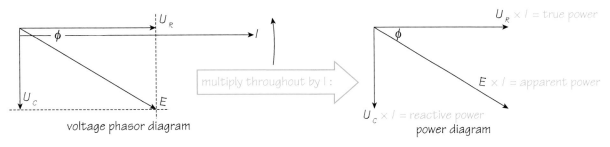

Figure 27.20

**Step 2**: *Multiply* throughout by the reference phasor, $\overline{I}$ (Figure 27.20).

So, by simply *multiplying the voltage phasor diagram by* $\overline{I}$, we have created a **power diagram**, from which the following equations have been generated:

$$\boxed{\text{apparent power} = \overline{I}\,\overline{E}} \quad \boxed{\text{true power} = \overline{I}\,\overline{U}_R}$$

$$\boxed{\text{reactive power} = \overline{I}\,\overline{U}_C}$$

If we know *any two* out of these three, then *we can find the third*, by applying *Pythagoras's Theorem* (Figure 27.21). For example:

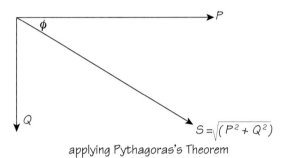

applying Pythagoras's Theorem

**Figure 27.21**

$$(\text{apparent power})^2 = (\text{true power})^2$$
$$+ (\text{reactive power})^2$$

$$\boxed{S = \sqrt{P^2 + Q^2}}$$

So, once again, the ability to draw a power diagram *saves you having to remember numerous equations*, as they can all be generated from the power diagram, by applying simple geometry (Pythagoras's Theorem) or simple trigonometry (the cosine ratio)! Let's look at what else we can derive from the power diagram.

For example, if we know the *apparent power* and the *true power*, then we could rearrange Pythagoras' Theorem, as follows:

$$\boxed{\begin{array}{l}\text{reactive power} \\ = \sqrt{(\text{apparent power})^2 - (\text{true power})^2}\end{array}}$$

To find the **power factor**, we simply need to find the cosine of the phase angle:

$$\boxed{\cos\phi = \frac{\text{adjacent}}{\text{hypotenuse}} = \frac{\text{true power}}{\text{apparent power}}}$$

This time, we *always* describe the power factor as **leading**, because, by definition, the *current leads the supply voltage*.

Again, we can manipulate this equation, to find another equation for the true power of a circuit:

$$\text{since } \cos\phi = \frac{\text{true power}}{\text{apparent power}}$$

$$\text{then true power} = \text{apparent power} \times \cos\phi$$

This gives us a very important equation:

$$\boxed{\text{true power} = (\overline{E}\,\overline{I})\cos\phi}$$

As was the case for the series *R-L* circuit, we can also convert an *R-C* impedance diagram into a power diagram, by simply multiplying throughout by $I^2$ (Figure 27.22).

Thus generating the following additional equations for power:

$$\boxed{\text{apparent power} = \overline{I}^2 Z} \quad \boxed{\text{true power} = \overline{I}^2 R}$$

$$\boxed{\text{reactive power} = \overline{I}^2 X_C}$$

> **Note!** If you learn *nothing else* from this chapter, you *must* learn **the importance of being able to draw a voltage phasor diagram and to be able to convert it into a power diagram**, for this will generate *all* the equations you will *ever* need to know in order to solve a.c. power problems.
>
> This will save you from *ever* needing to memorise these equations!

impedance diagram          power diagram

**Figure 27.22**

**Worked example 1**  An inductor, of inductance 16 mH and resistance 2 Ω, is connected across a 24-V, 50 Hz supply. Calculate each of the following:

a  inductive reactance
b  impedance
c  current
d  apparent power
e  true power
f  reactive power
g  power factor.

**Solution**  As with any problem, always start by sketching the circuit and inserting all the values given in the problem (Figure 27.23). You notice that this circuit consists of an inductor; you are given its inductance and resistance, so you can assume that it is equivalent to a *series R-L circuit*.

**Figure 27.23**

Next, sketch the corresponding voltage phasor diagram (Figure 27.24).

Next, convert this into an impedance diagram, by dividing throughout by $\bar{I}$ (Figure 27.25).

a  None of the generated equations allow us to determine the inductive reactance, so we must

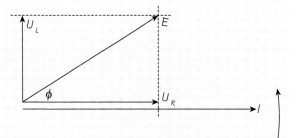

**Figure 27.24**

fall back on the basic equation that we have committed to memory:

$$X_L = 2\pi f = 2\pi \times 50 \times (16 \times 10^{-3}) = 5.03\ \Omega$$
(Answer a.)

b  $Z = \sqrt{R^2 + X_L^2} = \sqrt{2^2 \times 5.03^2} = \sqrt{29.30} = 5.41\ \Omega$
(Answer b.)

c  To find the current, we can use the following equation generated by the impedance diagram:

$$\bar{I} = \frac{E}{Z} = \frac{24}{5.41} = 4.44\ \text{A (Answer c.)}$$

For the rest of this problem, redraw the impedance diagram, and multiply by $\bar{I}^2$ to create a power diagram. Then, we can utilise the equations generated by the power diagram to solve the remaining parts of the problem (Figure 27.26).

d.  apparent power $= \bar{I}^2 Z$

$$= 4.44^2 \times 5.41$$

$$= 106.65\ \text{V} \cdot \text{A (Answer d.)}$$

e.  true power $= \bar{I}^2 R = 4.44^2 \times 2 = 39.43\ \text{W}$
(Answer e.)

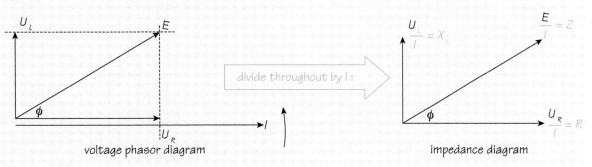

voltage phasor diagram          divide throughout by I :          impedance diagram

**Figure 27.25**

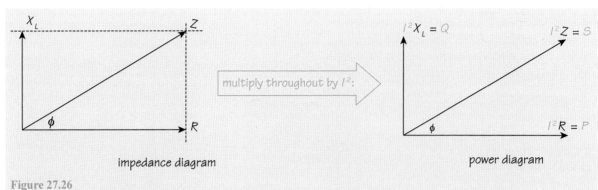

impedance diagram

**Figure 27.26**

power diagram

f.  reactive power $= \overline{I}^2 X_L$

$$= 4.44^2 \times 5.02$$

$$= 98.96 \text{ var (Answer f.)}$$

g.  power factor $= \dfrac{\text{true power}}{\text{apparent power}}$

$$= \dfrac{39.43}{106.65}$$

$$= 0.3697 \text{ lagging (Answer g.)}$$

('Lagging', because it is an **inductive** circuit, and current always lags the supply voltage in an inductive circuit.)

## Review your learning

Now that we've completed this chapter, we need to examine the **objectives** listed at its start. Placing '*Can I…*' at the beginning, and a question mark at the end, of each objective turns that objective into a **test item**. If we can answer each of those test items, then we've met the objectives of this chapter.

**Online resources**

The companion website to this book contains further resources relating to this chapter. The website can be accessed via the following link:
www.routledge.com/cw/waygood

# Chapter 28

# Parallel alternating-current circuits

## Objectives

On completion of this chapter, you should be able to:

1. draw a current phasor diagram for a parallel
   a. resistive-inductive *(R-L)* circuit
   b. resistive-capacitive *(R-C)* circuit
   c. resistive-inductive-capacitive *(R-L-C)* circuit.
2. derive an admittance diagram, and derive expressions for conductance, susceptance, admittance and phase angle for a parallel
   a. resistive-inductive *(R-L)* circuit
   b. resistive-capacitive *(R-C)* circuit
   c. resistive-inductive-capacitive *(R-L-C)* circuit.
3. derive a power diagram, and derive expressions for true power, reactive power, apparent power and power factor for a parallel
   a. resistive-inductive *(R-L)* circuit
   b. resistive-capacitive *(R-C)* circuit
   c. resistive-inductive-capacitive *(R-L-C)* circuit.
4. solve problems on parallel *R-L*, *R-C* and *R-L-C* circuits.

## Introduction

As it was with the earlier chapters on *series a.c. circuits* and *power in a.c. circuits*, the key to understanding and solving **parallel a.c. circuits** is the ability to be able to draw phasor diagrams from which *practically every equation you will ever need can be generated – providing you know how to use*

*Pythagoras's Theorem and basic trigonometry*. So you are urged *not* to waste your time trying to memorise all the various equations!

We have learnt how all 'real' a.c. circuits exhibit combinations of **resistance** (symbol: $R$), **inductance** (symbol: $L$) and **capacitance** (symbol: $C$).

We also learnt that 'real' circuits are relatively complicated because they contain *combinations of resistance, inductance and capacitance* and, in order to understand the behaviour of a.c. circuits, it is necessary to start by first considering how 'ideal' circuits would behave. We described these 'ideal' circuits as **purely resistive, purely inductive** and **purely capacitive**, and we discovered that

a. in a **purely resistive** circuit, *the current and voltage are in phase* with each other.
b. in a **purely inductive** circuit, *the current lags the voltage by 90°*.
c. in a **purely capacitive** circuit, *the current leads the voltage by 90°*.

To help us remember these *very* important relationships, we can use the mnemonic, '**CIVIL**', in which '**C**' stands for 'capacitive circuit', and '**L**' stands for '**inductive** circuit'.

Finally, we discovered that the opposition to current in a purely resistive circuit is called **resistance** $(R)$, the opposition to current in a purely inductive circuit is called **inductive reactance** $(X_L)$ and that the opposition to current in a purely capacitive circuit is called **capacitive reactance** $(X_C)$, where:

$$X_C = \frac{1}{2\pi f C}$$

An Introduction to Electrical Science, Waygood, ISBN 9780815391821, 2019. © Taylor & Francis

in                                     in

capacitive (**C**) circuits:      inductive (**L**) circuits:
**I** *before* ('leads') **V**          **I** *after* ('lags') **V**

Figure 28.1

$$X_L = 2\pi f L$$

where:

$X_L$ = inductive reactance (ohms)

$X_C$ = capacitive reactance (ohms)

$f$ = supply frequency (hertz)

$L$ = inductance (henrys)

$C$ = capacitance (farads)

Unlike *all* the other equations that we will need to use for solving a.c. circuits, these two equations have to be committed to memory, as their derivation requires a knowledge of calculus, which is beyond the scope of this text.

*All* other equations can be derived by learning to represent circuits using phasor diagrams and their derivatives (e.g. impedance diagrams and power diagrams). So if we learn how to draw these diagrams, then we will have absolutely no need to remember all the other various equations.

In this chapter we will be examining **parallel a.c. circuits**, specifically, **parallel *R-L*, parallel *R-C* and parallel *R-L-C*** circuits.

Again, a reminder that the circuit symbols used throughout this chapter represent the *quantities* resist*ance*, induct*ance*, and capacit*ance* – **not** resist*ors*, induct*ors* and capacit*ors*.

## Parallel R-L circuits

When a potential difference $(\overline{E})$ is applied across a parallel circuit, *that potential difference is common to each branch*. For an *R-L* parallel circuit, the resistive branch will then draw a current $\overline{I}_R$, and the inductive branch will draw a current $\overline{I}_L$. In accordance with Kirchhoff's Current Law, the supply current $(\overline{I})$ will then be the sum of the two branch currents. However, as we learnt in the chapter on *a.c. series circuits*, because these currents are not in phase with each other, *we must add them vectorially*.

So, let's examine this a little more closely, by drawing

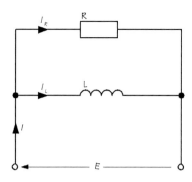

Figure 28.2

the phasor diagram for the circuit. In the following series of diagrams, the step being described is shown in **blue**, while previous steps are shown in black.

### Drawing the phasor diagram

#### *Step 1*

In a parallel circuit, the **supply voltage** $(\overline{E})$ is common to each branch and, so, *voltage is always chosen as the* **reference phasor**. The reference phasor is *always drawn first, and always along the horizontal positive axis* (i.e. horizontally and to the right), and it's also normally drawn fairly long in order to distinguish it from the others. In Figure 28.3, we have also given the reference phasor an outline, rather than a solid, arrow head although this is not really necessary. The small, curved, arrow head is there to remind us that phasors 'rotate' counterclockwise.

Figure 28.3

#### *Step 2*

The current, $\overline{I}_R$, flowing in the resistive branch is *in phase with the supply voltage* and, so, is also drawn along the horizontal positive axis (Figure 28.4).

Figure 28.4

## Step 3

The current, $\overline{I}_L$, flowing in the inductive branch *lags* the voltage by 90° (remember **CIVIL**) so, as phasors 'rotate' counterclockwise, it is drawn 90° clockwise from the reference phasor (Figure 28.5).

Figure 28.5

## Step 4

We know, from Kirchhoff's Current Law that, in a parallel circuit, the supply current $(\overline{I})$ is the sum of the individual branch currents. However, because, in this case, the two currents, $\overline{I}_R$ and $\overline{I}_L$, lie at right angles to each other, we have to add them *vectorially* (Figure 28.6).

Figure 28.6

From the completed phasor diagram we can see that the supply current, $\overline{I}$, is the **phasor sum** (or **vector sum**) of $\overline{I}_R$ and $\overline{I}_L$, which can be found by simply applying Pythagoras's Theorem:

$$\overline{I} = \sqrt{\overline{I}_R^2 + \overline{I}_L^2}$$

To find the phase angle, we can use the *cosine ratio*:

$$\cos\phi = \frac{adjacent}{hypotenuse} = \frac{\overline{I}_R}{\overline{I}}$$

so

$$\angle\phi = \cos^{-1}\frac{\overline{I}_R}{\overline{I}}$$

and, of course, because this is an inductive circuit, it is **lagging**.

We *could* have used the sine or tangent ratios, too, but by using the cosine ratio, we are able to work out the circuit's **power factor** (which is the cosine of the phase angle) at the same time! That is, you are 'killing two birds with one stone'!

**Worked example 1**    The current drawn by the resistive branch of a parallel *R-L* circuit is 15 A, and the current drawn by the inductive branch is 20 A. What is the value of the supply current? Also, what is the circuit's phase angle?

**Solution**    *Always* start by sketching the circuit diagram, and inserting all the values given to you in the question (Figure 28.7).

Figure 28.7

Next, sketch the phasor diagram, following the steps described above. You *don't* have to draw the phasor diagram to scale (Figure 28.8).

Figure 28.8

Now, you can apply Kirchhoff's Current Law, using Pythagoras's Theorem to solve the problem:

$$\overline{I} = \sqrt{\overline{I}_R^2 + \overline{I}_L^2}$$
$$= \sqrt{15^2 + 20^2}$$
$$= \sqrt{625} = 25 \text{ A (Answer)}$$

Since this is an inductive circuit, and the current *lags* the supply voltage, then the phase angle will be *lagging*:

$$\angle\phi = \cos^{-1}\frac{\overline{I}_R}{\overline{I}} = \cos^{-1}\frac{15}{25} = \cos^{-1}0.6$$
$$= 53.13° \text{ lagging (Answer)}$$

## Admittance diagram

You will recall that, in a **series *R-L*** diagram, we changed the voltage phasor diagram into an *impedance diagram* simply by dividing throughout by the reference phasor which, in that case, happened to be the supply current, $\overline{I}$.

Well, let's follow exactly the same rule and, again: *divide the phasor diagram by the reference phasor*. For our parallel circuit, however, the reference phasor is the supply voltage, $\overline{E}$, so let's see what happens this time.

## Drawing the admittance diagram

### Step 1

We start by drawing the circuit's phasor diagram, following the steps already explained (Figure 28.9).

Figure 28.9

### Step 2

Next, we *divide each of the current phasors by the reference phasor* ($\overline{E}$) (Figure 28.10).

As you can see, what we now have are expressions for the *reciprocals* of resistance, inductive reactance and impedance. We call these conductance, inductive susceptance and admittance – hence the term, '**admittance diagram**'.

Figure 28.11 shows exactly the same admittance diagram, but expressed directly in terms of **conductance** *(G)*, **inductive susceptance** *(B$_L$)* and **admittance** *(Y)*, each expressed in siemens (S).

As you can see, we have ended up with the *reciprocals* of resistance, inductive reactance and impedance. The resulting diagram is called an **admittance diagram** (sometimes called an *'admittance triangle'*), and is useful because it generates the following equations:

$$\boxed{\frac{\overline{I}_R}{\overline{E}} = \frac{1}{R}}\quad\boxed{\frac{\overline{I}_L}{\overline{E}} = \frac{1}{X_L}}\quad\boxed{\frac{\overline{I}}{\overline{E}} = \frac{1}{Z}}$$

No doubt you will have noticed that each of these equations can be turned upside down to give us the more familiar equations:

$$\boxed{\frac{\overline{E}}{\overline{I}_R} = R,}\quad\boxed{\frac{\overline{E}}{\overline{I}_L} = X_L,}\quad\text{and}\quad\boxed{\frac{\overline{E}}{\overline{I}} = Z}$$

... so you might be wondering why bother to express them in the way we have (i.e. as reciprocals)! Well, by applying Pythagoras's Theorem to them, as reciprocals, you can also use the following relationship:

$$\left(\frac{1}{Z}\right)^2 = \left(\frac{1}{R}\right)^2 + \left(\frac{1}{X_L}\right)^2$$

... which is exactly equivalent to the following equation that we are already familiar with for calculating the total resistance of a parallel d.c. circuit:

$$\frac{1}{R} = \frac{1}{R_1} + \frac{1}{R_2}$$

Manipulating the 'admittance' equation, then gives us:

$$\boxed{\frac{1}{Z} = \sqrt{\frac{1}{R^2} + \frac{1}{X_L^2}}}$$

... from which we can find the **impedance** of the parallel circuit.

We can also find the circuit's **phase angle**, using basic trigonometry, utilising either the *sine*, *cosine* or *tangent* ratios. In practice, for a reason we'll see shortly, the best choice is always to use the *cosine*:

$$\cos\phi = \frac{\text{adjacent}}{\text{hypotenuse}} = \frac{\left(\dfrac{1}{R}\right)}{\left(\dfrac{1}{Z}\right)} = \frac{Z}{R}$$

$$\boxed{\angle\phi = \cos^{-1}\frac{Z}{R}}$$

Figure 28.10

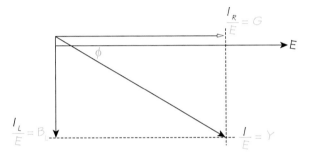

Figure 28.11

You will notice that this is rather different from the corresponding equation for a series *R-L* circuit! This is because the effective impedance is smaller than either the resistance or the inductive reactance (in just the same way that the effective resistance of a d.c. parallel circuit is less than either of the branch resistances).

**Important!** Dividing a current phasor diagram by voltage produces an admittance diagram which generates each of the equations shown above. So you don't have to learn *any* of these equations – they can all be generated *provided you learn how to draw the phasor and impedance diagrams!*

However, *if you prefer*, you can rewrite the above equations directly in terms of conductance, inductive susceptance and admittance, as follows:

$$\frac{\overline{I}_R}{\overline{E}} = G \qquad \frac{\overline{I}_L}{\overline{E}} = B_L \qquad \frac{\overline{I}}{\overline{E}} = Y$$

By applying Pythagoras's Theorem, you can also find the following relationship:

$$Y^2 = G^2 + B_L^2$$

$$Y = \sqrt{G^2 + B_C^2}$$

We can also find the circuit's **phase angle**, using basic trigonometry, utilising either the *sine*, *cosine* or *tangent* ratios. As usual, the best choice is always to use the *cosine*, because it also tells you what the circuit's power factor happens to be, should you need to know:

$$\cos\phi = \frac{\text{adjacent}}{\text{hypotenuse}} = \frac{G}{Y}$$

$$\angle\phi = \cos^{-1}\frac{G}{Y}$$

In many respects, it's far more convenient to work in terms of conductance, inductive susceptance and admittance, as it avoids the complications of having to work with reciprocals (fractions). But it's entirely up to you!

**Worked example 2**   A parallel circuit of resistance 5 Ω in parallel with an inductance 0.02 H is connected across a 230-V, 50 Hz a.c. supply. Calculate each of the following:

a   inductive reactance
b   impedance

c   current through the resistive branch
d   current through the inductive branch
e   supply current
f   phase angle of the circuit.

**Solution**   As always, the first step in solving *any* circuit problem is to sketch the circuit diagram, and label it with all values supplied in the problem (Figure 28.12).

**Figure 28.12**

The next step is to sketch the current phasor diagram, following the steps described earlier (Figure 28.13).

**Figure 28.13**

Next, we must convert the current phasor diagram into an admittance diagram (Figure 28.14), by dividing throughout by the reference quantity – i.e. the supply voltage. This generates all the equations that we need to solve the problem (without you having to remember them!).

a   To find the **inductive reactance** ($X_L$) of the circuit, we start by looking at the equations generated when we constructed the admittance diagram. There's only one equation with inductive reactance shown, $\dfrac{\overline{I_L}}{\overline{E}} = \dfrac{1}{X_L}$.
Unfortunately, we don't know the value of $\overline{I_L}$ so we can't use this formula. So we must fall back

on the basic equation for inductive reactance, as follows:

$$X_L = 2\pi f L = 2\pi \times 50 \times 0.02 = 6.28\ \Omega \text{ (Answer a.)}$$

b   To find the impedance, we can now use the equation generated by the admittance diagram:

$$\frac{1}{Z} = \sqrt{\frac{1}{R^2} + \frac{1}{X_L^2}}$$

$$\frac{1}{Z} = \sqrt{\frac{1}{5^2} + \frac{1}{6.28^2}} = \sqrt{\frac{1}{25} + \frac{1}{39.44}} = \sqrt{\frac{39.44 + 25}{25 \times 39.44}}$$

$$\frac{1}{Z} = \sqrt{\frac{64.44}{986}} = \sqrt{0.0654} = 0.2557$$

$$Z = \frac{1}{0.2557} \approx 3.91\ \Omega \text{ (Answer b.)}$$

This answer makes sense, because, in a parallel circuit, the impedance must be lower than either the resistance or the reactance (in exactly the same way that the equivalent resistance of a d.c. parallel circuit must be less than the resistance of any branch).

---

The alternative, and somewhat simpler, way of solving this part of the question, is to work directly with *conductance*, *inductive susceptance* and *admittance*.

First, find the **conductance** and the **inductive susceptance**:

$$G = \frac{1}{R} = \frac{1}{5} = 0.20\ \text{S} \quad \text{and} \quad B_L = \frac{1}{X_L} = \frac{1}{6.28} = 0.16\ \text{S}$$

Next, apply the Pythagoras's Theorem equation for **admittance**:

$$Y = \sqrt{G^2 + B_L^2} = \sqrt{0.2^2 + 0.16^2} = \sqrt{0.0656} = 0.256\ \text{S}$$

Finally, the **impedance** is the inverse of the admittance:

$$Z = \frac{1}{Y} = \frac{1}{0.256} \approx 3.91\ \Omega \text{ (Answer b.)}$$

**Figure 28.14**

c   To find the current through the resistive branch, we use the following equation generated by the admittance diagram:

$$\frac{1}{R} = \frac{\overline{I}_R}{\overline{E}}$$

rearranging, $\overline{I}_R = \dfrac{\overline{E}}{R} = \dfrac{230}{5} = 46$ A (Answer c.)

Again, if you prefer, you can use *conductance*, rather than resistance, to work out the current through the resistive branch:

$$\frac{\overline{I}_R}{\overline{E}} = G$$

rearranging, $\overline{I}_R = \overline{E}\,G = 230 \times 0.2 = 46$ A (Answer c.)

d   Again, using the equation generated by the admittance diagram:

$$\frac{1}{X_L} = \frac{\overline{I}_L}{\overline{E}}$$

rearranging, $\overline{I}_L = \dfrac{\overline{E}}{X_L} = \dfrac{230}{6.28} \approx 36.62$ A (Answer d.)

Again, if you prefer, you can use *conductance*, rather than resistance, to work out the current through the resistive branch:

$$\frac{\overline{I}_L}{\overline{E}} = B_L$$

rearranging, $\overline{I}_R = \overline{E}\,B_L = 230 \times 0.16$
$\approx 36.8$ A (Answer d.)

*(minor difference in answer due to rounding up/down)*

e   To find the supply current, we can apply Pythagoras' Theorem to the current phasor diagram,

$$\overline{I} = \sqrt{\overline{I}_R^2 + \overline{I}_L^2} = \sqrt{46^2 + 36.62^2}$$
$$= \sqrt{2116 + 1341} \approx 58.78 \text{ A (Answer e.)}$$

f   Using the cosine function:

$$\angle\phi = \cos^{-1}\frac{Z}{R} = \cos^{-1}\frac{3.911}{5}$$
$$= \cos^{-1}0.7822 \approx 38.54° \text{ lagging (Answer f.)}$$

(*'Lagging'*, because the *supply current lags the supply voltage* in an inductive circuit.)

Once again, if you prefer to work in terms of *admittance* and *conductance*, then:

$$\angle\phi = \cos^{-1}\frac{G}{Y} = \cos^{-1}\frac{0.2}{0.256} = \cos^{-1}0.7813$$
$$\approx 38.62° \text{ lagging (Answer f.)}$$

*(minor difference in answer, due to rounding up/down)*

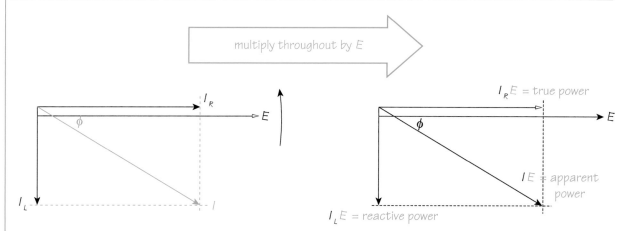

**Figure 28.15**

## Power diagram

To create a **power diagram** (or '**power triangle**') for a parallel *R-L* circuit we simply *multiply* throughout by the reference phasor (Figure 28.15).

This action generates the exact same power equations as we have already seen for a *series R-L* circuit. As always, if you can draw the circuit's current phasor diagram, and convert it into a power diagram, *then you can **derive** all the following equations, without having to remember them:*

$$\boxed{\overline{I}_R\overline{E} = \text{true power}} \quad \boxed{\overline{I}_L\overline{E} = \text{reactive power}}$$

$$\boxed{\overline{I}\ \overline{E} = \text{apparent power}}$$

By applying Pythagoras's Theorem, we can obtain an equation for apparent power (in volt amperes) in terms of true power (in watts) and reactive power (in reactive volt amperes):

$$(\text{apparent power})^2 = (\text{true power})^2 + (\text{reactive power})^2$$

Finally, to find the **power factor** of the circuit, we simply need to find the cosine of the phase angle, which is:

$$\boxed{\cos\phi = \frac{\text{adjacent}}{\text{hypotenuse}} = \frac{\text{true power}}{\text{apparent power}}}$$

## Parallel R-C circuits

We know that in a purely resistive circuit, the current and voltage are in phase with each other; and, in a purely capacitive circuit, the current *leads* the voltage by 90°.

So, in a *parallel R-C* circuit (Figure 28.26), the *current must lead the voltage by some angle between 0° and 90°* – this angle is called the circuit's **phase angle** (symbol: $\phi$, pronounced 'phi').

**Figure 28.16**

When a potential difference ($\overline{E}$) is applied across a **parallel *R-C* circuit**, a current, $\overline{I}_R$, will flow through the resistive branch, and a current, $\overline{I}_C$, will flow through the capacitive branch.

We needn't go through the step-by-step process of creating the current phasor diagram for this circuit, as it is practically the same as the procedure we have already gone through to create the phasor diagram for an *R-L* circuit. The important difference, of course, is that, for an *R-C* circuit, the current through the capacitive branch, $\overline{I}_C$, *leads* the supply voltage.

So, the finished current phasor diagram for a parallel *R-C* circuit will look like Figure 28.17.

Figure 28.17

From the completed phasor diagram, we can see that the supply current, $\overline{I}$, is the **phasor sum** (or **vector sum**) of $\overline{I}_R$ and $\overline{I}_C$, which can be found using Pythagoras's Theorem:

$$\overline{I} = \sqrt{\overline{I}_R^2 + \overline{I}_C^2}$$

To find the phase angle, we can use the *cosine ratio*:

$$\cos\phi = \frac{\text{adjacent}}{\text{hypotenuse}} = \frac{\overline{I}_R}{\overline{I}}$$

so

$$\angle\phi = \cos^{-1}\frac{\overline{I}_R}{\overline{I}}$$

and, of course, because this is an inductive circuit, it is **leading**.

## Admittance diagram

Again, by dividing *the phasor diagram by the reference phasor*, $\overline{E}$, we can change the phasor diagram into an admittance diagram (Figure 28.18). Remember, 'admittance' is the *reciprocal* of impedance.

The same admittance diagram, expressed directly in terms of **conductance (G)**, **capacitive susceptance ($B_C$)** and **admittance (Y)** is shown in Figure 28.19.

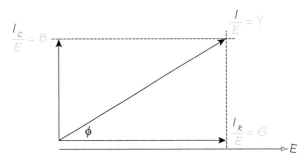

Figure 28.19

The resulting diagram is an **admittance diagram** (or *'admittance triangle'*), and is useful because it generates the following equations:

$$\frac{\overline{I}_R}{\overline{E}} = \frac{1}{R} \qquad \frac{\overline{I}_C}{\overline{E}} = \frac{1}{X_C} \qquad \frac{\overline{I}}{\overline{E}} = \frac{1}{Z}$$

Also from the admittance diagram, you can see that the admittance can be calculated by applying Pythagoras's Theorem:

$$\frac{1}{Z} = \sqrt{\left(\frac{1}{R}\right)^2 + \left(\frac{1}{X_C}\right)^2}$$

We can also find the circuit's **phase angle**, using basic trigonometry, utilising the *cosine* ratio.

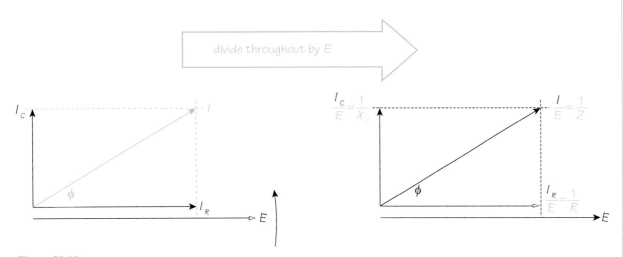

Figure 28.18

$$\cos\phi = \frac{\text{adjacent}}{\text{hypotenuse}} = \frac{\left(\dfrac{1}{R}\right)}{\left(\dfrac{1}{Z}\right)} = \frac{Z}{R}$$

$$\boxed{\angle\phi = \cos^{-1}\frac{Z}{R}}$$

**Important!** Dividing a current phasor diagram by voltage produces an admittance diagram which generates each of the equations shown above. So you don't have to learn *any* of these equations – they can all be generated *provided you learn how to draw the phasor and admittance diagrams!*

Again, you may prefer to work in terms of conductance, capacitive susceptance and admittance. In which case,

$$\boxed{\frac{\overline{I}_R}{\overline{E}} = G} \qquad \boxed{\frac{\overline{I}_C}{\overline{E}} = B_C} \qquad \boxed{\frac{\overline{I}}{\overline{E}} = Y}$$

Also from the admittance diagram, you can also see that the admittance can be calculated by applying Pythagoras's Theorem:

$$\boxed{Y = \sqrt{G^2 + B_C{}^2}}$$

We can also find the circuit's **phase angle**, using basic trigonometry, utilising the *cosine* ratio.

$$\cos\phi = \frac{\text{adjacent}}{\text{hypotenuse}} = \frac{G}{Y}$$

$$\boxed{\angle\phi = \cos^{-1}\frac{G}{Y}}$$

Once again, in many respects, it's more convenient to work in terms of conductance, inductive susceptance and admittance, as it avoids the complications of having to work with fractions. But it's entirely up to you!

## Power diagram

To create a **power diagram** (or '**power triangle**') for a parallel *R-C* circuit, we simply multiply throughout by the reference phasor (Figure 28.20).

This action generates the same power equations as we have already seen for a *series R-L* circuit. As always, if you can draw the circuit's current phasor diagram, and convert it into a power diagram, *then you can **derive** all the following equations, without having to remember them:*

$$\boxed{\overline{I}_R\overline{E} = \text{true power}} \quad \boxed{\overline{I}_C\overline{E} = \text{reactive power}}$$

$$\boxed{\overline{I}\,\overline{E} = \text{apparent power}}$$

By applying Pythagoras's Theorem, we can obtain an equation for apparent power (in volt amperes) in terms of true power (in watts) and reactive power (in reactive volt amperes):

$$\boxed{(\text{apparent power})^2 = (\text{true power})^2 + (\text{reactive power})^2}$$

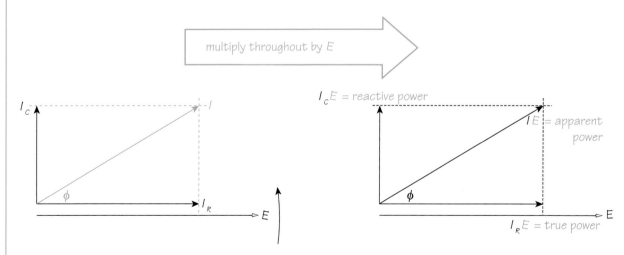

**Figure 28.20**

Finally, to find the **power factor** of the circuit, we simply need to find the cosine of the phase angle, which is:

$$\cos\phi = \frac{\text{adjacent}}{\text{hypotenuse}} = \frac{\text{true power}}{\text{apparent power}}$$

**Worked example 3**    A parallel *R-C* circuit comprises a 30-Ω resistor in parallel with a capacitor of capacitive reactance 20 Ω, connected across a 120-V supply. Calculate the circuit's (a) true power, (b) reactive power, (c) apparent power and (d) power factor.

**Solution**    As always, the first step is to sketch a circuit diagram, with all the supplied information inserted (Figure 28.21).

**Figure 28.21**

Before we can work out the true, reactive and apparent power, *we need to work out the current*

*flowing in each branch.* So we must sketch the current phasor diagram for the circuit, and divide throughout by the reference phasor $(\bar{E})$ to produce the **admittance diagram** (Figure 28.22). *All* the equations you will need are then generated without you having to remember them.

To find the current through the resistor, $\bar{I}_R$:

$$\frac{\bar{I}_R}{\bar{E}} = \frac{1}{R} \quad \text{rearranging:} \quad \bar{I}_R = \frac{\bar{E}}{R} = \frac{120}{30} = 4\ \text{A}$$

To find the current through the capacitor, $\bar{I}_C$:

$$\frac{\bar{I}_C}{\bar{E}} = \frac{1}{X_C} \quad \text{rearranging:} \quad \bar{I}_C = \frac{\bar{E}}{X_C} = \frac{120}{20} = 6\ \text{A}$$

To find the supply current, since we don't know the circuit's impedance, we apply Pythagoras's Theorem:

$$I = \sqrt{I_R^2 + I_C^2} = \sqrt{4^2 + 6^2} = 7.21\ \text{A}$$

Now we can convert the current phasor diagram into a power diagram (Figure 28.23), by multiplying throughout by the reference phasor $(\bar{E})$.

To find the true power:

$$\text{true power} = \bar{I}_R\bar{E} = 4 \times 120 = 480\ \text{W (Answer a.)}$$

To find the reactive power:

$$\text{reactive power} = \bar{I}_C\bar{E} = 6 \times 120 = 720\ \text{var (Answer b.)}$$

divide throughout by E

**Figure 28.22**

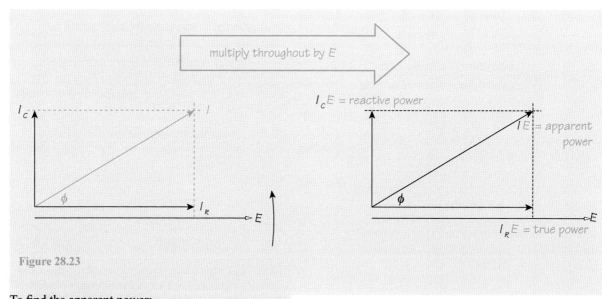

Figure 28.23

To find the apparent power:

apparent power $= \overline{IE} = 7.21 \times 120 \approx 865\ \text{V} \cdot \text{A}$
(Answer c.)

> An alternative method would be to apply Pythagoras's Theorem, as follows:
>
> apparent power $= \sqrt{\text{true power}^2 + \text{reactive power}^2}$
>
> $= \sqrt{480^2 + 720^2} \approx 865\text{V} \cdot \text{A}$

Finally, to find the power factor, we use the cosine ratio:

power factor $= \cos\phi = \dfrac{\text{adjacent}}{\text{hypotenuse}}$

$= \dfrac{\text{true power}}{\text{apparent power}}$

$= \dfrac{480}{865} = 0.555\ \text{leading (Answer d.)}$

'Leading' because, for an R-C circuit, the load current leads the supply voltage.

## Parallel R-L-C circuits

When a potential difference $(\overline{E})$ is applied across a **parallel R-L-C circuit** (Figure 28.24), a current, $\overline{I}_R$, will flow through the resistive branch of the circuit, a current, $\overline{I}_L$, will flow through the inductive branch, and a current, $\overline{I}_C$, will flow through the capacitive branch.

Figure 28.24

## Drawing the phasor diagram

### Step 1

In a parallel circuit, the **supply voltage $(\overline{E})$** is common to each branch and, so, this is chosen as the **reference phasor**. As we have learnt, the reference phasor is *always drawn along the horizontal positive axis*, and it's also normally drawn fairly long in order to distinguish it from the others (Figure 28.25).

Figure 28.25

### Step 2

The current, $\bar{I}_R$, flowing through the resistive branch, is *in phase with the supply voltage* and, so, is also drawn along the horizontal positive axis (Figure 28.26).

Figure 28.26

### Step 3

The current, $\bar{I}_L$, flowing through the inductive branch *lags the supply voltage by 90°* (remember C**I**V**IL**), so is drawn 90° clockwise from the reference phasor (Figure 28.27).

Figure 28.27

### Step 4

The current, $\bar{I}_C$, flowing through the capacitive branch *leads the supply voltage by 90°* (remember C**IV**IL), so is drawn 90° counterclockwise from the reference phasor (Figure 28.28). *Always* draw $\bar{I}_C$ shorter than $\bar{I}_L$ (or *vice versa*), or we will end up with a unique condition called **resonance** – just as we did in the *series R-L-C* circuit!

Figure 28.28

### Step 5

We know from Kirchhoff's Current Law that, in a parallel circuit, the supply current is the sum of the individual branch currents and, for a.c. circuits, we have to add them *vectorially*.

As always, it's a little more difficult to add *three* phasors. As $\bar{I}_L$ and $\bar{I}_C$ lie in opposite directions, the simplest thing to do is to start by subtracting them and, then, add the difference to phasor $\bar{I}_R$.

The snag is, of course, that we might not know whether $\bar{I}_L$ is bigger than $\bar{I}_C$, or *vice versa*! Fortunately, *it doesn't matter!* Once again, the purpose of the phasor diagram is to *generate equations*, not to accurately represent the *actual* conditions in the circuit to scale! And the phasor diagram will *always* generate the correct equations, whether or not $\bar{I}_L$ is bigger than $\bar{I}_C$!

So, the simplest solution is to get into the habit of *always drawing $\bar{I}_L$ longer than $\bar{I}_C$ – but whatever you do,* **never draw them the same length***!!*

So, as explained, start by subtracting $\bar{I}_C$ from $\bar{I}_L$, and *then* vectorially add the resultant to $\bar{I}_R$ to give the completed phasor diagram, as shown in Figure 28.29.

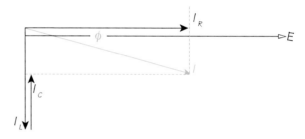

Figure 28.29

From the completed phasor diagram, we can see that the supply current, $\bar{I}$, is the **phasor sum** (or **vector sum**) of $\bar{I}_R$, $\bar{I}_L$ and $\bar{I}_C$, which can then be found using Pythagoras's Theorem:

$$\bar{I} = \sqrt{\bar{I}_R^2 + (\bar{I}_L - \bar{I}_C)^2}$$

To find the phase angle, we can use the *cosine ratio*:

$$\cos\phi = \frac{\text{adjacent}}{\text{hypotenuse}} = \frac{\bar{I}_R}{\bar{I}}$$

If $\bar{I}_L$ really is larger than $\bar{I}_C$, then the circuit is predominantly inductive and, so, the resulting phase angle will be **lagging**. On the other hand, if $\bar{I}_C$ happens to be larger than $\bar{I}_L$, then the circuit will be predominantly capacitive and, so, the resulting phase angle will be **leading**.

### Admittance diagram

How can we now proceed to find out further equations for solving a parallel *R-L-C* circuit?

Again, the answer is by means of an **impedance diagram**.

## Drawing the admittance diagram

### Step 1

We start by drawing the circuit's phasor diagram, following the steps already explained (Figure 28.30).

Figure 28.30

### Step 2

Next, we *divide each of the voltage phasors by the reference phasor (Ē)* (Figure 28.31).

The resulting diagram is an **admittance diagram** (or *'admittance triangle'*), and is useful because it generates the following equations:

$$\frac{\bar{I}_R}{\bar{E}} = \frac{1}{R} \qquad \frac{\bar{I}_L}{\bar{E}} = \frac{1}{X_L} \qquad \frac{\bar{I}_C}{\bar{E}} = \frac{1}{X_C}$$

$$\frac{\bar{I}}{\bar{E}} = \frac{1}{Z}$$

Also from the admittance diagram, you can see that the admittance is the *vector sum of conductance (1/R) and capacitive susceptance (1/X_C)*, which can be calculated by applying Pythagoras' Theorem:

$$\left(\frac{1}{Z}\right)^2 = \left(\frac{1}{R}\right)^2 + \left(\frac{1}{X_L} - \frac{1}{X_C}\right)^2$$

$$\frac{1}{Z} = \sqrt{\left(\frac{1}{R}\right)^2 + \left(\frac{1}{X_L} - \frac{1}{X_C}\right)^2}$$

We can also find the circuit's **phase angle**, using basic trigonometry, utilising the *cosine* ratio:

$$\cos\phi = \frac{\text{adjacent}}{\text{hypotenuse}} = \frac{\left(\frac{1}{R}\right)}{\left(\frac{1}{Z}\right)}$$

$$\angle\phi = \cos^{-1}\frac{Z}{R}$$

**Important!** Dividing a current phasor diagram by voltage produces an admittance diagram which generates each of the equations shown above. So you don't have to learn *any* of these equations – they can all be generated *provided you learn how to draw the phasor and impedance diagrams!*

Once again, you may prefer to work in terms of conductance, inductive susceptance, capacitive susceptance and admittance (after all, it is easier!), in which case the following equations apply (Figure 28.32).

$$\frac{\bar{I}_R}{\bar{E}} = G \qquad \frac{\bar{I}_L}{\bar{E}} = B_L \qquad \frac{\bar{I}_C}{\bar{E}} = B_C \qquad \frac{\bar{I}}{\bar{E}} = Y$$

Also from the admittance diagram, you can see that the admittance is the *vector sum of conductance (G), inductive susceptance (B_L) and capacitive susceptance (B_C)*, which can be calculated by applying Pythagoras's Theorem:

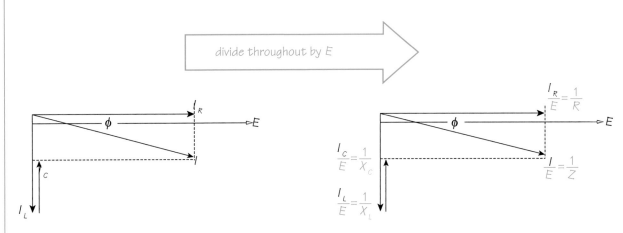

Figure 28.31

$$Y^2 = G^2 + (B_L - B_C)^2$$

$$Y = \sqrt{G^2 + (B_L - B_C)^2}$$

We can also find the circuit's **phase angle**, using basic trigonometry, utilising the *cosine* ratio:

$$\cos\phi = \frac{\text{adjacent}}{\text{hypotenuse}} = \frac{G}{Y}$$

$$\angle\phi = \cos^{-1}\frac{G}{Y}$$

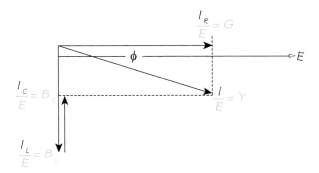

**Figure 28.32**

## Power diagram

To create a **power diagram** (or '**power triangle**') for a parallel *R-C* circuit, we simply multiply throughout by the reference phasor (Figure 28.33).

This action generates the same power equations as we have already seen for a *series R-L-C* circuit. As always, if you can draw the circuit's current phasor diagram, and convert it into a power diagram, *then you can* **derive** *all the following equations, without having to remember them:*

$$\overline{I_R}\,\overline{E} = \text{true power} \qquad \overline{I}\,\overline{E} = \text{apparent power}$$

$$(\overline{I_L} - \overline{I_C})\overline{E} = \text{reactive power}$$

By applying Pythagoras's Theorem, we can obtain an equation for apparent power (in volt amperes) in terms of true power (in watts) and reactive power (in reactive volt amperes):

$$(\text{apparent power})^2 = (\text{true power})^2 + (\text{reactive power})^2$$

Finally, to find the **power factor** of the circuit, we simply need to find the cosine of the phase angle, which is:

$$\cos\phi = \frac{\text{adjacent}}{\text{hypotenuse}} = \frac{\text{true power}}{\text{apparent power}}$$

**Worked example 4**   A parallel *R-L-C* circuit comprises a 6-Ω purely resistive branch, a 19-mH purely inductive branch and a 66-µF purely capacitive branch, supplied by a 48-V, 100-Hz supply. Calculate:

a   each of the branch currents
b   the supply current
c   the impedance
d   the true power of the complete circuit.

**Solution**   As always, the first step is to sketch the circuit, with all the supplied information inserted (Figure 28.34).

Before proceeding any further, we should determine the inductive reactance and capacitive reactance of the inductive and capacitive branches, as we will need to use these values:

$$X_L = 2\pi f L = 2 \times \pi \times 100 \times (19 \times 10^{-3}) \simeq 12\ \Omega$$

**Figure 28.33**

**Figure 28.34**

$$X_C = \frac{1}{2\pi fC} = \frac{1}{2\pi \times 100 \times (66 \times 10^{-6})} \approx 24\,\Omega$$

The next step is to sketch the current phasor diagram, and convert this into an admittance diagram, by dividing throughout by the reference phasor $(\overline{E})$ (Figure 28.35).

To find the current through the **resistive** branch:

$$\frac{\overline{I}_R}{\overline{E}} = \frac{1}{R} \quad \text{from which:} \quad \overline{I}_R$$

$$= \frac{\overline{E}}{R} = \frac{48}{6} = 8\,\text{A (Answer a.)}$$

To find the current through the **inductive** branch:

$$\frac{\overline{I}_L}{\overline{E}} = \frac{1}{X_L} \quad \text{from which:}$$

$$\overline{I}_L = \frac{\overline{E}}{X_L} = \frac{48}{12} = 4\,\text{A (Answer a.)}$$

To find the current through the **capacitive** branch:

$$\frac{\overline{I}_C}{\overline{E}} = \frac{1}{X_C} \quad \text{from which:}$$

$$\overline{I}_C = \frac{\overline{E}}{X_C} = \frac{48}{24} = 2\,\text{A (Answer a.)}$$

We now need to refer back to the current phasor diagram, and apply Pythagoras's Theorem, to determine the supply current:

$$I = \sqrt{I_R{}^2 + (I_L - I_C)^2} = \sqrt{8^2 + (4-2)^2}$$

$$= \sqrt{68} = 8.25\,\text{A (Answer b.)}$$

To find the impedance of the circuit, we can use the appropriate equation generated by the admittance diagram:

$$\frac{\overline{I}}{\overline{E}} = \frac{1}{Z} \quad \text{from which:}$$

$$Z = \frac{\overline{E}}{\overline{I}} = \frac{48}{8.25} = 5.82\,\Omega \text{ (Answer c.)}$$

To find the true power of the circuit, we must convert the current phasor diagram into a power diagram, by multiplying throughout by the reference phasor $(\overline{E})$ (Figure 28.36).

From the various equations generated, we can use the following:

$$\text{true power} = I_R E = 8 \times 48 = 384\,\text{W (Answer d.)}$$

**Figure 28.35**

**Figure 28.36**

## Review your learning

Now that we've completed this chapter, we need to examine the **objectives** listed at its start. Placing '*Can I...*' at the beginning, and a question mark at the end, of each objective turns that objective into a **test item**. If we can answer each of those test items, then we've met the objectives of this chapter.

> **Online resources**
> The companion website to this book contains further resources relating to this chapter. The website can be accessed via the following link:
> www.routledge.com/cw/waygood

# Chapter 29

# Summary of equations derived from phasor diagrams

## Objectives

On completion of this chapter, you should be able to:

1  sketch a current phasor diagram, representing a
   a  series $R$-$L$ circuit
   b  series $R$-$C$ circuit
   c  series $R$-$L$-$C$ circuit.
2  convert a series circuit phasor diagram into an impedance diagram.
3  convert a series circuit phasor diagram into a power diagram.
4  apply basic geometry and/or trigonometry, to derive equations for
   a  impedance
   b  resistance
   c  reactance
   d  apparent power
   e  true power
   f  reactive power
   g  phase angle
   h  power factor.
5  sketch a voltage phasor diagram, representing a
   a  parallel $R$-$L$ circuit
   b  parallel $R$-$C$ circuit
   c  parallel $R$-$L$-$C$ circuit.
6  convert a parallel circuit phasor diagram into an admittance diagram.
7  convert a parallel circuit phasor diagram into a power diagram.
8  apply basic geometry and/or trigonometry, to derive equations for
   a  admittance
   b  conductance
   c  susceptance
   d  impedance
   e  resistance
   f  reactance
   g  apparent power
   h  true power
   j  reactive power
   k  phase angle
   l  power factor.

## Generating equations for series a.c. circuits

During an earlier chapter, we learnt how to construct a phasor diagram for a **series a.c. circuit**. By constructing a phasor diagram, solving an 'electrical' problem becomes a simple matter of applying either *basic geometry* (Pythagoras's Theorem) or *basic trigonometry* (sine, cosine or tangent ratios) to simple right-angled triangles.

*Understanding this concept is of vital importance in understanding the behaviour of alternating current.* So much so that, in this chapter, we are *not* going to learn anything new, but we are going to summarise all that we – hopefully – will have learnt from the preceding chapters.

An Introduction to Electrical Science, Waygood, ISBN 9780815391821, 2019. © Taylor & Francis

For example, by applying **Pythagoras's Theorem** to each of the following voltage phasor diagrams, which are really just simple **right-angled triangles**, the following relationships are revealed:

**Series *R-L* circuit:**

Figure 29.1

$$\overline{E}^2 = \overline{U}_R^2 + \overline{U}_L^2$$

**Series *R-C* circuit:**

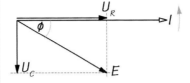

Figure 29.2

$$\overline{E}^2 = \overline{U}_R^2 + \overline{U}_C^2$$

**Series *R-L-C* circuit:**

Figure 29.3

$$\overline{E}^2 = \overline{U}_R^2 + \left(\overline{U}_L - \overline{U}_C\right)^2$$

From which we can derive the following equations:

$$\overline{E} = \sqrt{\overline{U}_R^2 + \overline{U}_L^2}$$
$$\overline{U}_R = \sqrt{\overline{E}^2 - \overline{U}_L^2}$$
$$\overline{U}_L = \sqrt{\overline{E}^2 - \overline{U}_R^2}$$

$$\overline{E} = \sqrt{\overline{U}_R^2 + \overline{U}_C^2}$$
$$\overline{U}_R = \sqrt{\overline{E}^2 - \overline{U}_C^2}$$
$$\overline{U}_C = \sqrt{\overline{E}^2 - \overline{U}_R^2}$$

$$\overline{E} = \sqrt{\overline{U}_R^2 + \left(\overline{U}_L - \overline{U}_C\right)^2}$$
$$\overline{U}_R = \sqrt{\overline{E} - \left(\overline{U}_L - \overline{U}_C\right)^2}$$
$$\left(\overline{U}_L - \overline{U}_C\right) = \sqrt{\overline{E} - \overline{U}_R^2}$$

We can also determine the phase angle for each circuit, as follows:

$$\phi = \cos^{-1}\left(\frac{\overline{U}_R}{\overline{E}}\right)$$

$$\phi = \cos^{-1}\left(\frac{\overline{U}_R}{\overline{E}}\right)$$

$$\phi = \cos^{-1}\left(\frac{\overline{U}_R}{\overline{E}}\right)$$

## Impedance diagrams

We can convert each of the above *voltage* phasor diagrams into a corresponding **impedance diagram**, by simply *dividing throughout by the reference phasor,* $\overline{I}$ , as follows:

**Series *R-L* circuit:**

Figure 29.4

**Series *R-C* circuit:**

Figure 29.5

**Series *R-L-C* circuit:**

Figure 29.6

From which we get the following equations:

$$Z = \frac{\overline{E}}{\overline{I}}$$

$$R = \frac{\overline{U}_R}{\overline{I}}$$

$$X_L = \frac{\overline{U}_L}{\overline{I}}$$

$$Z = \frac{\overline{E}}{\overline{I}}$$

$$R = \frac{\overline{U}_R}{\overline{I}}$$

$$X_C = \frac{\overline{U}_C}{\overline{I}}$$

$$Z = \frac{\overline{E}}{\overline{I}}$$

$$R = \frac{\overline{U}_R}{\overline{I}}$$

$$X_L = \frac{\overline{U}_L}{\overline{I}}$$

$$X_C = \frac{\overline{U}_C}{\overline{I}}$$

$$X = \frac{\left(\overline{U}_L - \overline{U}_C\right)}{\overline{I}}$$

Applying **Pythagoras's Theorem** to each impedance diagram:

$$Z^2 = R^2 + X_L^2$$

$$Z^2 = R^2 + X_C^2$$

$$Z^2 = R^2 + \left(X_L - X_C\right)^2$$

From which we can find the values of $Z$, $R$, $X_L$, and $X_C$:

$$Z = \sqrt{R^2 + X_L^2}$$

$$R = \sqrt{Z^2 - X_L^2}$$

$$X_L = \sqrt{Z^2 - R^2}$$

$$Z = \sqrt{R^2 + X_C^2}$$

$$R = \sqrt{Z^2 - X_C^2}$$

$$X_C = \sqrt{Z^2 - R^2}$$

$$Z = \sqrt{R^2 + \left(X_L - X_C\right)^2}$$

$$R = \sqrt{Z^2 + \left(X_L - X_C\right)^2}$$

$$\left(X_L - X_C\right) = \sqrt{Z^2 - R^2}$$

We can also determine the phase angle for each circuit, as follows:

$$\phi = \cos^{-1}\left(\frac{R}{Z}\right)$$

$$\phi = \cos^{-1}\left(\frac{R}{Z}\right)$$

$$\phi = \cos^{-1}\left(\frac{R}{Z}\right)$$

## Power diagrams

We can convert any *voltage* phasor diagram into a corresponding **power diagram**, simply by *multiplying throughout by the reference phasor,* $\overline{I}$ , as follows.

Remember, '**apparent power**' *(S)* is expressed in volt amperes, '**true power**' *(P)* in watts and reactive power *(Q)* in reactive volt amperes.

**Series *R-L* circuit:**

**Series *R-C* circuit:**

**Series *R-L-C* circuit:**

Figure 29.7

Figure 29.8

Figure 29.9

From which we get the following equations:

| | | |
|---|---|---|
| apparent power, $S = \overline{E}\,\overline{I}$ | apparent power, $S = \overline{E}\,\overline{I}$ | apparent power, $S = \overline{E}\,\overline{I}$ |
| true power, $P = \overline{U}_R\,\overline{I}$ | true power, $P = \overline{U}_R\,\overline{I}$ | true power, $P = \overline{U}_R\,\overline{I}$ |
| reactive power, $Q = \overline{U}_L\,\overline{I}$ | reactive power, $Q = \overline{U}_C\,\overline{I}$ | $Q = \left(\overline{U}_L - \overline{U}_C\right)\overline{I}$ |

Applying **Pythagoras's Theorem** to each power diagram:

| | | |
|---|---|---|
| $S^2 = P^2 + Q^2$ | $S^2 = P^2 + Q^2$ | $S^2 = P^2 + Q^2$ |

From which, we can find the values of apparent power *(S)*, true power *(P)* and reactive power *(Q)*:

| | | |
|---|---|---|
| $S = \sqrt{P^2 + Q^2}$ | $S = \sqrt{P^2 + Q^2}$ | $S = \sqrt{P^2 + Q^2}$ |
| $P = \sqrt{S^2 - Q^2}$ | $P = \sqrt{S^2 - Q^2}$ | $P = \sqrt{S^2 - Q^2}$ |
| $Q = \sqrt{S^2 - P^2}$ | $Q = \sqrt{S^2 - P^2}$ | $Q = \sqrt{S^2 - P^2}$ |

We can also convert any *impedance* diagram into a **power diagram**, by *multiplying throughout by the square of the current $(\overline{I}^2)$*, as follows.

**Series *R-L* circuit:**

Figure 29.10

**Series *R-C* circuit:**

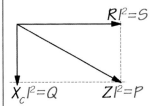

Figure 29.11

**Series *R-C-L* circuit:**

Figure 29.12

From which we get the following equations:

| | | |
|---|---|---|
| reactive power, $S = \overline{I}^2 Z$ | reactive power, $S = \overline{I}^2 Z$ | reactive power, $S = \overline{I}^2 Z$ |
| true power, $P = \overline{I}^2 R$ | true power, $P = \overline{I}^2 R$ | true power, $P = \overline{I}^2 R$ |
| reactive power, $Q = \overline{I}^2 X_L$ | reactive power, $Q = \overline{I}^2 X_C$ | $Q = \overline{I}^2\left(X_L - X_C\right)$ |

## Generating equations for parallel a.c. circuits

During an earlier chapter, we learnt how to construct a phasor diagram for a **parallel a.c. circuit**. Once we have learnt to construct a phasor diagram for a **parallel a.c. circuit**, solving an 'electrical' problem becomes a simple matter of applying either *basic geometry* (Pythagoras's Theorem) or *basic trigonometry* (sine, cosine or tangent ratios), where the phasor diagrams are simple, right-angled triangles!

Applying **Pythagoras's Theorem** to each of these current phasor diagrams will give us the following relationships:

**Parallel *R-L* circuit:**

Figure 29.13

$$\overline{I}^2 = \overline{I}_R^2 + \overline{I}_L^2$$

**Parallel *R-C* circuit:**

Figure 29.14

$$\overline{I}^2 = \overline{I}_R^2 + \overline{I}_C^2$$

**Parallel *R-L-C* circuit:**

Figure 29.15

$$\overline{I}^2 = \overline{I}_R^2 + \left(\overline{I}_L - \overline{I}_C\right)^2$$

From which we can derive the following equations:

$$\overline{I} = \sqrt{\overline{I}_R^2 + \overline{I}_L^2}$$

$$\overline{I}_R = \sqrt{\overline{I}^2 - \overline{I}_L^2}$$

$$\overline{I}_L = \sqrt{\overline{I}^2 - \overline{I}_R^2}$$

$$\overline{I} = \sqrt{\overline{I}_R^2 + \overline{I}_C^2}$$

$$\overline{I}_R = \sqrt{\overline{I}^2 - \overline{I}_C^2}$$

$$\overline{I}_C = \sqrt{\overline{I}^2 - \overline{I}_C^2}$$

$$\overline{I} = \sqrt{\overline{I}_R^2 + \left(\overline{I}_L - \overline{I}_C\right)^2}$$

$$\overline{I}_R = \sqrt{\overline{I} - \left(\overline{I}_L - \overline{I}_C\right)^2}$$

$$\left(\overline{I}_L - \overline{I}_C\right) = \sqrt{\overline{I} - \overline{I}_R^2}$$

We can also determine the phase angle for each circuit, as follows:

$$\phi = \cos^{-1}\left(\frac{\overline{I}_R}{\overline{I}}\right)$$

$$\phi = \cos^{-1}\left(\frac{\overline{I}_R}{\overline{I}}\right)$$

$$\phi = \cos^{-1}\left(\frac{\overline{I}_R}{\overline{I}}\right)$$

## Admittance diagrams

You will recall that, for a **parallel d.c. circuit**, the total resistance is found as follows:

$$\frac{1}{R} = \frac{1}{R_1} + \frac{1}{R_2} + \frac{1}{R_3} + \text{etc.}$$

The *reciprocal* of resistance is termed '**conductance**' (symbol: *G*), expressed in **siemens** (symbol: **S**). So, we could rewrite the above equation in terms of conductances, that is:

$$G = G_1 + G_2 + G_3 + \text{etc.}$$

In a.c. circuits, the reciprocal of impedance is called '**admittance**' (symbol: *Y*), and the reciprocal of reactance is called '**susceptance**' (symbol: *B*). Specifically, the reciprocal of inductive reactance is called '**inductive susceptance**' (symbol: $B_L$) and the reciprocal of capacitive reactance is called '**capacitive susceptance**' (symbol: $B_C$).

Generally, in a.c. parallel circuits, it's usually *much* easier to work with admittance, conductance and susceptance, than it is to work with impedance, resistance and reactance.

So, to create an **admittance diagram** we *divide the current phasors by the reference phasor* – i.e. by the supply voltage:

**Parallel *R-L* circuit:**

**Parallel *R-C* circuit:**

**Parallel *R-L-C* circuit:**

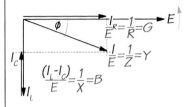

Figure 29.16                    Figure 29.17                    Figure 29.18

From which we get the following equations:

$$Y = \frac{1}{Z} = \frac{\overline{I}}{\overline{E}}$$

$$G = \frac{1}{R} = \frac{\overline{I}_R}{\overline{E}}$$

$$B_L = \frac{1}{X_L} = \frac{\overline{I}_L}{\overline{E}}$$

$$Y = \frac{1}{Z} = \frac{\overline{I}}{\overline{E}}$$

$$G = \frac{1}{R} = \frac{\overline{I}_R}{\overline{E}}$$

$$B_C = \frac{1}{X_C} = \frac{\overline{I}_C}{\overline{E}}$$

$$Y = \frac{1}{Z} = \frac{\overline{I}}{\overline{E}}$$

$$G = \frac{1}{R} = \frac{\overline{I}_R}{\overline{E}}$$

$$B_C = \frac{1}{X_C} = \frac{\left(\overline{I}_L - \overline{I}_C\right)}{\overline{E}}$$

Applying **Pythagoras's Theorem** to each impedance diagram:

$$Y^2 = G^2 + B_L^2$$

or, if you prefer:

$$\left(\frac{1}{Z}\right)^2 = \left(\frac{1}{R}\right)^2 + \left(\frac{1}{X_L}\right)^2$$

$$Y^2 = G^2 + B_C^2$$

or, if you prefer:

$$\left(\frac{1}{Z}\right)^2 = \left(\frac{1}{R}\right)^2 + \left(\frac{1}{X_C}\right)^2$$

$$Y^2 = G^2 + \left(B_L - B_C\right)^2$$

or, if you prefer:

$$\left(\frac{1}{Z}\right)^2 = \left(\frac{1}{R}\right)^2 + \left(\frac{1}{\left(X_L - X_C\right)}\right)^2$$

From which we can find the values of $Y$, $G$, $B_L$ and $B_C$:

$$Y = \sqrt{G^2 + B_L^2}$$

$$G = \sqrt{Y^2 - B_L^2}$$

$$B_L = \sqrt{Y^2 - G^2}$$

$$Y = \sqrt{G^2 + B_C^2}$$

$$G = \sqrt{Y^2 - B_C^2}$$

$$B_C = \sqrt{Y^2 - G^2}$$

$$Y = \sqrt{G^2 + \left(B_L - B_C\right)^2}$$

$$G = \sqrt{Y^2 - \left(B_L - B_C\right)^2}$$

$$\left(B_L - B_C\right) = \sqrt{Y^2 - G^2}$$

If you prefer, you could *still* work in terms of impedance, resistance and reactance:

$$\frac{1}{Z} = \sqrt{\left(\frac{1}{R}\right)^2 + \left(\frac{1}{X_L}\right)^2}$$

$$\frac{1}{R} = \sqrt{\left(\frac{1}{Z}\right)^2 - \left(\frac{1}{X_L}\right)^2}$$

$$\frac{1}{X_L} = \sqrt{\left(\frac{1}{Z}\right)^2 - \left(\frac{1}{R}\right)^2}$$

$$\frac{1}{R} = \sqrt{\left(\frac{1}{Z}\right)^2 + \left(\frac{1}{X_L}\right)^2}$$

$$\frac{1}{R} = \sqrt{\left(\frac{1}{Z}\right)^2 - \left(\frac{1}{X_C}\right)^2}$$

$$\frac{1}{X_C} = \sqrt{\left(\frac{1}{Z}\right)^2 - \left(\frac{1}{R}\right)^2}$$

$$\frac{1}{Z} = \sqrt{\left(\frac{1}{R}\right)^2 + \left(\frac{1}{X_L - X_C}\right)^2}$$

$$\frac{1}{R} = \sqrt{\left(\frac{1}{Z}\right)^2 - \left(\frac{1}{X_L - X_C}\right)^2}$$

$$\frac{1}{X_L - X_C} = \sqrt{\left(\frac{1}{Z}\right)^2 - \left(\frac{1}{R}\right)^2}$$

We can also determine the phase angle for each circuit, as follows:

$$\phi = \cos^{-1}\left(\frac{G}{Y}\right)$$

$$\phi = \cos^{-1}\left(\frac{G}{Y}\right)$$

$$\phi = \cos^{-1}\left(\frac{G}{Y}\right)$$

Or again, if you prefer: in terms of impedance and resistance:

$$\phi = \cos^{-1}\left(\frac{(1/R)}{(1/Z)}\right) = \frac{Z}{R} \qquad \phi = \cos^{-1}\left(\frac{(1/R)}{(1/Z)}\right) = \frac{Z}{R} \qquad \phi = \cos^{-1}\left(\frac{(1/R)}{(1/Z)}\right) = \frac{Z}{R}$$

## Power diagrams

We can convert any of the above *current* phasor diagrams into a **power diagram**, by *multiplying throughout by the reference phasor,* , as follows. Remember, '**apparent power**' *(S)* is expressed in volt amperes, '**true power**' *(P)* in watts and reactive power *(Q)* in reactive volt amperes.

**Parallel *R-L* circuit:**

Figure 29.19

**Parallel *R-C* circuit:**

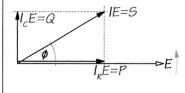

Figure 29.20

**Parallel *R-L-C* circuit:**

Figure 29.21

From which we get the following equations:

$$S = \overline{E}I$$
$$P = \overline{E}I_R$$
$$Q = \overline{E}I_L$$

$$S = \overline{E}I$$
$$P = \overline{E}I_R$$
$$Q = \overline{E}I_C$$

$$S = \overline{E}I$$
$$P = \overline{E}I_R$$
$$Q = \overline{E}\left(\overline{I}_L - \overline{I}_C\right)$$

Applying **Pythagoras's Theorem** to each power diagram:

$$S^2 = P^2 + Q^2 \qquad\qquad S^2 = P^2 + Q^2 \qquad\qquad S^2 = P^2 + Q^2$$

From which, we can find the values of apparent power *(S)*, true power *(P)* and reactive power *(Q):*

$$S = \sqrt{P^2 + Q^2}$$
$$P = \sqrt{S^2 - Q^2}$$
$$Q = \sqrt{S^2 - P^2}$$

$$S = \sqrt{P^2 + Q^2}$$
$$P = \sqrt{S^2 - Q^2}$$
$$Q = \sqrt{S^2 - P^2}$$

$$S = \sqrt{P^2 + Q^2}$$
$$P = \sqrt{S^2 - Q^2}$$
$$Q = \sqrt{S^2 - P^2}$$

Hopefully, what should have been made abundantly clear from the above is that *every single equation* is derived by applying Pythagoras's Theorem, or the sine, cosine or tangent ratios to phasor diagrams *which are nothing more than right-angled triangles!*

Accordingly, *there is absolutely no need whatsoever to commit any of these equations to memory!*

You will learn **far** more about the behaviour of a.c. circuits by (1) **learning how to construct phasor diagrams** and (2) using Pythagoras's Theorem and basic trigonometry to **derive all these equations**, than you will **ever** learn by trying to commit them to memory!

## Review your learning

Now that we've completed this chapter, we need to examine the **objectives** listed at its start. Placing *'Can I ... '* at the beginning, and a question mark at the end, of each objective turns that objective into a **test item**. If we can answer each of those test items, then we've met the objectives of this chapter.

# Chapter 30

# Power factor improvement

## Objectives

On completion of this chapter, you should be able to:

1 define the terms, 'power factor', 'lagging' power factor and 'leading' power factor.
2 explain why increasing a load's power factor towards unity will reduce the current supplied to that load.
3 outline the main advantages of improving the power factor of a load.
4 explain why a power factor capacitor is rated in reactive volt amperes, rather than in farads.
5 solve problems on power factor improvement.

## Need for power factor improvement

As we have learned, '**power factor**' is defined as *the ratio of true power to apparent power of a load*, and is expressed as the *cosine of a circuit's phase angle* – i.e. the cosine of the angle by which the load's *supply current* lags or leads its supply voltage. We *always* express phase angle in terms of the position of the load current, relative to the supply voltage, not the other way around. So,

$$\text{power factor} = \cos\phi = \frac{\text{true power}}{\text{apparent power}}$$

Power factor can be expressed as 'per unit', e.g. '0.75 lagging', or as a 'percentage', e.g. '75% lagging'. The 'per unit' method is more widely used, and expressing it as a percentage is now considered rather old fashioned.

It's important to specify whether a power factor is 'lagging' or 'leading'. For inductive circuits, where the load current *lags* the supply voltage, we use the term '**lagging power factor**' (Figure 30.1); for capacitive circuits, where the load current *leads* the supply voltage, we use the term '**leading power factor**' (Figure 30.2) (remember that phasors 'rotate' counterclockwise).

Figure 30.1 circuit with a lagging power factor.

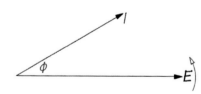

Figure 30.2 circuit with a leading power factor.

Most domestic, commercial, and industrial loads are **resistive-inductive** (**R-L**) comprising resistive loads (i.e. incandescent lighting and heating loads) and inductive loads (i.e. motors, welding sets, flourescent lamps etc.). The majority of loads, therefore, have **lagging power factors**.

The circuit diagram in Figure 30.3 represents an equivalent circuit for an *R-L* load, with the resistive and inductive branches taking currents, $\overline{I}_R$ and $\overline{I}_L$ respectively.

An Introduction to Electrical Science, Waygood, ISBN 9780815391821, 2019. © Taylor & Francis

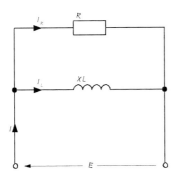

Figure 30.3

The phasor diagram for this circuit is shown in Figure 30.4. As you see, the supply current is the phasor sum of the two branch currents.

$$\overline{I} = \overline{I}_R + \overline{I}_L$$

And, of course, this current is *lagging* the supply voltage by some angle, $\phi$, and, so, the circuit is said to have a 'lagging power factor'.

Figure 30.4

Now, let's see what happens if we add a capacitor in parallel with the circuit's existing branches (Figure 30.5).

As we already know, with a parallel circuit, *adding another branch will have absolutely no effect whatsoever on the behaviour or operation of the existing branches* – they will continue to operate normally and draw exactly the same amounts of current *regardless of how many additional branches are added*.

However, adding an additional branch *will*, of course, affect the value of the load current the circuit draws from the supply.

Figure 30.5

By applying Kirchhoff's Current Law, we can express the supply current in terms of the branch currents, as follows:

$$\overline{I} = \overline{I}_R + \overline{I}_L + \overline{I}_C$$

You would, perhaps, be forgiven for assuming that, by adding another branch current $(\overline{I}_C)$, then the load current is *bound* to increase due to the additional branch current.

However, this is *not* the case!

Let's redraw the phasor diagram, to see what effect the capacitive-branch current will have on the supply current (Figure 30.6).

Figure 30.6

As you can see, because $\overline{I}_C$ leads the supply voltage, it is in *antiphase* (acts in the opposite sense) to $\overline{I}_L$ and therefore acts to reduce the circuit's overall reactive current (i.e. $\overline{I}_L - \overline{I}_C$) which, of course, reduces the phase angle and, therefore, reduces the value of the supply current – in this case, from $\overline{I}$ to $\overline{I}'$.

If we use a capacitor that draws an even bigger leading current, then the phase angle will reduce even further – reducing the supply current from $\overline{I}'$ to $\overline{I}''$ (Figure 30.7).

Figure 30.7

So, while adding a capacitive branch has no effect whatsoever on the behaviour of the currents in the other two branches, it acts *to reduce the circuit's phase angle which, in turn, reduces the value of the supply current.* The more leading current the capacitive branch draws, the more it acts to reduce the phase angle and the corresponding supply current. Eventually, the capacitive branch will draw enough current to cause the load current to become 'leading', rather than 'lagging', and start to increase in value again. However, it is both unnecessary and, in fact, undesirable to reduce the phase angle to zero, for reasons we will discuss shortly.

The practice of adding capacitors in parallel with a load in order to adjust the lagging power factor towards unity is termed **power factor improvement**, or **power factor correction**.

The result of **power factor improvement**, then, is to reduce the size of the load current and, thus, allowing the volume of copper in the electricity network company's supply equipment, such as cables, switchgear, transformers, etc., to be lower and, therefore, less expensive than would be the case if the power factor had *not* been improved. A lower supply current also means a reduction in the voltage drop along the supply cables and, therefore, a higher terminal voltage at the load. The overall effect of all this is to improve system capacity (the ability to supply energy).

So, what is considered to be a 'good power factor' and what is considered to be a 'poor power factor'? Well, **EDF Energy**, one of the UK's largest energy-supply companies, considers a power factor *above* 0.95 to be 'good' and a power factor *below* 0.95 to be 'bad'.

To summarise, then. From the electricity network company's point of view, an improved power factor results in

- reduced system losses
- improved system voltages
- improved system capacity.

But power-factor improvement also offers advantages to the industrial or large commercial users too. Unlike residential consumers, these larger consumers not only have to pay for the number of **kilowatt hours** they consume, but they must also pay a penalty based on any excessive **reactive volt ampere hour** 'consumption' which increases with poor values of power factor.

To minimise such penalties, it would seem sensible to increase the installation's power factor to unity (i.e. reduce the phase angle to zero), by adding as much capacitance in parallel with a load as is necessary, until this is achieved. Perhaps, surprisingly, however, this is *not* the case, because the savings achieved by doing so are offset by the costs incurred in purchasing, installing, and maintaining the required capacitors and their associated control equipment! *So, it is rarely financially worthwhile to try to achieve a unity power factor.* In practice, it is usual to increase the load's power factor beyond 0.95.

It is possible to determine the optimum power factor for any given load, however, this is well beyond the scope of this chapter.

But power factor improvement also offers advantages to the industrial or large commercial users too. Unlike residential consumers, these larger consumers not only have to pay for the **energy** they consume, but they must also pay an *additional fee* based on their installation's **demand** (i.e. its apparent power, expressed in kilovolt amperes) usually together with a **power factor surcharge** if the power factor of their load is too low.

To minimise these additional charges, it would seem sensible to improve the installation's power factor to unity (i.e. reduce the phase angle to zero), by adding as much capacitance in parallel with a load as is necessary, until this is achieved. Perhaps, surprisingly, however, this is *not* the case, because the savings achieved by doing so are offset by the costs incurred in purchasing, installing and maintaining the required capacitors and their associated control equipment! *So, it is rarely financially worthwhile to try to achieve a unity power factor.* It is possible to determine the optimum power factor for any given load, however, this is well beyond the scope of this chapter.

To further complicate matters, practical loads vary continuously: lighting and heating are always being switched on and off, and motors are being started, their loads adjusted, and switched off. So, even if it were financially sensible to achieve a unity power factor, any temporary or instantaneous reduction in load would then result in a *leading* power factor – i.e. an increasingly leading load current would result! For reasons beyond the scope of this chapter, *leading power factors are highly undesirable,* as they lead to system instability and, so, must be avoided at all cost.

In practice, for financial reasons, it is desirable for an installation's lagging power factor to approach, *but to never achieve,* unity.

It should be understood that the advantages of power factor improvement *do **not** apply to residential loads.* Residential consumers (you and me, in other words) are charged simply for the **energy** that they consume. Residential energy meters monitor the supply voltage and in-phase component of the load current, so their readings are based on the true power of the load. And, of course, there is no surcharge imposed on residential consumers for 'poor' power factor.

There are a number of online companies selling 'energy-reduction capacitors' which, they promise, 'will substantially reduce your electricity bill'! Some of these are even demonstrated on *YouTube,* showing how they 'reduce the current' drawn by a residential load. And they do, indeed, reduce the load current. But, of course, they fail to mention that residences are *not* charged for the amount of current they draw from the supply, but for the energy they consume! These devices are a complete and utter waste of money, and the whole business is a scam!

## Power factor improvement in practice

At the beginning of this chapter, we said that power factor is defined as *the ratio of **true power** to **apparent power**.* Let's remind ourselves what a power diagram for an inductive load looks like (Figure 30.8).

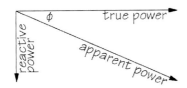

Figure 30.8

In order to reduce ('improve') the power factor, we must reduce the phase angle. We can do this by *reducing the circuit's overall reactive power.* How can we do this in practice? Well, there are a number of ways including through the use of synchronous machines but, in general, the least expensive way is, as already explained, through the use of **individal capacitors** or **capacitor banks** (sometimes called 'static capacitors').

There are *three* approaches to power factor correction using capacitors:

1   Motors in excess of around 35 kW can have their individual power factor improved by connecting capacitors directly across their terminals.
2   Installing capacitor banks at motor control centres will improve the power factor of several motors.
3   Installing capacitor banks at the installation's service entrance.

Of these three approaches, the third is usually the least expensive. However, while this method will decrease the supply current in the distribution company's feeder, it will have absolutely no effect on the current between the service entrance (where the network company's cables are terminated) and the bulk of the load, whereas the other two methods will.

Capacitors intended for power factor improvement must be capable of withstanding the peak voltages of the installation's supply voltage – this information, together with their reactive power ratings is shown on their information nameplates. Power factor capacitors are normally oil-impregnated paper dielectric types, sealed within oil-filled metal containers.

Overcorrecting power factor can be avoided through the use of automatic switching, in which a number of individual capacitors are added or removed, to meet changes in the load's power factor.

So, power factor improvement can be achieved by adding a capacitor bank with an appropriate value of leading reactive power which, of course, acts in the *opposite sense* to the load's lagging reactive power (Figure 30.9).

To achieve any desired value of power factor, we need to know what size of capacitor to place in parallel with the load. In this case, a power factor improvement capacitor's 'size' is expressed in **reactive volt amperes (var)**, and *not* in terms of its capacitance.

Knowing the *capacitance* of a power factor improvement capacitor is of no practical use, and is of academic interest only.

**Figure 30.9**

Power-factor capacitors are usually oil-impregnated paper dielectric capacitors, with their plates enclosed in oil-filled metal tanks. These capacitors are always rated in **reactive volt amperes (var)**, *never* in farads. Banks of capacitors enable individual capacitors to be brought on line, automatically or manually, to ensure optimum power factor improvement for variations in the load's reactive power.

Let's illustrate this with a worked example.

**Worked example 1**  A single-phase a.c. supply of 400 V at 50 Hz, supplies a 50-kW load having a power factor of 0.5 lagging. Find the reactive power of a capacitor bank necessary to bring the power factor to 0.85.

**Solution**  The first step is to determine the load's phase angle, given its power factor:

$$\cos\phi = 0.5$$
$$\phi = \cos^{-1}0.5 = 60°$$

The next step is to find out what the load's *existing reactive power* is. We can do that by applying basic trigonometry to its power diagram (Figure 30.10).

**Figure 30.10**

$$\tan\phi = \frac{\text{opposite}}{\text{adjacent}} = \frac{\text{reactive power}}{\text{true power}}$$

so, reactive power = true power × tan $\phi$

$$= 50 \times \tan 60°$$
$$= 50 \times 1.732$$
$$= 86.60 \text{ kvar}$$

Now, we need to find out *what value of reactive power would be required* to achieve a power factor of 0.85:

$$\phi' = \cos^{-1}0.85 = 31.79°$$
reactive power = true power × tan $\phi$
$$= 50 \times \tan 31.79°$$
$$= 50 \times 0.62$$
$$= 31.00 \text{ kvar}$$

**Figure 30.11**

To reduce the load's existing reactive power of 86.60 kvar down to 31.00 kvar, we must supply the circuit with:

$$86.60 - 31.00 = 55.60 \text{ kvar}$$

. . . which must act in *the opposite sense*. In other words, the necessary capacitor bank must be rated at **55.60 kvar** (Answer).

## Review your learning

Now that we've completed this chapter, we need to examine the **objectives** listed at its start. Placing '*Can I...*' at the beginning, and a question mark at the end, of each objective turns that objective into a **test item**. If we can answer each of those test items, then we've met the objectives of this chapter.

**Online resources**
The companion website to this book contains further resources relating to this chapter. The website can be accessed via the following link:
www.routledge.com/cw/waygood

# Chapter 31

# Balanced three-phase a.c. systems

On completion of this chapter, you should be able to:

1 briefly explain the major advantages of three-phase a.c. systems, compared with single-phase a.c. systems.
2 identify delta- and star-connected systems.
3 distinguish between 'phases' and 'lines'.
4 briefly explain how lines are identified.
5 sketch a balanced, three-phase, delta-connected load, showing all voltage and current 'sense arrows'.
6 construct a three-phase phasor diagram for a balanced star- and delta-connected load, showing all line and phase voltages and currents.
7 for a star-connected balanced load, define the relationships between its line and phase voltages and currents.
8 for a delta-connected balanced load, define the relationships between its line and phase voltages and currents.
9 express the total power of a balanced star- or delta-connected three-phase load, in terms of both their phase values and their line values.
10 Solve three-phase problems for balanced loads.

## Introduction

Almost without exception, alternating current is generated, transmitted and largely distributed using **three-phase** systems which use *three* conductors rather than the *two* required by single-phase systems. The main reason for this is **economy**, because it can be shown that, for a given load, and despite the additional conductor, the overall volume of copper required by a three-phase system is approximately 75 per cent of that required by an equivalent single-phase system.

But there are other excellent reasons for using three-phase a.c. systems, which include:

- 'smoother' energy delivery (see below), compared with single-phase a.c.
- three-phase machines are self-starting; single-phase machines are not.
- three-phase machines are simpler, smaller, lighter and more efficient than single-phase machines of equivalent rating.
- three-phase machines run more smoothly and quietly than single-phase machines.

The waveforms shown in Figure 31.1 compare the rates at which energy is supplied (i.e. the 'power') to a load by a single-phase and by a three-phase supply. The grey areas below each waveform represent the power 'supplied' by each system (in these examples, we are assuming the load is purely resistive in each case): for the three-phase system, the power supplied is close to being constant whereas, for the single-phase system, the power is supplied as a series of pulses.

 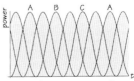

single-phase power waveform    three-phase power waveform

Figure 31.1

An Introduction to Electrical Science, Waygood, ISBN 9780815391821, 2019. © Taylor & Francis

## Generating three-phase voltages

A three-phase alternator's armature consists of three independent windings that are physically displaced from each other by 120°. Accordingly, the voltages (labelled **A**, **B** and **C** in the diagram) which are induced into each winding, each reach their peak values, in sequence, 120 electrical degrees apart – as illustrated in Figure 31.2.

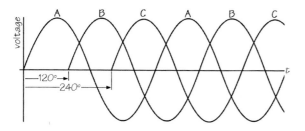

**Figure 31.2**

The sequence in which each voltage reaches peak value (**A-B-C**) is called the '**phase sequence**' of the system. Should the voltages reach their peak values in *reverse* sequence (**A-C-B**) – e.g. should the alternator run backwards – then we describe the system as having a 'negative phase sequence', but this topic is beyond the scope of this text.

As already explained, practically all high-voltage a.c. transmission and distribution systems are three-phase systems. For low-voltage a.c. distribution systems, single-phase supplies are required for residences while three-phase supplies may be required for small workshops, etc. This is achieved as illustrated in Figure 31.3 which shows a typical low-voltage, three-phase, distribution system used in most European countries. The three conductors, labelled **a**, **b** and **c**, are each energised at a nominal potential of 230 V with respect to the fourth, neutral, conductor, with each potential displaced by 120°. As we shall learn, the potential differences between conductors **a-b**, **b-c** and **c-a** are 400 V.

**Figure 31.3**

The electricity network company tries to 'balance' their loads (i.e. apply a similar electrical load to each line) as far as possible, by supplying adjacent consumers from alternate lines, as illustrated in Figure 31.3. The significance of these connections will become clear later in this chapter.

This chapter is intended to supply us with an overview of three-phase a.c. systems, and is *not* intended to cover the subject in-depth.

## Three-phase connections and terminology

### Delta and star connections

The two most common three-phase connections are known as '**delta**' and '**star**'. In the United States and Canada, the 'star' connection is more commonly known as a '**wye**' connection.

A **delta** connection is shown in Figure 31.4, for a supply and for a load. Essentially, the three supply windings (as well as three identical load impedances) are connected is series with each other, and the external connections are made from points **A**, **B** and **C**, between each supply winding (e.g. alternator windings or transformer secondary windings) or load impedance.

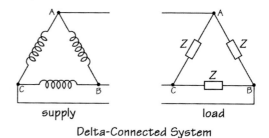

Delta-Connected System

**Figure 31.4**

A **star** connection is shown in Figure 31.5, for a supply and for a load made up of three identical impedances. The common point of connection for a star connection is called its '**star point**'; at the supply, the star point is normally connected to earth for the purpose of stabilising the three voltages **A-N**, **B-N** and **C-N**, and it is from this point that a neutral connection is made.

There is absolutely no reason why a delta-connected supply cannot feed a star-connected load, or why a star-connected supply cannot feed a delta-connected load, providing the voltage requirements are met.

Figure 31.5

There are various other three-phase connections, but these are all beyond the scope of this text.

**Delta-connected** supplies are connected to their loads using *three* conductors, and are known as '**three-phase, three-wire systems**', which accounts for why high-voltage pole-mounted distribution lines carry three conductors, and high-voltage transmission towers carry multiples of three conductors – as shown in Figure 31.6).

Figure 31.6

The **132-kV distribution tower**, illustrated, actually carries two, entirely-separate, three-phase circuits suspended on opposite sides of the same tower. Each circuit consists of three single, steel-cored aluminium, conductors, arranged vertically, and supported from 'chains' of toughened-glass 'dish' insulators.

> The operating voltages of transmission towers can be identified by the number of conductors suspended from each insulator chain. **275-kV towers** have a pair of conductors, while **400-kV towers** have a 'bundle' of *four*.

**Star-connected** supplies are connected to a common point, called a 'star point', which is normally then connected to earth, providing the **neutral point** of the system. Star-connected supplies are connected to their loads using *four* conductors, one of which is a neutral conductor, and are known as '**three-phase, four-wire, systems**'. Typically, in the UK at least, low-voltage distribution systems are three-phase, four-wire systems, which is why you will typically see four vertically mounted conductors on low-voltage distribution poles (any additional conductors are used for supplying street lighting, etc.).

> Delta and star configurations not only apply to alternator windings, but also to a three-phase transformer's primary and secondary windings, and to the way in which three-phase loads are connected.

## Phase and line

The individual windings of a delta- or star-connected alternator or transformer, and their corresponding load impedances (as illustrated in Figures 31.7 and 31.8) are called '**phases**', whereas the conductors that interconnect three-phase supplies and their loads are called '**lines**'.

Figure 31.7

Figure 31.8

• In a three-phase, three-wire system, there are three '**line conductors**'.

- In a three-phase, four-wire system, there are three '**line conductors**' and a '**neutral**' conductor.

Generally speaking, the **phases** of any three-phase machine, including transformers, are inaccessible, whereas the lines and their terminals are easily accessible. This is because the phases are normally enclosed within the machine or transformer tank.

Figures 31.7 and 31.8 showing three-phase loads should make this clear. Don't become confused over the 'crossed over' line conductors. The only reason for doing this is to make the vertical sequence **A-B-C** match the clockwise sequence **A-B-C**.

---

**Vitally important!**

Unfortunately, in the field (as well as in some textbooks), the term 'phase' is frequently misused which leads to confusion. For example, 'line conductors' are sometimes referred to as 'phase conductors' or, simply, as 'phases'. *This is completely incorrect*, and you **must** learn to use the correct terminology if you want to avoid becoming confused!

---

A single-phase supply can be obtained from a three-phase system very easily, simply by connecting the single-phase load between any pair of line conductors or, alternatively, between any line conductor and a neutral conductor – depending on the level of voltage required for the single-phase load.

### Phase and line voltages and currents

The potential difference appearing across any phase is called a '**phase voltage**' ($E_p$), and the current flowing through any phase is called a '**phase current**' ($I_p$).

The potential difference appearing between any pair of line conductors is called a '**line voltage**' ($E_L$), and the current flowing through a line conductor is called a '**line current**' ($I_L$).

### Identifying line conductors and terminals

Three-phase **line** conductors and terminals (*not* 'phases'!) are identified in various ways, according to

relevant national electrical standards. At the time of writing this textbook, in the UK's electricity supply industry, lines are still identified using the colours **red-yellow-blue**. For commercial and industrial installations these colours have been replaced by **brown-black-grey**, in accordance with EU harmonisation requirements. Alternative systems of identifying lines include the use of *letters* (e.g. **A-B-C**), *numerals* (e.g. **1-2-3**), or a *combination of the two* (e.g. $L_1$-$L_2$-$L_3$).

Whichever system is used, the order, or sequence, in which the colours, letters or numerals are listed match the phase sequence of the three-phase system. For example, the letters, **A-B-C**, would indicate normal, or 'positive', phase sequence, whereas the letters, **A-C-B**, would indicate reverse, or 'negative', phase sequence. We will *not* be considering a negative phase sequence situation in this chapter.

In this chapter, we will be using the letters **A-B-C** to identify line conductors or line terminals. Upper-case letters can then be used to indicate high voltages, while lower-case letters can be used to indicate lower voltages, whenever we need to differentiate between the two (such as for transformers). This is also in accordance with the relevant British Standard for the marking of transformer connections.

---

It's important to understand that this system of colours, letters or numbers is used to identify **line conductors** (or **line terminals**), *not* phases. The phases themselves, are not individually identified although, if necessary, they can be identified in terms of their line terminals; for example, 'phase A-B', 'phase B-C' and 'phase C-A'.

Many textbooks talk about '**phase A**', '**phase B**' and '**phase C**'. This makes no sense whatsoever, and simply leads to confusion.

---

### Sense arrows and double-subscript notation

Line and phase voltages can be identified using 'sense arrows', which we met in an earlier chapter on *series, parallel and series-parallel circuits*, combined with what is termed '**double-subscript notation**'. This system helps us determine the sense, or direction, in which a voltage or a current is acting *at any given instant*, and allows us to easily construct three-phase phasor diagrams.

The directions of voltage and current sense arrows are always drawn as illustrated in the schematic diagrams in this section. You will notice that sense arrows that represent line currents *always* point towards the load.

Double-subscript notation takes the following forms:

$$\bar{E}_{AB} \text{ or } \bar{E}_{AN}$$

These examples are read as: *'the potential at point A with respect to point B'*, and as: *'the potential at point A with respect to the neutral'*. So the first subscript letter defines the line whose potential is being measured, and the second subscript defines the point of reference.

Double-subscript notation applies to both **line voltages** and **phase voltages**, and they *always* follow the phase sequence, A-B-C-A, so they're written as:

$$\bar{E}_{AB}, \bar{E}_{BC}, \text{and } \bar{E}_{CA}$$
$$\bar{E}_{AN}, \bar{E}_{BN}, \text{and } \bar{E}_{CN}$$

Double-subscript notation can also be used with **phase currents**, and takes the following form:

$$\bar{I}_{AB} \text{ or } \bar{I}_{AN}$$

Double-subscript notation for currents only applies to *phase* currents, and the above examples are read as *'the current flowing through the load, from terminal A towards terminal B'* or as *'the current flowing through the load, from terminal A towards the neutral terminal'*.

For currents, too, double-subscript notation *always* follows the phase sequence, A-B-C-A, so they're always written:

$$\bar{I}_{AB}, \bar{I}_{BC}, \text{and } \bar{I}_{CA}$$
$$\bar{I}_{AN}, \bar{I}_{BN}, \text{and } \bar{I}_{CN}$$

For **line currents**, we use '**single-subscript notation**', which takes the following form:

$$\bar{I}_{A}$$

This example is always read as, *'the current flowing in line A from the supply towards the load'*.

Figure 31.9 illustrates how this system of sense arrows and double-/single-subscript notation is used to identify the potential differences and currents for a four-wire, star-connected load.

Figure 31.9

Figure 31.10 illustrates how sense arrows and double-/single-subscript notation is used to identify the potential differences and currents for a three-wire, delta-connected load.

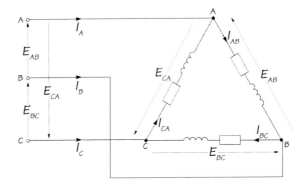

Figure 31.10

Do not get confused by the way in which the line conductors have been drawn, in Figures 31.9 and 31.10, with line B crossing over line C. This has been done for no other reason than neatness and consistency! The reason for doing this is simply to ensure that the vertical sequence of the line conductors follow the order A-B-C, and that each terminal of the star or delta connection is labelled A-B-C in a clockwise sequence.

## Three-phase phasor diagrams

There are two '**Golden Rules**' which we must *always* follow whenever we want to construct a three-phase **phasor diagram**. These are:

1  *Always draw a phasor diagram as it applies to the* **load**, *never to the supply*.
2  *Always draw the phasor diagram in the following order:*
   *a  phase voltages*
   *b  line voltages*
   *c  phase currents*
   *d  line currents.*

### Balanced star-connected load

So, let's draw a phasor diagram for the following star-connected load. We will assume a 'balanced' load: that is, each phase has an identical impedance and phase angle and, so, draws an identical phase current. *Unbalanced loads are well beyond the scope of this text.*

Of course, the phase current could either lag or lead its corresponding phase voltage, depending on whether the load is resistive-inductive or resistive-capacitive but, in the following example, we'll assume that each load is resistive-inductive, with the phase current lagging by φ degrees.

The first thing we must do, then, is to draw a schematic diagram of a star-connected load, and insert the **sense arrows**, labelled using **double-/single-subscript notation**, for *all* voltages and currents. Don't forget, the sequence of letters used in double-subscript notation *must* follow in the sequence: A-B-C-A (Figure 31.11).

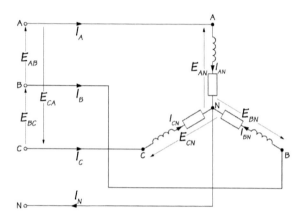

Figure 31.11

So, following the second of our two 'golden rules', we start by drawing the **phase voltages**, and it makes sense to choose $\bar{E}_{AN}$ as the reference phasor which, as usual, is drawn along the horizontal, positive, axis (Figure 31.12).

Figure 31.12

Next, we draw the second phase voltage, $\bar{E}_{BN}$, lagging the reference phasor, by 120° (Figure 31.13).

Figure 31.13

To complete drawing the phase voltages, we draw $\bar{E}_{CN}$, lagging the reference phasor by 240° (Figure 31.14).

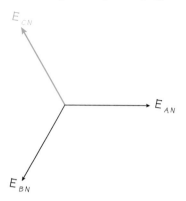

Figure 31.14

Having now drawn all the phase voltages, the next step is to construct each of the line voltages: $\bar{E}_{AB}, \bar{E}_{BC}$ and $\bar{E}_{CA}$. We'll start by drawing line voltage, $\bar{E}_{AB}$. So let's refer back to the schematic diagram to find out how this line voltage relates to the phase voltages (Figure 31.15).

Figure 31.15

Line voltage, $\bar{E}_{AB}$, is the voltage between terminal **A** and terminal **B**. So it's made up of the phase voltage acting *from* terminal **A** *to* terminal **N** and the phase voltage acting *from* terminal **N** *to* terminal **B** – i.e. phase voltage $\bar{E}_{AN}$ and phase voltage $\bar{E}_{NB}$ (which is in the *opposite* direction to sense arrow, $\bar{E}_{BN}$). These, of course, are *phasor* quantities, so they must be added vectorially.

Imagine you are *walking* from point A to point B. Your route will take you from point A to point N and, then, from point N to point B. The 'distance' you have walked is the sum of A to N *plus* N to B. You (vectorially) add the voltages in exactly the same sequence!

If we now refer back to the partly completed phasor diagram, we can readily identify voltage phasor $\bar{E}_{AN}$; *but there is no phasor labelled $\bar{E}_{NB}$!* However, we can easily obtain it, simply by *reversing the voltage phasor $\bar{E}_{BN}$* (Figure 31.16).

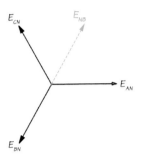

Figure 31.16

So, now, all we have to do is to **vectorially-add** $\bar{E}_{AN}$ and $\bar{E}_{NB}$ to obtain line voltage $\bar{E}_{AB}$ (Figure 31.17):

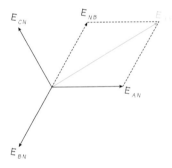

Figure 31.17

We can repeat this exercise line voltage, $\bar{E}_{BC}$, where (referring back to our schematic diagram and its sense arrows) (Figure 31.18):

$\bar{E}_{BC} = \bar{E}_{BN} + \bar{E}_{NC}$ (where $\bar{E}_{NC}$ is simply phase voltage $\bar{E}_{CN}$ reversed)

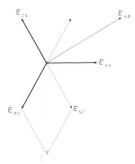

Figure 31.18

... and, again, for line voltage, $\bar{E}_{CA}$ (Figure 31.19), where:

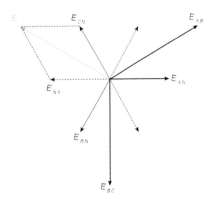

Figure 31.19

$$\bar{E}_{CA} = \bar{E}_{CN} + \bar{E}_{NA}$$
(where $\bar{E}_{NA}$ is simply phase voltage $\bar{E}_{AN}$, reversed).

Let's clean up the phasor diagram, by removing the 'reversed' phase voltages, which we only needed to construct the line voltages (Figure 31.20).

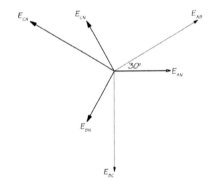

Figure 31.20

It's obvious, from the voltage phasor diagram, that *the line voltages are larger than the phase voltages*, and they *lead the phase voltages by 30°*. But *how much larger are the line voltages*, compared to the phase voltages? Well, we'll worry about that in a moment!

In the meantime, let's now finish the phasor diagram, by constructing the **current phasors**.

In a star-connected system, the phase and line currents are identical, so when we sketch the phase currents, we are *also* sketching the line currents. As with the previous example, we'll assume that the phase currents lag the phase voltage by $\phi$ degrees, so we start by drawing the phase current, $\bar{I}_{AN}$ (which is exactly the same as the line current $\bar{I}_A$), lagging (clockwise) the phase voltage, $\bar{E}_{AN}$, by a phase angle of $\phi$ degrees, behind its corresponding phase-voltage, $\bar{E}_{AN}$ (Figure 31.21).

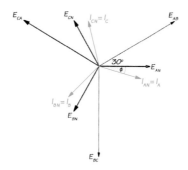

**Figure 31.21**

To complete the phasor diagram, we simply repeat this process for phase/line currents, $\bar{I}_{BN}$ ($\bar{I}_B$) and $\bar{I}_{CN}$ ($\bar{I}_C$) (Figure 31.22).

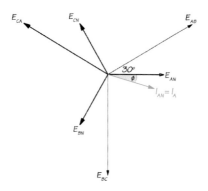

**Figure 31.22**

## Relationship between line and phase voltages

By simple observation, it's quite obvious from the completed phasor diagram that, for a **star-connected system**, the *line voltages are larger than the phase voltages* – but the question is, *how much larger?* Of course, we *could* draw the phasor diagram to scale and, then, measure and compare the two (in fact, you might like to do that with the phasor diagram in Figure 31.22 which has been drawn to scale).

However, we can, instead, apply simple trigonometry to the problem, as follows.

Referring back to the completed phasor diagam, let's compare the length of phase voltage $\bar{E}_{AN}$ to the line voltage $\bar{E}_{AB}$ (Figure 31.23).

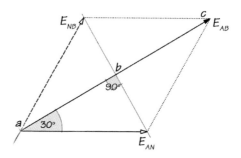

**Figure 31.23**

If we draw a line between the tips of the phase voltage phasors $\bar{E}_{AN}$ and $\bar{E}_{NB}$, it will bisect the line voltage phasor, $\bar{E}_{AB}$, at point **b**. The phase voltage $\bar{E}_{AN}$ and the distance **a–b** then form a right-angled triangle, so we can find the length **a–b** as follows:

$$\cos 30° = \frac{\text{adjacent}}{\text{hypotenuse}} = \frac{\text{distance } ab}{\bar{E}_{AN}}$$

$$\text{distance } ab = \bar{E}_{AN} \cos 30°$$

The line voltage, $\bar{E}_{AN}$, is equal to the distance **a–c**, or *twice* the distance **a–b**, so:

$$\bar{E}_{AB} = \text{distance } ac = 2 \times \text{distance } ab = 2 \times \bar{E}_{AN} \cos 30°$$
$$= 2 \times 0.866 \, \bar{E}_{AN}$$
$$= 1.732 \, \bar{E}_{AN}$$

By an amazing coincidence (!), **1.732** just happens to be the *square root of 3!* So, for a **wye-connected** system:

$$\boxed{\bar{E}_L = \sqrt{3} \, \bar{E}_P}$$

As the **supply voltages** are independent of any load, this important relationship *always* holds true for voltages in a star-connected system.

---

Worked example 1    What is the line voltage for a three-phase, four-wire system having a phase voltage of 230 V?

Solution

$$\bar{E}_L = \sqrt{3} \, \bar{E}_P = \sqrt{3} \times 230 \approx 400 \text{ V (Answer)}$$

The worked example, above, explains what we mean when we describe the UK's low-voltage supply as being a '400/230-V system'. The '400 V' refers to its nominal line voltage, while the '230 V' refers to its nominal phase voltage.

## Balanced delta-connected load

Now let's draw a phasor diagram for the following delta-connected load. Again, we will assume a 'balanced' load: that is, each phase has an identical impedance and phase angle and, so, draws an identical phase current. *Unbalanced loads are well beyond the scope of this text.*

We'll again assume that each load is resistive-inductive, this time with a phase angle of $\phi$ degrees, lagging – i.e. each phase current will lag its corresponding phase voltage by $\phi$ degrees (any angle between 0° and –90°).

As for the previous, star-connected load, the first thing we must do is to sketch a schematic diagram of the delta-connected load, and insert **sense arrows** into the schematic diagram for all voltages and currents, and to **label these sense arrows** using double-/single-subscript notation (Figure 31.24).

Figure 31.24

So, following the second of our two 'golden rules', we start by drawing the **phase voltages** and, this time, it makes sense to choose $\bar{E}_{AB}$ as the reference phasor which, as usual, is drawn along the horizontal, positive, axis. To save time (now that we know how to do so), we'll also draw the other two phase voltages, $\bar{E}_{BC}$ and $\bar{E}_{CA}$, which lag the reference phasor by 120° and 240°, respectively (Figure 31.25).

For a delta-connected system, of course, the line voltages are identical to the phase voltages. So the phasor diagram in Figure 31.25 represents both the phase *and* the line voltages.

Figure 31.25

Next, we'll start to construct the phase and line currents; starting, as always, with the phase currents $\bar{I}_{AB}, \bar{I}_{BC}$, and $\bar{I}_{CA}$.

The first step, then, is to draw the phase current, $\bar{I}_{AB}$, which we are told to assume lags the corresponding phase current by an angle of $\phi$ degrees (Figure 31.26).

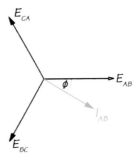

Figure 31.26

Again, we can save time by drawing the other two phase currents, $\bar{I}_{BC}$ and $\bar{I}_{CA}$, at 120° and 240°, respectively, from the first (Figure 31.27).

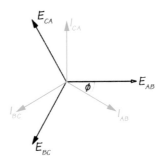

Figure 31.27

The next step is to construct the line currents, starting with $\bar{I}_A$. To understand how to do this, we must first look at junction **A** in our schematic diagram (Figure 31.28).

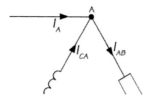

Figure 31.28

If we apply Kirchhoff's Current Law to junction **A**, it can be expressed as follows:

*The phasor sum of the currents approaching the juction equals the phasor sum of the currents leaving that junction. That is:*

$$\overline{I}_A + \overline{I}_{CA} = \overline{I}_{AB}$$

Rearranging this equation, in terms of $\overline{I}_A$:

$$\overline{I}_A = \overline{I}_{AB} - \overline{I}_{CA}$$

Which is exactly the same thing as

$$\overline{I}_A = \overline{I}_{AB} + \overline{I}_{AC}$$

If we now examine the phasor diagram, we see that we already have a phase current, $\overline{I}_{AB}$, but unfortunately we *don't* have a phase current $\overline{I}_{AC}$! However, that's easily rectified, by simply *reversing phasor $\overline{I}_{CA}$* (Figure 31.29).

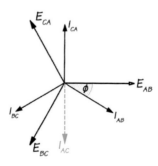

Figure 31.29

We now simply vectorially add the phase currents, $\overline{I}_{AB}$ and $\overline{I}_{AC}$, to obtain the line current, $\overline{I}_A$ (Figure 31.30).

As you can see, the line current, $\overline{I}_A$, *lags* the phase current, $\overline{I}_{AB}$, by 30°. The line current is also greater than the phase current which, of course, we already knew simply by inspection. If we were to conduct a similar analysis to that we did for the line and phase voltages

in a star-connected system, we would find that the line current is $\sqrt{3}$ larger than the phase current.

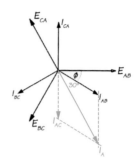

Figure 31.30

To complete drawing the line voltages, we apply exactly the same procedure (Figure 31.31) – i.e. apply Kirchhoff's Current Law to junctions **B** and **C** of the schematic diagram:

$$\overline{I}_B = \overline{I}_{BC} + \overline{I}_{BA} \quad \text{(where } \overline{I}_{BA} \text{ is simply } \overline{I}_{AB}, \text{ reversed).}$$

$$\overline{I}_C = \overline{I}_{CA} + \overline{I}_{CB} \quad \text{(where } \overline{I}_{CB} \text{ is simply } \overline{I}_{BC}, \text{ reversed).}$$

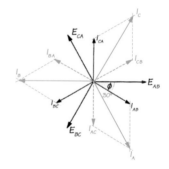

Figure 31.31

Finally, let's clear up the phasor diagram, by removing all the construction lines (Figure 31.32).

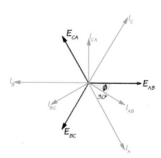

Figure 31.32

In this particular example, we were told that the phase current *lagged* the corresponding phase voltage by $\phi$ degrees (as indicated by the shaded sector). The *actual* phase angle, of course, depends entirely on the ratio of resistance to impedance of the loads and, of course, it could either lead or lag the corresponding phase voltage, depending on whether the actual load is an *R-L* load or an *R-C* load.

So, for a **delta-connected** system:

$$\overline{I}_L = \sqrt{3}\,\overline{I}_P$$

## Summary for balanced star and delta systems

Three-phase system supplying a balanced star-connected load:

$$\boxed{\overline{E}_L = \sqrt{3}\,\overline{E}_P} \quad \boxed{\overline{I}_L = \overline{I}_P}$$

Three-phase system supplying a balanced delta-connected load:

$$\boxed{\overline{E}_L = \overline{E}_P} \quad \boxed{\overline{I}_L = \sqrt{3}\,\overline{I}_P}$$

## Behaviour of currents in balanced three-phase systems

In this chapter, we are only concerned with **balanced loads**. A 'balanced' load is one in which each phase is identical in all respects: each having the same impedance and phase angle.

### Neutral current in a balanced star load?

In a star-connected load, the three phase currents, $\overline{I}_{AN}, \overline{I}_{BN}$ and $\overline{I}_{CN}$, each converge on the star point, and return to the supply, via the neutral conductor. So we can say that the neutral current $(\overline{I}_N)$ must be the phasor sum of the three phase currents:

$$\overline{I}_N = \overline{I}_{AN} + \overline{I}_{BN} + \overline{I}_{CN}$$

Now, let's see what happens if we vectorially add the three phase currents in a **balanced** star-connected system. To add *three* phasors, we start by adding *any two* and, then, *add the resultant to the third*. In the example in Figure 31.33, $\overline{I}_{NC}$ is the phasor sum of $\overline{I}_{AN}$ and $\overline{I}_{BN}$, and is *equal and opposite* (and will, therefore, *cancel*) $\overline{I}_{CN}$.

So, the phasor sum of the three phase currents is **zero**! That is:

$$\boxed{\overline{I}_N = \overline{I}_{AN} + \overline{I}_{BN} + \overline{I}_{CN} = 0}$$

**Figure 31.33**

So, *what's the point in having a neutral conductor?*

Well, in practise, star-connected loads are **rarely perfectly balanced** and, if the three-phase load is unbalanced, even slightly, there *will* be a neutral current – albeit a small one (in comparison with the phase currents).

*Real* three-phase loads are rarely perfectly balanced, so there is practically *always* a neutral return current and, therefore, the need for a neutral. An exception to this are three-phase loads such as motors, which are *always* balanced because each of their phase windings is identical; so, if a motor has a star-connected field winding, there is *no* requirement for a neutral.

A failure of the neutral conductor supplying an unbalanced load will result in any neutral current having to return through the line conductors, resulting in *unbalanced phase voltages* (*not* line voltages, they are independent of load) appearing across the load – however, any further discussion of this topic is beyond the scope of this chapter.

### Where is the return current in a balanced delta load?

It's probably instinctive to understand why no neutral current should flow back to the supply from a balanced, **star-connected** load. However, many students are confused as to how three-line conductors can supply current *to* a **delta-connected** load, if there is no *return* conductor (equivalent to a neutral) for current to flow back to the supply!

The answer, of course, is that these line currents are *not* constant-value currents; they are both *out of phase* with each other, and *vary in value and direction*. Hopefully, the following explanation should make this clear.

(fig. a)

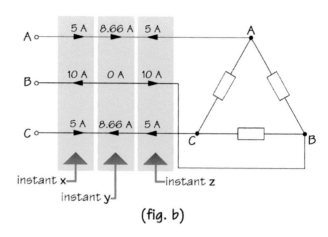

(fig. b)

Figure 31.34

In Figure 31.34a, we show the waveforms representing the three line currents, each displaced by 120°, and flowing through lines **A**, **B** and **C**. We have randomly chosen three instants in time, **x**, **y** and **z**, in order to examine what is happening to the currents at each of these particular moments.

- At **instant x** (occurring at a displacement angle of 30°), the current in lines **A** and **C** are both +5 A. The positive sign indicates that these currents are acting in the positive direction (i.e. flowing *towards* the load), while the current in line **B** is –10 A. The negative sign indicates that this current is acting in the negative direction (*away* from the load, *back towards the supply*).
- At **instant y** (occurring at a displacement angle of 120°), the current in line **A** is +8.66 A, so is flowing *towards* the load. The current in line **B** is passing through the horizontal axis and, so, is zero. The current in line **C** is –8.66 A, so is flowing *away* from the load.
- At **instant z** (occurring at a displacement angle of 210°), the currents in line **A** and **C** are both –5 A and, so, are both flowing *away* from the load. The current in line **B** is +10 A, so is flowing *towards* the load.

If we now transfer this data to the schematic diagram (Figure 31.34b), you can see that:

- At **instant x**, a total of 10 A is flowing along lines **A** and **C** *towards the load*, while a total of 10 A is flowing *back to the supply*, along line **B**.
- At **instant y**, a current of 8.66 A is flowing along line **A** *towards the load*, while a current of 8.66 A is flowing along **C** *back to the supply*. No current is flowing through line **B**.

- At instant **z**, a current of 10 A is flowing along line **B** *towards the load*, while a total of 10 A is flowing through lines **A** and **B**, *back to the supply*.

## Power in three-phase systems

You will recall that power in a single-phase circuit is given by:

$$P = \bar{E}\,\bar{I}\cos\phi$$

where:

$$P = \text{power (watts)}$$
$$\bar{E} = \text{voltage (volts)}$$
$$\bar{I} = \text{current (amperes)}$$
$$\cos\phi = \text{power factor}$$

As already explained, in this chapter we are only going to consider **balanced loads** (i.e. with each phase being identical in all respects).

For a three-phase load, the **total power** will be *the sum of the power developed by each phase*. So, the total power in a three-phase *balanced* load is given by:

$$P = 3 \times (\bar{E}_P \bar{I}_P \cos\phi)$$

Unfortunately, in practise, we *cannot* measure phase voltages or phase currents *directly* because, usually, the phases are not easily accessible. For example, we cannot measure the phase voltages or currents for a three-phase motor because its phase windings are *inside* the casing of the machine (Figure 31.35) and, therefore, *inaccessible*. However, we can *always* measure *line* voltages and *line* currents because they are 'outside' the machine and, therefore, easily accessible.

**Figure 31.35**

So, it would be much more practical if we could develop an equation that gave us the total power of a three-phase system *in terms of line values* rather than in terms of phase values.

The explanation is as follows:

| **For a delta-connected system:** | **For a star-connected system** |
|---|---|
| $\bar{E}_P = \dfrac{\bar{E}_L}{\sqrt{3}}$   and   $\bar{I}_P = \bar{I}_L$ | $\bar{E}_P = \bar{E}_L$   and   $\bar{I}_P = \dfrac{\bar{I}_L}{\sqrt{3}}$ |
| $P = 3\bar{E}_P\bar{I}_P\cos\phi$ | $P = 3\bar{E}_P\bar{I}_P\cos\phi$ |
| $= 3\left(\dfrac{\bar{E}_L}{\sqrt{3}}\right) \times \bar{I}_L \times \cos\phi$ | $= 3\bar{E}_L \times \left(\dfrac{\bar{I}_L}{\sqrt{3}}\right) \times \cos\phi$ |
| $= 1.732\,\bar{E}_L\bar{I}_L\cos\phi$ | $= 1.732\,\bar{E}_L\bar{I}_L\cos\phi$ |
| $= \sqrt{3}\,\bar{E}_L\bar{I}_L\cos\phi$ | $= \sqrt{3}\,\bar{E}_L\bar{I}_L\cos\phi$ |

So, as you can see, it doesn't really matter whether you are dealing with a balanced star-, or a balanced delta-connected, exactly the same equation applies:

$$\boxed{P = \sqrt{3}\,\bar{E}_L\,\bar{I}_L\cos\phi}$$

where:

$$P = \text{power (watts)}$$
$$\bar{E}_L = \text{line voltage (volts)}$$
$$\bar{I}_L = \text{line current (amperes)}$$
$$\cos\phi = \text{power factor}$$

> Once again, a reminder that the equations used in this section only apply for balanced loads.

By extension, we can show the equations for **apparent power** (in volt amperes) and **reactive power** (in reactive volt amperes):

$$\text{apparent power} = \sqrt{3}\,\bar{E}_L\,\bar{I}_L$$
$$\text{reactive power} = \sqrt{3}\,\bar{E}_L\,\bar{I}_L\sin\phi$$

And, of course, the same relationship exists between these three quantities in three-phase circuits as it did for single-phase circuits, i.e.

$$(\text{apparent power})^2 = (\text{true power})^2$$
$$+ (\text{reactive power})^2$$

## Summary of important three-phase relationships

**Figure 31.36**

For a **star-connected** system (Figure 31.36):

$$\boxed{\bar{E}_{line} = \sqrt{3}\,\bar{E}_{phase}}\quad \text{and} \quad \boxed{\bar{I}_{line} = \bar{I}_{phase}}$$

For a **balanced** load:

$$\boxed{\bar{I}_{neutral} = 0}$$

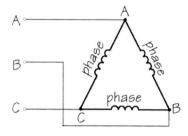

**Figure 31.37**

For a **delta-connected** system (Figure 31.37):

$$\boxed{\bar{E}_{line} = \bar{E}_{phase}} \quad \text{and} \quad \boxed{\bar{I}_{line} = \sqrt{3}\,\bar{I}_{phase}}$$

For both **star** *and* **delta** connections, with a *balanced* load:

$$\text{true power} = 3\,\bar{E}_p\bar{I}_p\cos\phi \text{ or}$$
$$\text{true power} = \sqrt{3}\,\bar{E}_L\bar{I}_L\cos\phi$$

and

$$\text{apparent power} = 3\,\bar{E}_p\bar{I}_p \text{ or}$$
$$\text{apparent power} = \sqrt{3}\,\bar{E}_L\bar{I}_L$$

(**true power** is measured in **watts**; **apparent power** in **volt amperes**)

**Worked example 2** Complete the missing data in the schematic diagram shown as Figure 31.38.

**Figure 31.38**

**Solution**

**Figure 31.39**

**Worked example 3**

**Figure 31.40**

In the schematic shown in Figure 31.40, calculate each of the following:

a   line voltage
b   line current
c   transformer's phase current
d   load's phase voltage
e   the load's apparent power.

**Solution**

a   We know the supply transformer's secondary
    phase voltage $(\bar{E}_P)$ is 230 V so, for a star
    connection:

$$\bar{E}_L = \sqrt{3}\ \bar{E}_S = \sqrt{3} \times 230 \approx 400 \text{ V (Answer a.)}$$

b   We know that the load's phase current is 15 A
    so, for a delta connection:

$$\bar{I}_L = \sqrt{3}\ \bar{I}_P = \sqrt{3} \times 15 \approx 26 \text{ A (Answer b.)}$$

c   For a star connection, the phase current is the
    same as the line current, so the supply's phase
    current:

$$= \text{line current} = 26 \text{ A (Answer c.)}$$

d   For a delta connection, the phase voltage is
    the same as the line voltage, so the load's line
    voltage:

$$= \text{phase voltage} = 400 \text{ V (Answer d.)}$$

e   apparent power $= \sqrt{3}\ \bar{E}_L\bar{I}_L = \sqrt{3} \times 400 \times 26$

$$\approx 18 \text{ kV} \cdot \text{A (Answer e.)}$$

**Worked example 4**   In the above example, what
is the power of the load, assuming that it has a
power factor of (a) 0.5, and (b) 0.75?

**Solution**

a   For a power factor of 0.5:

$$P = \sqrt{3}\ \bar{E}_L\bar{I}_L \cos\phi$$
$$= \sqrt{3} \times 400 \times 26 \times 0.5$$
$$\approx 9000 \text{ W or 9 kW (Answer a.)}$$

b   For a power factor of 0.75,

$$P = \sqrt{3}\ \bar{E}_L\bar{I}_L \cos\phi$$
$$= \sqrt{3} \times 400 \times 26 \times 0.75$$
$$\approx 13\,500 \text{ W or 13.50 kW (Answer b.)}$$

$$\bar{I}_L = \sqrt{3}\ \bar{I}_L = \sqrt{3} \times 31 \approx 53.7 \text{ A}$$

## Review your learning

Now that we've completed this chapter, we need to
examine the **objectives** listed at its start. Placing '*Can
I...*' at the beginning, and a question mark at the end, of
each objective turns that objective into a **test item**. If
we can answer each of those test items, then we've met
the objectives of this chapter.

**Online resources**

The companion website to this book contains further resources relating to this chapter. The website can be
accessed via the following link:
www.routledge.com/cw/waygood